Architectural Material & Detail Structure

建筑材料与细部结构

（荷）尼尔斯·凡·麦里恩博尔 编　常文心 译

Masonry 砖石

辽宁科学技术出版社

Preface 前言

Morden interpretation of traditional material

古老材料的现代演绎

Masonry may be best understood as the building of structures from individual units, laid and jointed together by mortar. Alongside timber, it is perhaps mankind's most ancient building means of construction.

Stone hewn from the earth has formed some the grandest and most enduring structures, from the pyramids of Egypt and the Parthenon in Athens to the Great Wall of China and the Taj Mahal in India.

The humble brick, made from clay set in moulds and dried in the sun or pressed and baked in kilns has provided simple, durable, relatively cheap building material for centuries. Standardised, repetitive in nature but enduringly acceptable in most societies as bringing a tangible human scale to even the largest of buildings, the small unit size of bricks lends them great versatility. Careful positioning in the kiln with its temperature gradients means that the fired colours of brick will vary considerably despite a consistent batch mix. The surface hues and textural variation have offered great opportunity to create pattern and diversity of colour, whereas tight selection can result in an almost uniform consistency.

Brickwork's high qualities of fire resistance, good load bearing capacity, and the fact that it is capable of being structurally reinforced with steel, have all been key to its longevity. Architects and engineers have been able to lift this basic building block to create the grandest of civic works. From fine villas and viaducts to defensive military structures, stadia such as the Roman coliseum, 19th century railways stations and for much of Europe's and North America's urban housing stock the brick has been a central element in the creation of our cities. Extraordinary buildings such as the 17th century Børsen in Copenhagen and the much later Larkin Administration Building 1906 by Frank Lloyd Wright, in different ways use brick to achieve monumental form coupled with decorative detail, making majestic the architecture of commerce. The simple brick has risen far above its very humble origins.

Eladio Dieste (1917-2000), a Uruguayan engineer and architect, made his reputation by building a range of structures from grain silos, factory sheds, markets and churches, in his home country. Using bricks in extraordinary ways he created the sense that masonry structures could shed their weight and instead of bearing heavily upon the ground would seem to hover or take on the appearance of fabric floating in the wind. Buildings in Uruguay such as Northern Soft Drinks, Inc. Salto, Uruguay 1980 and Iglesia de Estación in Atlántida illustrate plainly that Dieste achieved similar limits of possibility with brick masonry to those reached by Pier Luigi Nervi (1891-1971) in his masterful use of concrete.

Though production is now as mechanised as possible, stone masonry, to a considerable extent still requires hundreds of hours of human labour from skilled tradespeople to assemble and lay the components according to tightly controlled standards and skill derived from centuries of practice.

The Valetta City Gate project by Renzo Piano Building Workshop completed in 2015, takes in the complete reorganisation of the principal entrance to the Maltese capital of Valletta. The project comprises four parts: the Valletta City Gate and its site immediately outside the city walls, the design for an open-air theatre "machine" within the ruins of the former Royal opera house, the construction of a new Parliament building and the landscaping of the ditch. The parliament's façades are finished in solid stone. This stone has been sculpted as though eroded by the direction of the sun and the views around it, creating a fully functional device that filters solar radiation while allowing natural daylight inside,

砖石建筑是由独立个体通过叠加和灰浆黏合而成的建筑结构。除了木材以外，它可能是人类最古老的建筑形式。

来自大地的石材能够形成最宏大、最持久的建筑结构，从埃及的金字塔、雅典的帕台农神庙、中国的长城到印度的泰姬陵。

低调的砖块先由黏土在模具中成形，然后在阳光下晒干或在窑炉内煅烧，是一种简单、耐用、相对廉价的建筑材料。它尺寸标准，易于重复，适应性强，即使在大型建筑中也具有良好的质感，小而统一的尺寸赋予了砖块极好的灵活性。窑炉里的温度渐变意味着即使采用统一的配料比例，煅烧的砖块也会呈现出不同的色彩。表面色彩和纹理的变化让砖块可以组成各种图案和色彩的组合，而严格的筛选又能实现高度的统一。

砖砌结构具有良好的防火和承重能力，并可以通过钢筋加固，具有极长的使用寿命。建筑师和工程师可以利用这种基本的建筑构件来创造伟大的市政工程。从精美的别墅、高架桥到坚固的军事设施，例如罗马竞技场、19世纪的火车站以及欧洲和北美的城市住宅，砖块已经成为我们创造城市的核心元素之一。17世纪哥本哈根的旧股票交易中心、弗兰克·劳埃德在1906年设计的拉金行政楼等建筑都选择用砖块来实现宏大的效果，配以装饰性细节，成就伟大的建筑。简单砖块的价值已经远远超过了它卑微的出身。

乌拉圭工程师、建筑师埃拉迪欧·迪斯特（1917-2000）以他在乌拉圭所建造的各种谷仓、工厂厂房、市场和教堂而闻名于世。他通过特有的方式让砖石结构超脱了自身的重量，呈现出盘旋或飘浮在空中的模样。北方软饮料公司（乌拉圭萨尔托，1980）、阿特兰蒂达车站教堂等建筑向世人展示了埃拉迪欧·迪斯特的卓越成就，他超越了砖石结构的极限，正如皮埃尔·奈尔维（1891-1971）超越了混凝土的极限一样。

尽管机械化生产已经实现了常规化，石材在很大程度上仍然依赖经验丰富的技术工人花费数百小时根据数百年来流传下来的严格标准和技巧来进行装配和铺装。

由伦佐·皮亚诺建筑工作室设计的瓦莱塔城市之门项目（2015）帮马耳他首都瓦莱塔的主要门户实现了全面的重组。项目由四个部分构成：瓦莱塔城市之门和它紧靠外城墙的地块、建在皇家歌剧院遗址上的露天剧院、新

figure1~ figure 3: Luan Gallery, Ireland
图1——图3：位于爱尔兰的卢安美术馆

all the while maintaining views from the building. Each of these blocks of façade has been sculpted by a numerically controlled machine. The result is a stone architecture that is fitting for its historic context but also the product of cutting-edge technology*. (source Archdaily*)

Stone masonry, for centuries split and carved by hand, later cut by machine, is clearly part of the 21st century technological revolution bringing architecture new opportunities for exploration of form and surface. What interests me is how to use material as a means of expressing more effectively the architectural pre-occupation of my projects.

My firm's buildings are shaped by my core belief that architecture is primarily concerned with simple components such as light, form, space, scale, material, and context. It is to the architect to use his or her skills to determine the way in which these things are combined that will have the capability to enthral and enrich the experience of people who are touched by them.

At the Luan Gallery, Ireland, my firm converted and extended the brick built Father Mathew Hall, to create a new contemporary art gallery. Pale buff coloured limestone was used in both rough split and smooth plane to form the new gallery wing. The horizontal plinth of random strip stone underpins visually the smooth clad limestone gallery above. The textural subtleties in the different use of the same material were crucial to the project's elevational composition and the project's materiality. (see figure 1,2)

At the Unicorn Theatre in London, we used sheer walls of dark blue/grey brick to create a seemingly taught skin of massive unbroken scale. Compositionally the brickwork forms a background plane in the elevational order of the building into which the glazing was incised, let flush into the brickwork, and through which key elements such as the glazed theatre "green room" project.

In Architectural Material and Detail Structure: Masonry we see a number of trends continuing the exploration and invention in architecture through an examination of form, surface and material. What this book demonstrates in accomplished fashion is that the very best architects in exploring new aesthetic direction be it through radical challenge of form or precise refinement of materiality continue to invent and reinterpret ancient building materials in the context of the new technological age.

Texe by Keith R Williams (member of RIBA, MRIAI, FRSA and founder + director at Keith Williams Architects: London)

国会大厦以及沟渠的景观美化。国会大厦的外墙上装饰着坚固的石材。石材经过雕刻，就像被阳光直射所侵蚀过一样，具有过滤日照辐射，引入自然光照的作用，同时又能保证建筑内部的良好视野。建筑外墙的砌块由数控机器进行切割雕刻。这座石砌建筑既切合它的历史环境，又是尖端技术的产物。（来源：Archdaily*）

数百年来，石材一直由手工切割雕刻，后来则采用机械切割，它充分体现了21世纪技术革命为建筑带来的新机遇，让我们可以充分探索新的形式和表面装饰。我所在意的是如何利用材料来为我的建筑项目带来更好的视觉表达。

我司的建筑全部以我的核心理念来打造，即建筑首先要考虑的是光、造型、空间、比例、材料、环境等简单元素。建筑师需要利用他们的技能来决定这些元素的组合方式，以实现丰富使用者体验的目标。

在爱尔兰的卢安美术馆项目中，我们改造并扩建了砖砌的马修教父厅，形成了一个全新的现代美术馆。新美术馆的翼楼选用了粗糙和光滑的浅黄色石灰岩作为建筑材料。由条状石材构成的水平底座支撑着由光滑石灰岩包覆的美术馆。同种材料在纹理上的细微差异对项目的立面构成和质感呈现起到了重要的作用。（见图1,2）

在伦敦的独角兽剧院，我们采用深蓝灰色的砖墙来打造连续的表皮。砖砌结构形成了一片背景，上面嵌入了玻璃装配，让砖块也变得鲜活起来，建成了一座"玻璃温室"剧院。

在本书中，我们将看到建筑师针对各种建筑形式、表皮和材料所进行的一系列探索和尝试。在本书所收录的案例中，优秀的建筑师为了追求新的美学方向，有的选择激进的造型，有的选择精致的材料。在新技术时代的背景下，他们不断创新和重新诠释砖石这种古老的建筑材料。

本文由基斯·威廉姆斯（英国皇家建筑师协会会员、爱尔兰皇家建筑师协会会员、英国皇家艺术协会会员，伦敦KWA建筑事务所的创始人兼设计总监）撰写

Contents 目录

9	**Overview**	
	概述	
15	**Chapter 1 Brick & Ceramic**	
	第一章 砖与瓷砖	
22	Lingang New City Planning Exhibition Hall	
	临港新城展览馆	
26	Paasitorni Hotel	
	帕斯托米酒店	
30	Ofenhalle	
	高炉车间办公楼	
32	The West Village Building	
	西村楼	
36	Building for Laboratories and Administrative Offices	
	实验楼与行政楼	
42	Lyric Theatre	
	歌词剧院	
46	Sport & Cultural Centre NIARA	
	尼亚拉体育文化中心	
50	Primary School ZV Zavelput	
	ZV小学	
54	Fort Cortina	
	科尔蒂纳堡垒——公司办公楼	
60	Piazza Céramique	
	陶瓷广场	
64	The Curving House	
	曲线住宅	
72	Chengdu Skycourts	
	成都空中庭院	
78	23 Housing Blvd de Hollande Béthune	
	奥朗德贝蒂纳23户住宅楼	

82	Amstelveen College
	阿姆斯特尔芬学院
86	Low-Care and Senior Housing
	监护病房及老年公寓
90	56 Dwellings and Office Space at the Bloemsingel
	布罗伊姆辛格尔住宅与办公空间
94	Split View
	错层住宅
96	School Restaurant of Paul Bert and Léon Blum Scholar Group, Lille-Lomme
	保罗·伯特与莱昂·布鲁姆学校集团餐厅
100	O|2 Laboratory Building of VU Amsterdam University
	阿姆斯特丹自由大学O|2实验楼
108	PARQUE Kindergarten
	公园幼儿园
112	Salmtal Secondary School Canteen
	萨姆塔尔中学食堂
118	Kindergarten Parque Goya
	戈雅公园幼儿园
122	Corte Verona Apartment Building
	柯特维罗纳公寓楼
126	Blood Centre
	血液中心
130	Teacher Training Centre, Archbishop of Granada
	天主教格拉纳达总教区教师培训中心
136	Fire Station, Vilnius
	维尔纽斯消防站
140	Apartments for Young People, Nursery and Park in San Sebastián
	圣塞巴斯蒂安青年公寓、托儿所和公园
146	Agencia IDEA Headquarters
	IDEA公司总部

152	Na Vackove Residential Houses in Prague 纳瓦克夫住宅区
156	Building 18 of Getafe Campus 赫塔费校园18号楼
160	Centre of Air Navigation Services 空中导航服务中心
164	Małopolska Garden of Arts 小波兰艺术花园
168	Martinet Primary School 马丁内特小学
172	Mediatheque in Carballo 卡瓦略图书馆
178	**Chapter 2 Stone** 第二章　石材
184	Bajo Martin County Seat 巴乔马丁郡政府
188	NUK College of Humanities and Social Sciences 高雄大学人文社科学院
194	Rheinfels III, Chur 莱茵岩大厦
198	Sheikh Zayed Desert Learning Centre 扎伊德酋长沙漠学习中心
206	The Samaranch Memorial Museum 萨马兰奇纪念馆
214	Luan Gallery 卢安美术馆
218	Tainan Yuwen Library 台南裕文图书馆
224	County Council of Zamora 萨莫拉郡议会
230	Argul Weave 阿古尔波浪大楼
240	Head Office Archipel Habitat/Municipalpole of The Quarter 阿希佩尔置业总部办公楼/区域行政中心

244	Crescendo Maaswaard Elderly Housing	
	马斯沃德老年住宅	
248	Police Station and Civil Protection Headquarters	
	警察局和民众防护总部	
252	Social Housing for Mine-Workers	
	矿工社会福利住宅	
258	CULT	
	文化中心	
262	Health Centre	
	健康中心	
266	Churchyard Offices and Staff Housing in Gufunes Cemetery	
	古芬斯墓园办公楼和员工宿舍	
272	School Centre Paredes	
	帕雷德斯学校中心	
276	"La Grajera" Institutional Winery	
	格拉赫拉酒庄	
282	Hotel Hospes Palma	
	帕尔马霍斯佩斯酒店	
288	Apartment No.1	
	一号公寓	
292	Administrative Building of the Croatian Bishops Conference	
	克罗地亚主教会议行政楼	
298	Dingli Sculpture Art Museum	
	鼎立雕刻馆	
306	Centre for Interpretation of Jewish Culture Isaac Cardoso	
	犹太文化讲解中心	
312	Cangzhou Merchant Mansion	
	沧州市招商大厦	
318	**Index**	
	索引	

Overview

概述

As primitive construction materials, bricks, ceramics and stone have played an important role in traditional architecture history. Masonry architecture contains high historic values and emotional factors. It implies the spirit of the time. Today, even with the extensive use of industrial construction methods and new construction materials, traditional masonry materials still maintain strong vitality. Nowadays, the concepts of architecture and environment have changed a lot. Ecology, environmental protection, health, energy saving and sustainable development have become consensus. Architects are trying to explore new application of masonry materials in the construction of contemporary architecture and have developed some innovative ideas which refresh our traditional understanding. Masonry materials are ancient and contemporary as well. (See Figure 1 to Figure 3)

砖石作为原始的建筑材料,在传统建筑发展史上发挥了重要的作用。砖石建筑本身蕴含了很高的历史价值和情感因素,体现时代精神。在大量采用工业化建造手段以及各种新型建筑材料的今天,传统的砖石材料以及其建造技艺依旧具有强大的生命力。如今,关于建筑和环境的认识理念发生了很大变化,生态、环保、健康、节能、可持续发展等成为共识,建筑师对砖石材料在现代建筑中的应用进行不断探索,造就了不少创新的构思,大大拓展了人们对其的传统认识。它们既是古老的,也是现代的。(见图1~图3)

Figure 1 to Figure 3 Different façades made from brick, ceramic and stone
图1~图3 砖、瓷砖、石材建筑立面，各具特色

1 Types of Materials

·Brick

A brick is a block or a single unit of a kneaded clay-bearing soil, sand and lime, or concrete material, fire hardened or air dried, used in masonry construction. Brick are produced in numerous types, materials, and sizes which vary with region and time period, and are produced in bulk quantities. Two most basic categories of brick are fired and non-fired brick. Fired brick are one of the longest lasting and strongest building materials sometimes referred to as artificial stone and have been used since circa 5000 BC. Air dried bricks have a history older than fired bricks, are known by the synonyms mud brick and adobe, and have an additional ingredient of a mechanical binder such as straw.

·Ceramic

A ceramic is an inorganic, nonmetallic solid comprising metal, nonmetal or metalloid atoms primarily held in ionic and covalent bonds. The earliest ceramics made by humans were pottery objects, including 27,000 year old figurines, made from clay, either by itself or mixed with other materials like silica, hardened, sintered, in fire. Later ceramics were glazed and fired to create smooth, coloured surfaces, decreasing porosity through the use of glassy, amorphous ceramic coatings on top of the crystalline ceramic substrates.[1] Ceramics now include domestic, industrial

1 砖石材料的类型

·砖

砖是一种建筑材料，由黏土、沙子、石灰或混凝土材料煅烧或晾晒而成，用于砌筑砖石建筑。根据地区和年代不同，砖的种类、制作材料、尺寸也各不相同。大体来讲，砖可分为烧结砖和非烧结砖两大类。烧结砖坚固结实、经久耐用，有时也指人造石，其历史可追溯到公元前约5000年。风干砖的历史比烧结砖更长，又被称为泥砖或土砖，内部含有稻草等其他成分作为黏合剂。

·瓷砖

陶瓷是一种无机非金属固体，由主要存在于离子和共价键的金属、非金属或类金属原子构成。人类制造的最早的陶瓷是陶器，其中包括具有27,000年历史的小雕像。这些小雕像由黏土制成，一些还混合了硅土等材料，在火中经过硬化烧结而成。后来的陶瓷表面经过了上釉，烧制形成光滑的彩色表面，通过在透明陶瓷基质上涂抹一层光亮的非晶质陶瓷涂层而减少了陶瓷制品的孔隙。[1]现在，陶瓷包含家用产品、工业用产

and building products, as well as a wide range of ceramic art. In the 20th century, new ceramic materials were developed for use in advanced ceramic engineering; for example, semiconductors.

·Stone

Stone is an expensive product in architecture industry. Natural stone can be generally categorised as granite, slate, sandstone, limestone, volcanic, etc. With the continuous development of technology, various artificial stone products have emerged with almost equal quality as natural stone, such as terrazzo and synthetic stone. With the development of architectural design, stone have become one of the most important materials for architecture, decoration, road and bridge constructions.

2 Development and Application
·Brick

Brick is one of the oldest masonry materials. With the requirement of sustainable development, industrial wastes such as coal gangue and coal ash have replaced clay to become brick's main raw materials. In the meanwhile, the types are transformed from solid brick to porous brick and hollow brick, and from fired brick to non-fired brick.

Today, based on traditional small piece of brick, brick blocks with various sizes are created, including concrete or decorative concrete hollow block, light-weight aggregate concrete hollow block, gypsum block, etc. These blocks are also used in load bearing wall or exposed exterior wall. Architectural structure has break through brick wall load bearing mode, and bricks are more and more used as maintaining material and finishing material. (See Figure 4)

·Ceramic

The history of ceramic can be dated back to B.C. when the Egyptians used ceramics to decorate their houses. People put clay bricks in the sun or baked them until they were dry. Then they painted the bricks with blue glaze extracted from copper. In Middle Ages, ceramic decorations achieved their peak in Persia. Later, ceramic prevailed over the world. In ancient times, ceramic products were all handmade. Today, mass manufacture has become the main stream and hands are just used to operate the equipments. As in the past, ceramic products are both used in the interior and exterior for decorations. (See Figure 5)

品、建筑用产品以及各种陶瓷艺术品。在20世纪，新的陶瓷材料开始被应用在先进的陶瓷工程中，例如半导体。

• 石材
石材是建筑装饰材料的高档产品，天然石材大体分为花岗岩、板岩、砂岩、石灰岩、火山岩等，随着科技的不断发展和进步，人造石的产品也不断日新月异，质量和美观已经不逊色天然石材，如水磨石和合成石等。随着建筑设计的发展，石材早已经成为建筑、装饰、道路、桥梁建设的重要原料之一。

2 砖石材料的发展和应用
• 砖
砖是最传统的砌体材料。出于可持续发展需求，已由黏土为主要原料逐步向利用煤矸石和粉煤灰等工业废料发展，同时由实心向多孔、空心发展，由烧结向非烧结发展。

如今，在传统小型砖块的基础上，人们创造出多种外形尺寸的砌块，如混凝土及装饰混凝土空心砌块、轻骨料混凝土空心砌块、石膏砌块等。这些砌块常用于不承重的框架填充墙体，对于建筑形式没有任何影响。此外，混凝土空心砌块还可用于承重墙体、清水外墙等。建筑结构形式突破了砖墙体承重模式，使得砖更多的作为维护材料和装饰材料。(见图4)

• 瓷砖
瓷砖的历史应该追溯到公元前，当时，埃及人已开始用瓷砖来装饰各种类型的房屋。人们将黏土砖在阳光下晒干或者通过烘焙的方法将其烘干，然后用从铜中提取出的蓝釉进行上色。在中世纪伊斯兰时期，所有瓷砖的装饰方法在波斯达到了顶峰。随后，瓷砖的运用逐渐盛行全世界。在古代，都是手工制作。如今，全世界范围内，是运用自动化的生产技术，人的手只是用来操作设备。与过去一样，室内室外都使用瓷砖进行装饰。(见图5)

·Stone

In recent 20 to 30 years, international stone industry has developed rapidly. Since 1990, the global stone output and trade volume has been increasing by the speed of 7.3% and 9.2% each year, much higher than other industries. Stone industry grows much faster than global economy. Before 1949, there was no stone industry in China. With the drive of market requirement, Chinese stone industry is growing with two-digit rates per year. In the next few years, the China's consumer demand of stone are still high, so the speed of development will maintain in double digit.

Different parts of the building use different types of stone as facing materials. Granite is apt to be used as exterior material because it doesn't contain carbonate and is resistant to water and weather. It is also a perfect choice for flooring of lobby or hall with high requirements for physical and chemical stableness. Marble with beautiful patterns and lower mechanical strength are usually used in wall facing and living floors.

Figure 4 The world's highest brick tower of St. Martin's Church in Landshut, Germany, completed in 1500 (from Wikipedia)
图4 全球最高的砖砌建筑——圣马丁大教堂（德国兰茨胡特，1500年完工）（来自维基百科）

Figure 5 Mid-16th century Ceramic Tilework on the Dome of the Rock, Jerusalem (from Wikipedia)
图5 建于16世纪中期的耶路撒冷圆顶清真寺的陶瓷装饰（来自维基百科）

·石材

近二三十年来，国际石材工业发展十分迅速。自1990年以来，全球石材生产量和贸易额每年分别以7.3%和9.2%的速度增长，比其他工业都高，整个石材行业的发展明显快于全球经济的发展。在新中国成立以前，中国没有石材工业，新中国成立后，中国的石材工业才开始起步。由于市场需求的推动，中国石材工业的发展每年以两位数的速度增长。今后几年国内石材消费需求依然旺盛，发展速度依然会保持在两位数。

由于使用天然饰面石材装饰的部位不同，所以选用的石材类型也不同。用于室外建筑物装饰时，需经受长期风吹雨淋日晒，花岗岩因为不含有碳酸盐，吸水率小，抗风化能力强，最好选用各种类型的花岗岩石材；用于厅堂地面装饰的饰面石材，要求其物理化学性能稳定，机械强度高，应首选花岗岩类石材；用于墙裙及家居卧室地面的装饰，机械强度稍差，用具有美丽图案的大理石。

Reference
参考资料
1. Carter, C. B. & Norton, M. G. (2007). Ceramic materials: Science and engineering. Springer. pp. 20 & 21. ISBN 978-0-387-46271-4.

Chapter 1
Brick & Ceramic

第一章 砖与瓷砖

With the improvement of modernisation level, more advanced materials and techniques emerge continuously. Even so, traditional masonry and ceramic still attract architects with their unique sense of reality and natural feeling and are extensively used in architectural field.

随着现代化水平的提高,更多的先进材料及工艺不断涌现。尽管如此,传统的砖砌体及瓷砖因其特有的真实感和自然气息重新打动了建筑师,被大量运用到建筑领域。

1.1 Brick

· Categories

According to architectural performance, bricks are categorised into load-bearing brick, non-load-bearing brick, engineering brick, insulating brick, acoustic brick, facing brick, patterned brick, etc.

According to raw materials, bricks are categorised into clay brick, shale brick, coal gangue brick, coal ash brick, concrete brick, etc.

According to appearance, bricks are categorised into solid brick (without holes or hollow ratio<25%), porous brick (hollow ratio≥25%; bricks with many small-sized holes are usually used in load-bearing area due to their high strength), hollow brick (hollow ratio≥40%; bricks with few large-sized holes are usually used in non-load-bearing area due to their low strength); normal brick and special-shaped brick.

According to manufacturing technique, bricks are categorised into fired brick (baked in fire) and non-fired brick (steam-pressing brick, steam-cured brick).

According to fired or not, bricks are categorised into baking-free brick (cement brick) and fired brick (red brick, black brick). (See Figure 1.1 to Figure 1.4)

1.1 砖

· 分类

根据建筑性能：承重砖、非承重砖、工程砖、保温砖、吸声砖、饰面砖、花板砖等。

根据材质：黏土砖、页岩砖、煤矸石砖、粉煤灰砖、灰砂砖、混凝土砖等。

根据外形：实心砖（无孔洞或孔洞小于25%的砖）、多孔砖（孔洞率等于或大于25%，孔的尺寸小而数量多的砖，常用于承重部位，强度等级较高）、空心砖（孔洞率等于或大于40%，孔的尺寸大而数量少的砖，常用于非承重部位，强度等级偏低）；普通砖和异型砖。

根据生产工艺：烧结砖（经焙烧而成的砖）、非烧结砖（蒸压砖、蒸养砖）；

根据烧结与否：免烧砖（水泥砖）和烧结砖（红砖、青砖）。（见图1.1~图1.4）

Figure 1.1 to Figure 1.4 Various types of brick (block) façade
图 1.1~图1.4 不同类型砖（砌块）建筑立面
Figure 1.5 to Figure 1.6 Building façade made of fired bricks
图1.5、图1.6 烧结砖建筑立面

·Features

With soft colours and original feelings, fired bricks are perfect high-end finishing materials for villas, office buildings, student apartments, dwellings, gardens and antique-style buildings.

Fire bricks' values are accretive. Although the initial cost of buildings built with fired bricks is high, the aesthetics won't fade with time and the bricks don't need frequent maintenance.

Fire bricks provide a comfortable built environment. Fired bricks absorb and exhaust humidity. The holes inside the brick exhaust humidity in the daytime and exhaust it at night. This absorb-and-exhaust property helps to keep local environment moist and avoid dry air caused by rapid evaporation and moisture condensation. In addition, fire bricks' heat-conduction property is extraordinary. Buildings built with fired bricks can adjust the temperature, creating a comfortable living environment.

Fired bricks are environment friendly. The composition of split brick is simple, with shale as its main component and without any glaze on surface. Fired in high temperature, the brick product is radiation-free and is a clean, safe and environment-friendly product.

Fired bricks are durable. They can resist fire and weathering. Their colour will never fade even exposed in UV radiation and last for centuries.

As a classic material for European-style buildings, fired bricks are culture-oriented. The warm colour palette, natural transition of colours and unique grains are not only naturally beautiful but also high-valued. As a nature-friendly material, fired bricks endow the building a cultural theme of eco friendly. (See Figure 1.5 and Figure 1.6)

·特性

烧结砖色泽柔和、返璞归真，是别墅、办公大楼、学生公寓、住宅、园林、仿古等建筑的高档装饰材料。

烧结砖的增值性：烧结砖的建筑虽然初期投资比较大，但是建筑的美感不会随时间的流逝而消失，也无需后期频繁的维护。

烧结砖的舒适性：烧结砖具有吸排湿机能，砖内分布的孔隙可以吸排外界的湿气，砖体白天释放湿气，夜晚吸收湿气。烧结砖的这种吸排湿机能有利于保持局部环境的湿润，可以避免水分迅速蒸发而造成的空气干燥，还可以避免结露，同时，烧结砖的导热性能是其他材料无法比拟的。烧结砖的建筑可以调节温度，

·Application in Architectural Field
Besides primary load bearing structure, bricks can be also used as exterior finishing material, interior wall loading structure or interior decoration with its decoration and weather resistance properties. Double wall system is a commonly used energy-saving wall system. There are two basic types divided by the construction principles: Back-infill masonry (similar to traditional wet hanging of stone plate) is stable but easy to crack due to the distortion of primary structure. The other type separates the two walls. The external wall is supported by the bearing element fixed on the primary structure and fixed to the inner wall by metal connecters. An air cavity is usually left between the two layers of walls. An insulation layer is often added to the inner wall to insulate heat and sound.

Double wall system, especially the one with air cavity and insulation, is a comparative ideal structure for energy saving. With concerns of sustainability, it deserves to be generalised on the premise that the stability of exterior wall is confirmed. (See Figure 1.7 to Figure 1.11)

The arrangement, composition and construction mode of bricks (blocks) vary a lot. We can choose apt methods according the specific condition of a building and create innovative and unique design. (See Figure 1.12 to Figure 1.14)

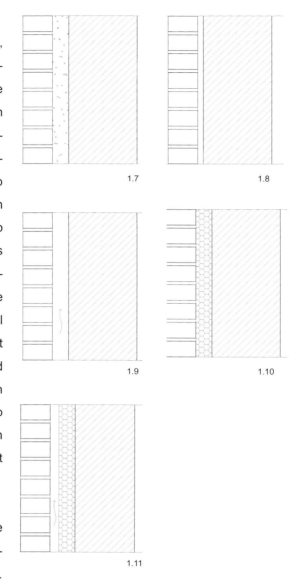

1.7　　　　　　　1.8

1.9　　　　　　　1.10

1.11

Figure 1.7 to Figure 1.11 Typical construction styles of double wall system: grouting double wall, double wall without air cavity or insulation, double wall with air cavity but no insulation, double wall with insulation but no air cavity, double wall with both air cavity and insulation

图1.7~图1.11 双层墙体的几种典型构造样式：灌浆式双层墙体、无空气层无保温层双层墙体、有空气层无保温层双层墙体、无空气层有保温层双层墙体、有空气层有保温层双层墙体

创造更加舒适的生活环境。

烧结砖的环保性：烧结砖的成分很单纯，主要以页岩为主，表面不施釉。产品经高温烧制，没有放射性，是清洁、安全、理想的环保产品。

烧结砖的耐久性：烧结砖可以防火阻燃，抵御风雨。即使紫外线照射自始自终永不褪色，可以历经百年。

烧结砖的文化性：烧结砖是表达欧式建筑风格的经典材料。其温暖的色调，过渡自然的色差以及特殊的外观质地，不仅具有自然美，更凝聚着厚重的价值感。烧结砖是最具有自然保护的材料，它的建筑同样也表现了自然环保的文化主题。（见图1.5、图1.6）

在建筑中的应用

除用作打造主体承重结构之外，砖的装饰性和耐候性决定其可作为外饰面材料、内层墙体承受主体结构或维护室内空间。双层墙体是建筑设计中比较常用的节能墙体结构。其从构造原理上主要包含两种基本做法：其一是背后灌浆式砖砌体（类似于传统石板的湿挂法），自身稳定性好，但易受主体结构变形影响出现开裂等现象。其二是两层墙体分开，通过固定在主体结构上的承重构件承受外墙重力，并通过金属连接件将外层固定于内层墙体上。内外墙体之间通常留有空气流通层，内层墙体外侧可以增添保温层，实现节能、隔音效果。

双层墙体，特别是带有空气流通层及保温层，是比较理想的节能结构，在提倡关注可持续发展的环境下，非常值得推广，但一定确保外墙结构的稳定性。（见图1.7~图1.11）

砖（砌块）排列、组合及构造方式多样，可以根据建筑的具体情况进行选择和创新，从而打造别具特色的样式。（见图1.12~图1.14）

1.2 Ceramic

·Categories

According to use, ceramics are categorised into exterior wall tile, interior wall tile, floor tile, plaza tile, industrial tile, etc. According to moulding principle, ceramics are categorised into dry pressing tile, extrusion moulding tile and plastic moulding tile.

According to component, ceramics are categorised into oxidation ceramic and reduction ceramic.

According to glazed or not, ceramics are categorised into glazed ceramic and unglazed ceramic.

According to water absorption rate, ceramics are categorised into porcelain tile, stoneware porcelain tile, fine stoneware tile, stoneware tile and ceramic tile.

According to manufacturing technique, ceramics are categorised into print tile, polished tile, brindled tile, crystal tile and unglazed tile.

With the continuous development of ceramic manufacturing techniques, more and more products have emerged, such as energy-saving tile, thin tile, photochromic glazed tile, light-weighted tile, permeable tile, etc. (See Figure 1.15 to Figure 1.17)

1.2 瓷砖

·分类

根据用途：外墙砖、内墙砖、地砖、广场砖、工业砖等。

根据成型原理：干压成型砖、挤压成型砖、可塑成型砖。

根据成分：氧化性瓷砖、还原性瓷砖。

根据施釉：有釉砖、无釉砖。

根据吸水率：瓷质砖、炻瓷砖、细炻砖、炻质砖、陶质砖。

根据生产工艺：印花砖、抛光砖、斑点砖、水晶砖、无釉砖。

瓷砖制造工艺不断发展，更多的产品将不断涌现，如节能砖、薄砖、变色釉瓷砖、轻质砖、透水砖等。（见图1.15~图1.17）

Figure 1.12 to Figure 1.14 Various arrangements of bricks (blocks)
图1.12~图1.14 砖（砌块）的不同排列组合方式

· Features

Size: Ceramic products are generally uniform sized, which help to shorten construction time and look neat and aesthetic.

Water absorption rate: The lower the water absorption rate, the higher the product's vitrification level is. With good chemical and physical properties is, ceramic products won't crack or peel off easily due to temperature changes. Ceramic products are water repellent and air permeable. They are light-weighted and flexible, resistant to acid, alkali, freezing, thawing, seismic and crack. They are also compatible to insulation system of exterior wall.

Smoothness: Ceramic products with smooth surface do not bend or rake angle, which facilitates the construction process.

Strength: Ceramic possesses high bending strength and is abrasive resistant. It is a durable material ideal for public places.

· Application in Architectural Field

Most ceramics feature good durability, water resistance, acid-base resistance, abrasive resistance and self-cleaning ability. Therefore, they are extensively used in buildings' exterior and interior walls and floors. The ceramic tiles' texture, colour, size, applied range, water absorption rate, rigidity, abrasive resistance and freezing-thawing frequency are closely related to their application in architecture. With

· 特性

尺寸：产品大小片尺寸齐一，可节省施工时间，而且整齐美观。

吸水率：吸水率越低，玻化程度越好，产品理化性能越好，越不易因气候变化热胀冷缩而产生龟裂或剥落。瓷砖拒水透气性强，自重轻，具有柔性，耐酸碱耐冻融，抗震，抗裂，与外墙外保温体系相容性很好。

平整性：平整性佳的瓷砖，表面不弯曲、不翘角、容易施工。

强度：瓷砖抗折强度高、耐磨性佳且抗重压、不易磨损、历久弥新，适合公共场所使用。

· 在建筑中的应用

多数瓷砖具有良好的耐久性、耐水性、耐酸碱性、耐磨性及自洁性等优点，因此被广泛运用在内外墙和地面饰面上。瓷砖的质感、色彩、规格、应用范围、吸水率、硬度、强度、耐磨性及冻融次数等与建筑应用具有密切的关系。其经过多年的发展，外形及内在品质都有了很大的进步。

瓷砖用于外墙饰面常用的方式为人工现场粘贴，施工方便，适合于人力成本较低的地区，但施工质量不易保证。在日本等国家，除了直接粘贴，还将瓷砖浇筑在混凝土板上，这样能使其与背板紧密结合，技术要求相对较高。考虑到节能要求，也可将瓷砖粘贴在外保温板材上，然后将板材固定于结构墙体。粘结瓷砖外墙设计需考虑墙面整体色调、质感、规格及比例、排列方式、勾缝宽度及色彩等。

Figure 1.15 to Figure 1.17 Building façade made of ceramics
图1.15~图1.17 面砖建筑立面
Figure 1.18 and Figure 1.19 Façade sunshade structure made of ceramic products
图1.18、1.19 面砖构成立面遮阳结构

decades of development, ceramics have progressed in both exterior looking and interior quality.

Ceramics are commonly stuck to exterior walls manually, which is convenient for construction and fit for areas with low labour cost. However, this method cannot ensure the construction quality. In Japan and some other countries, besides direct sticking, ceramics are cast in concrete slabs to ensure close connection with backing slab. Accordingly, this requires higher technique levels. In consideration with energy saving, we can also stick ceramics on exterior insulation panels and fix the panels on structural walls. The ceramic design on exterior wall should consider the wall's colour palette and tile's texture, size, arrangement mode, jointing width and colour comprehensively.

Sticking ceramic tiles are restricted in sizes because large sized tiles are easy to fall. Generally, ceramic tile in 250x250mm size can only be stuck to lower levels. The development and emergence of dry-hang ceramic tile and hollow ceramic plate provide new opportunities. The basic types include dry-hang large-scale plate tile (common construction techniques include slot dry hanging, plug dry hanging, back-bolt dry hanging and supporting dry hanging), dry-hang back-tie monolayer ceramic tile, dry-hang hollow ceramic plate and other dry-hang ceramic components (e.g. horizontal grid dry-hang screen, porous acoustic hollow ceramic plate). (See Figure 1.18 and Figure 1.19)

粘贴瓷砖受规格限制，太大容易掉落，增添危险，通常250x250mm规格的瓷砖粘贴时只能用于距地面不高的楼层。干挂瓷砖及空心陶板的研发出现则打开了新的局面，其类型一般为干挂大面积平板瓷砖（常用工艺为扣槽式干挂工艺、插销式干挂工艺、背栓式干挂工艺及支承式干挂工艺）、干挂背勒式单层陶土瓷砖、干挂式空心陶板及其他干挂式陶土构件（水平格栅状干挂遮阳板、孔状吸音空心陶板）。（见图1.18~图1.19）

Lingang New City Planning Exhibition Hall
临港新城展览馆

Location/ 地点：Jiangyin, China/ 中国，江阴
Architect/ 建筑师：Rong Zhaohui/ 荣朝晖
Photos/ 摄影：Yao Li/ 姚力
Site area/ 占地面积：2,451.6m²
Built area/ 建筑面积：4,481.2m²
Completion date/ 完成时间：2013

Key materials: Façade – terracotta brick, glass
Structure – steel, concrete
主要材料：立面——陶土砖、玻璃
结构——钢、混凝土

Overview
The site of Lingang New City Planning Exhibition Hall is located in a triangle zone of two crossed roads and the sunken north area of the site is naturally connected to the reserved access and the park to the north. A passage through the building and the open space on the first floor are two main spatial features of the building. The bricks which respond to the surrounding context achieve a light feeling which contrasts with the nature of brick through unique construction method. The transparent façade like a veil reflects the publicity of the exhibition hall.

Site
In city with rapid development, the whole new district has become scenery. The building is set back from the wide road, far away the mundane life and human feelings. The road has reserved an underground pedestrian passageway in its construction and the south entrance is exactly located in the site. South to the site is Jiangyin High School. With a group of red facing brick buildings arranged in a courtyard style, the school dominates the environment due to its large scale.

Function
Due to its governmental functions, the planning exhibition hall's exterior expression differs from general exhibition halls, for it's not completely open to public. The introversion in actual use and the openness in planning is a unique expression of the exhibition hall.

Detail and Materials
Structure
The main volume is a cube floating above two floors, looking peaceful and pure. However, it is difficult to build a 7.5-metre-high hollow-out brick wall above the structure. Restricted by the module of brick, if the construction leaves a cavity of half brick, the width of hole is destined to be less than the depth, which will impact the transparency greatly. This method is obviously against the design desire for translucency so

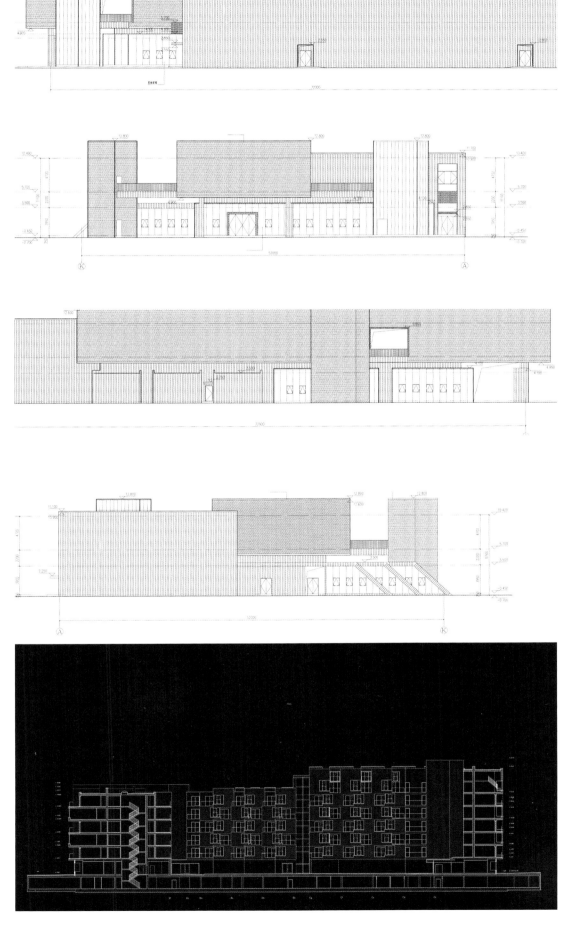

the architects decide to build the whole wall with half bricks. The reinforcement detailing uses standard reinforcing structural measure, i.e. to insert a transverse reinforcement into the mortar joint of several layers of brick. But this structural measure is unfavourable to earthquake proof with the width of half brick. Besides, this measure highly mortar-dependent and highly required on masonry technology, which is impossible to be realised by existing technological level. Therefore, the only solution is to link the bricks with reinforcing steel bars.

Material
A transparent envelop is the basis of openness of the public space. In the meanwhile, in order to look grand and dignified with the requirements of governmental functions, the building adopts semi-transparent envelope. Semi-transparency not only ensures the mass feeling but also adds ambiguity, which responds to the somewhat monotonous urban space. The semi-transparency also makes the interior open space attractive. From the starting point, the architects have decided to use red brick in collaboration with the neighbouring high school. Perforated plate can also achieve the semi-transparent effect of brick. However, as an open system, the two sides of the envelope will both be exposed to public space which requires a consistency of materiality between the two sides and the internal frame of meat plate cannot be hidden, so brick is left as the final choice. On the other hand, the relation between the holes on perforated plate and the mass is intermediated by the aluminium plate, while the relation between the brick and the mass more simple and pure. Brick is a heavy and warm material with great intimacy and is a favoured by many architects. How to use bricks to achieve semi-transparency is the main challenging of the architectural design.

After various experiments, it is time for mass construction. The first layer of bricks is the core of the wall design. Because the construction length is not a integral multiple of the brick module, the hole width of the first brick of the corner is not half brick but a size determined by sample practice. In other words, the first brick is a size adjuster of the whole wall, which is interesting. In the final results, the small hole in the corner is remained completely as the stitch of clothes' hemming. The bricks is layered with a vertical cavity every two layers. A coping ring beam is constructed after the masonry is finished. The texture of the façade is completed with half brick filling.

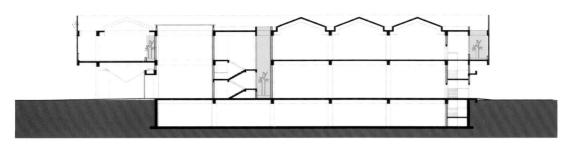

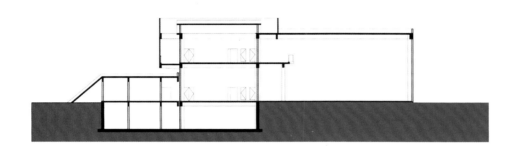

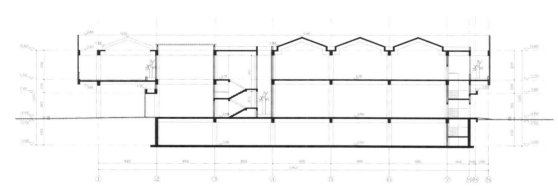

定制砖

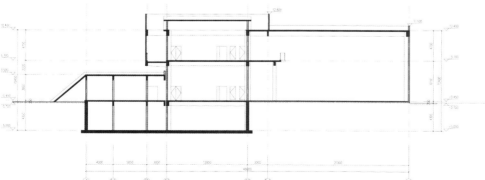

项目概况

临港新城展览馆基地位于两条道路交叉的三角地带，下沉北侧基地自然连接预留的通道和北侧公园。一条穿越建筑的通道和二层的开放空间是建筑的基本空间特点。顺应周边环境的砖通过特定的构造方式获得了与砖自身质感相反的轻灵感觉。立面通透如纱，这也正是展览建筑公共性的体现。

场地

过于快速发展的城市，使得整个新区都布景化了。建筑被远远的退在宽阔的道路后面，远离了世俗的生活和熟悉的人情冷暖。道路建设的时候，地下就已经预留了人行通道。而南端的开口就在建设场地内。基地南侧是江阴市高级中学，学校是一组院落式的红色面砖建筑群体，由于体量因素，在环境中占主导地位。

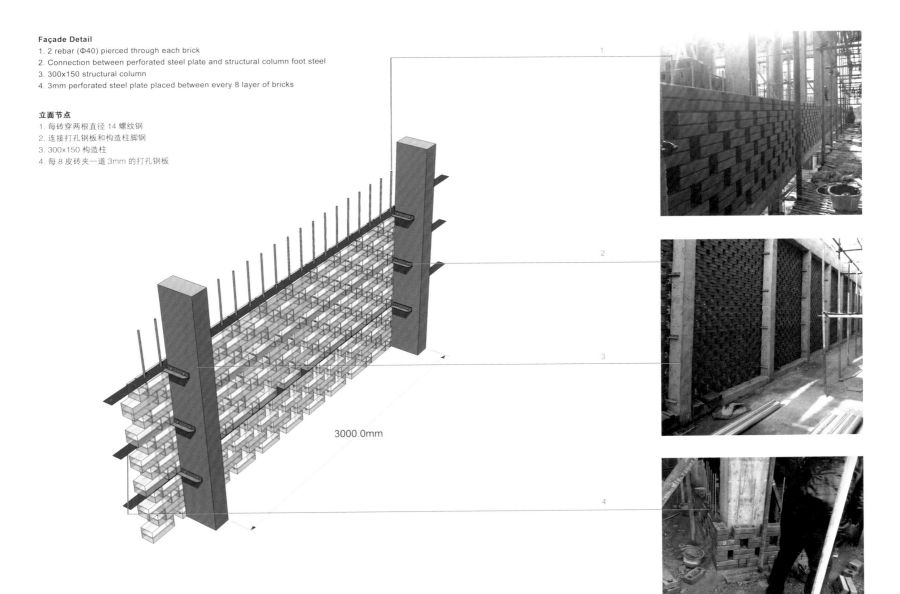

Façade Detail
1. 2 rebar (Φ40) pierced through each brick
2. Connection between perforated steel plate and structural column foot steel
3. 300x150 structural column
4. 3mm perforated steel plate placed between every 8 layer of bricks

立面节点
1. 每砖穿两根直径 14 螺纹钢
2. 连接打孔钢板和构造柱脚钢
3. 300x150 构造柱
4. 每 8 皮砖夹一道 3mm 的打孔钢板

功能
规划展览馆由于其实际的政府职能功能，其外在表现不同于一般的展览馆，它并不具有真正的公共性。以开放窗口为目的的建筑在其实际使用层面的自闭，是规划展示馆的特殊表现方式。

细部与材料
构造
建筑的主要体块是一个飘浮在二层的立方体，平静而纯粹。但高达 7.5 米的整面砖墙空砌在构造上却是十分困难的。由于砖的模数限制，在整砖砌筑下留半砖的空洞，孔的宽度肯定会小于深度。这样通透的感觉会大打折扣。这与半透明的设计愿望是不符合的，所以决定整面墙要用半砖来砌筑完成。配筋构造是标准的加强构造措施，也就是每几皮砖在灰缝里加一根横向钢筋。但在半砖的宽度尺寸下，这种构造方式显然对抗震不利。加上有空砌的要求，底面钢筋不可避免要暴露在外面。而且这种砌筑方式对砂浆严重依赖，对砌筑的工艺要求太高，按现有工匠的水准是无法高质量的达到设计要求的。所以用钢筋穿孔的方式把砖联结为一个整体是唯一的解决方案。

材料
一个透明的表皮是保证公共空间开放性的基础。同时为了满足展示馆作为政府职能功能所需的稳重与宏伟要求，建筑表皮采用一种折中的半透明策略。在保证建筑体量感的同时，半透明更多了一份暧昧。这也是对略显乏味的城市空间的一种回应。建筑因为半透明，使得内部的开放空间更具有吸引力。在半透明开放表皮的材料选择上，由于相邻学校的存在，很早就确定了红砖的使用。与砖相比，穿孔板同样可以达到半透明效果，但问题在于，展览馆的外表皮是一个全开放的体系，里外两个面将全部暴露在公共空间内，这就要求内外两个表皮需要完全一致，而金属板内侧的骨架是无法隐藏的，所以最终还是决定用砖。另一方面穿孔板的孔洞和整体之间还有一个铝板的模数关系，而砖就是简单的个体和整体的关系，因此逻辑上也就会显得更加纯粹。砖是一种很有重量感和温润感的材料，非常具有亲和力，这也是建筑师所喜欢使用的材料。而如何使砖获得半透明的开放性变成了这个建筑设计完成之后最花精力的地方。

施工
实验之后，可以大面积施工了！第一皮砖的定位是整片墙的核心。由于实际的建筑长度和砖的模数不是一个整倍数关系，所以在四个角部的第一块砖的孔洞宽度不是半砖，而是根据实际放样排砖后留下的大小。也就是说，第一块砖是整面墙的一个尺寸调节器。这是很有意思的一点。从最终结果看，角部的小孔作为施工痕迹如同衣服走边的针线活一样被完整的保留了下来。砖是以每两层留一个竖向空为单位来砌筑的，在整体砌筑完成后做好压顶圈梁。立面上的肌理则是用半砖填空的方式完成的。

Paasitorni Hotel
帕斯托米酒店

Location/ 地点: Helsinki, The Netherlands/ 荷兰，赫尔辛基
Architect/ 建筑师: Kimmo Lintula, Niko Sirola and Mikko Summanen/K2S Architects Ltd.
Built area/ 建筑面积: 11,400m²
Completion date/ 竣工时间: 2012

Key materials: Façade – brick
Structure – in-situ concrete
主要材料：立面——砖
结构——现浇混凝土

Overview
The granite Paasitorni fortress was originally built for the use of Helsinki labour movement. It was designed by Karl Lindahl and the first phase was completed 1908. Now the block functions as a successful conference centre. Paasitorni Hotel comprises of two existing buildings and a new wing hidden behind early 20-century and 1950s façades.

The architectural guide line has been to emphasise the architectures of four distinct eras that are present in the Paasitorni quarters – 1908, 1925, 1955 and 2012. The construction of the new additions are mostly cast-in-situ concrete.

The design philosophy concentrates in respecting the listed buildings and in adding a unique contemporary layer of architecture inside the existing framework. The restoration of listed façades and interiors has been made in close cooperation with the Helsinki City Museum and the National Board of Antiquities.

Detail and Materials
The geometry of the new wing is softly curving in order to allow flow of space and light. Façades are constructed of ivory white brick. On two sides of building the brick façade turns into brick "lace" which functions as a filtering layer between the rooms and the inner court yard. During the night the new wing glows like a snow lantern. A customised brick was developed for especially for the project. The brick has oval shaped holes in both ends to allow tolerance for the steel supports used to strengthen the wall. Also the lively surface texture of the brick was developed for this particular use.

项目概况
花岗岩建造的帕斯托米堡最初是为赫尔辛基工人运动使用所建。项目由卡尔·林达尔设计，第一阶段于1908年完工。现在它是一个成功的会议中心。帕斯托米酒店由两座旧楼和一座新建的翼楼构成，后者隐藏在建于20世纪初和50年代的立面之后。

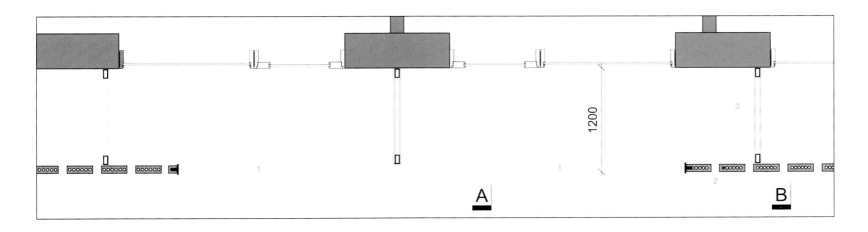

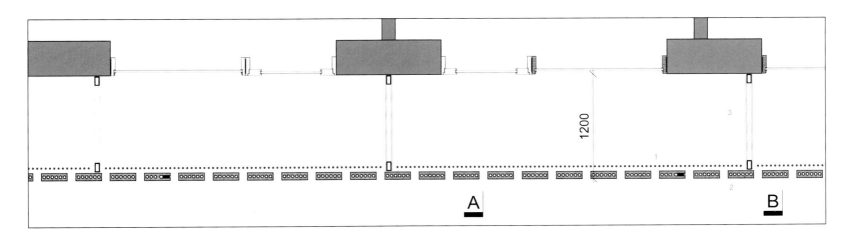

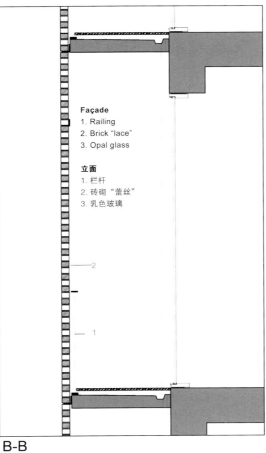

Façade
1. Railing
2. Brick "lace"
3. Opal glass

立面
1. 栏杆
2. 砖砌"蕾丝"
3. 乳色玻璃

A-A   B-B

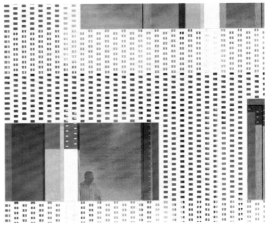

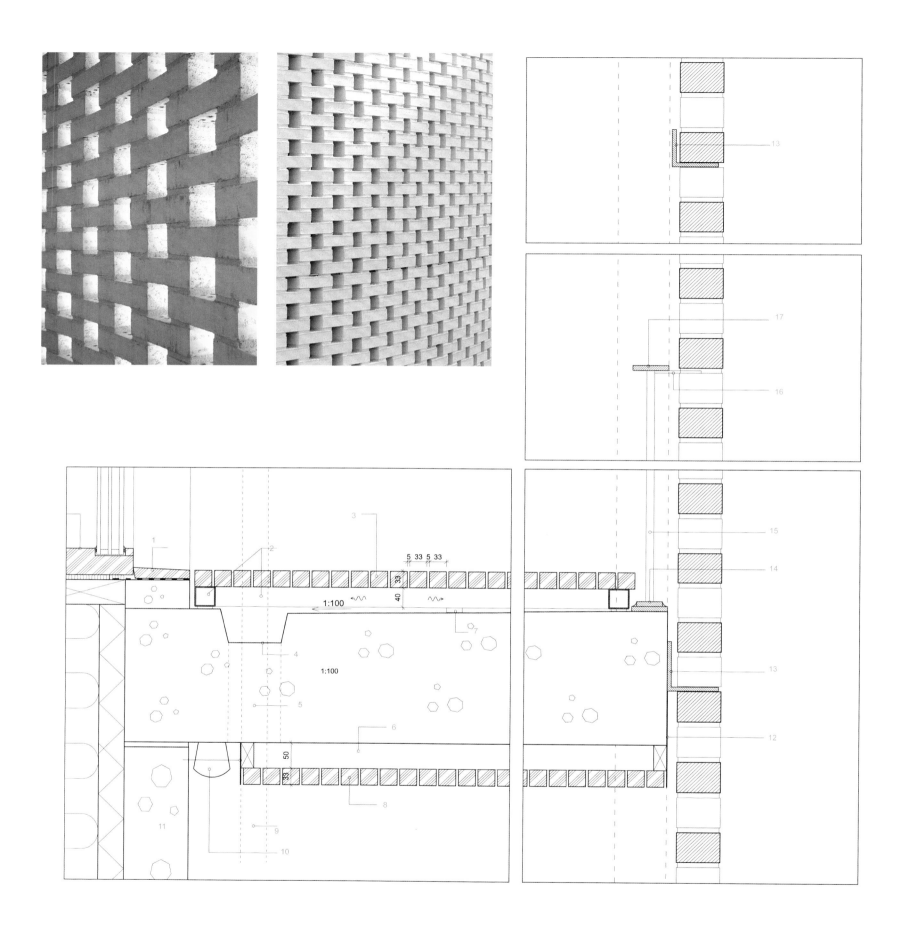

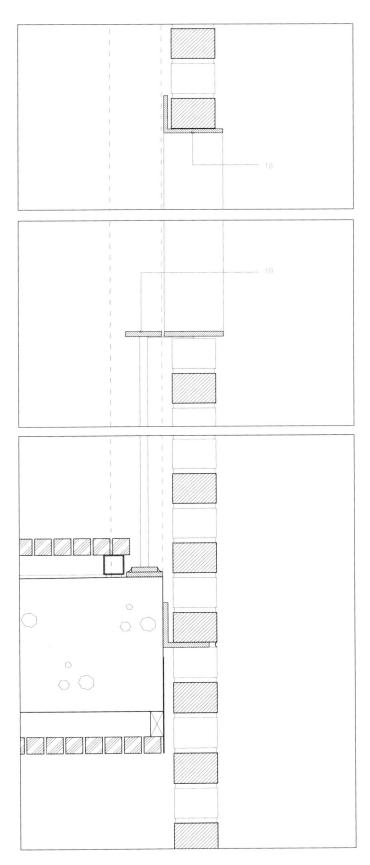

建筑的设计路线旨在突出四个独立区域的建筑特色，它们分别建于1908年、1925年、1955年和2012年。新建结构主要使用现浇混凝土进行建造。

设计将重点放在保护历史建筑上，力求在原有框架内添加一个现代的建筑层次。在历史建筑的外墙和室内空间的修复上，建筑师与赫尔辛基博物馆以及荷兰国家文物局进行了紧密的合作。

细部与材料

新翼楼的外观呈柔和的弧形，实现了空间和光线的流动。建筑立面由象牙白的砖块构造而成。在建筑的两面，砖砌外墙变身为砖砌"蕾丝"，成为了房间和内庭之间的过滤层。夜晚，新翼楼会像冰灯一样发出柔和的白光。项目所用的砖块是特别定制的，砖块的两端带有椭圆形孔洞，从而让钢条穿过，加固墙壁。砖墙活泼的纹理也具备这种功能。

Detail
1. Lacquered solid oak
2. Balcony terrace housing: stainless steel thin-walled tube 40x40x2mm
3. Balcony grating: larch lath 33x33mm
4. Uniform trough the balcony, slope 1:100
5. Balcony wells RPM 102//50/140 (RP-Systems)
6. Wooden joists, painted lower edge of the concrete tone
7. Rubber feet
8. Balcony roof zone: larch lath 33x33mm
9. Drainpipe
10. LED light string, the entire balcony façade measure all floors
11. Stainless steel plate 1mm
12. Matt polished stainless steel plate
13. Structural hst-angles
14. Balcony railing welded to the steel frame (hst)
15. The railing cannot crib
16. Crib railing supported side terrace brickwork
17. Crib railing 70x10mm (hst)
18. Structural angles of the frame (hst)
19. Crib railing 70x10mm (hst)

节点
1. 涂漆实心橡木
2. 阳台露台：不锈钢薄壁管 40x40x2mm
3. 阳台格栅：落叶松木板条 33x33mm
4. 阳台水槽，坡度 1:100
5. 阳台井 RPM 102//50/140
6. 木托梁，下缘涂成混凝土色
7. 橡胶垫
8. 阳台屋顶区：落叶松木板条 33x33mm
9. 排水管
10. LED 灯带，贯穿整个阳台立面
11. 不锈钢板 1mm
12. 亚光面不锈钢板
13. 结构角钢
14. 阳台栏杆，焊接在钢架上
15. 非框形栏杆
16. 框形栏杆，支撑侧平台砖砌结构
17. 框形栏杆 70x10mm（钢）
18. 框架结构角钢
19. 框形栏杆 70x10mm（钢）

Ofenhalle
高炉车间办公楼

Location/ 地点: Pfungen, Switzerland/ 瑞士，普丰根
Architect/ 建筑师: Gramazio & Kohler
Completion date/ 竣工时间: 2012

Key materials: Façade – brick
主要材料: 立面——砖

Façade material producer、:
外墙立面材料生产商：
Keller

Overview
The new front façade for the modification of a former production hall into the new headquarter of the brick manufacturer Keller in Pfungen is a freestanding, self-supporting construction of bricks in front of a steel-glass façade. The bricks are positioned and glued together by a robot. This new manufacturing process is called ROBmade (http://www.robmade.ch). The slight rotation of the bricks generates a vivid image of light and shadow and enhances the volumetric reading of the diagrid.

Detail and Materials
This façade with its delicate diagrid brick structure creates a strong identity for the new headquarter of the façade construction company Keller AG Ziegeleien. It combines craftsmanship with innovation and was finished in October 2012. In the very dense situation on the premises the façade works as a filtering element to the adjacent housing building. Digital production is expanding the traditional process of prefabrication in the building and construction industry. The production of specific elements with a high degree of differentiation is based on the methodical usage of computers for design and production.

The elements are composed of bricks with horizontal adhesive joints. This fabrication method allows completely new design possibilities for walls and façades through real 3D design by means of stone rotation and twist, open joints, vertical offsets and curvatures.

This digital process manages to develop projects from design through to programming and production with robot technology. The combination of architectural design-software and robot-based production processes result in non generic, highly differentiated walls and façades which may in the end consist of a variety of materials.

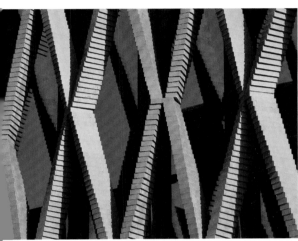

项目概况

项目将一个生产车间改造成了凯勒砖业制造公司的新总部办公楼。全新的立面设计采用独立自承式砖砌结构，里面是钢铁玻璃墙面。砖块的定位和黏合均有自动机械完成。这种全新的生产流程被称为 ROB 制作（详见 http://www.robmade.ch）。砖块的轻微旋转度形成了生动活泼的光影形象，同时也提升了斜肋构架的空间感。

细部与材料

斜肋构架砖砌结构立面为凯勒砖业新总部办公楼的设计增添了强烈的辨识度。它将工艺与创新理念结合起来，于2012年10月正式竣工。在密集的建筑环境中，这个立面将建筑与旁边的住宅隔开。十字花生产扩展了建筑业和建造业的传统预制作流程。高度分化的特定元件的生产制造以电脑设计和生产的系统化应用为基础。

建筑元件有水平砌筑的砖块组成。这种建造方式让墙面通过砖块的旋转和扭曲、开放接口、垂直错段、弯曲等三维设计实现了千变万化的外观。

数字化流程让项目设计、开发到自动机械化生产制造形成了一个流畅的流程。建筑设计软件和自动机械制造流程的结合实现了非常规、高度分化的墙壁和立面，可以让多种材料自由组合。

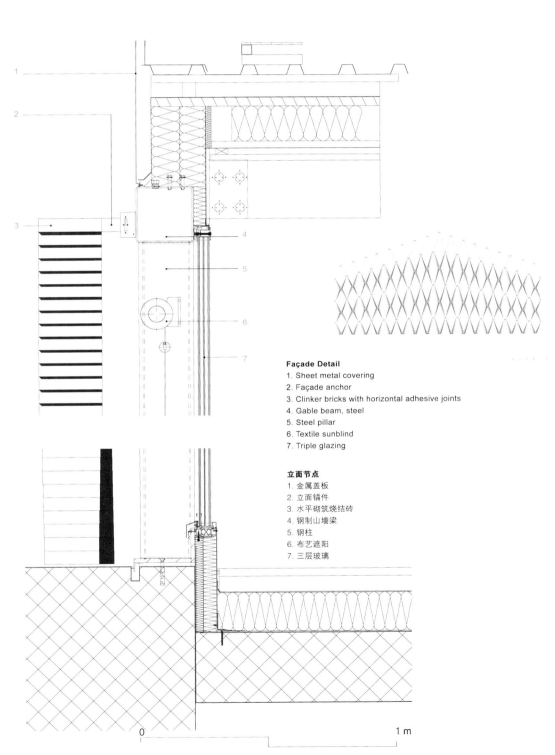

Façade Detail
1. Sheet metal covering
2. Façade anchor
3. Clinker bricks with horizontal adhesive joints
4. Gable beam, steel
5. Steel pillar
6. Textile sunblind
7. Triple glazing

立面节点
1. 金属盖板
2. 立面锚件
3. 水平砌筑烧结砖
4. 钢制山墙梁
5. 钢柱
6. 布艺遮阳
7. 三层玻璃

The West Village Building
西村楼

Location/ 地点：Seoul, Korea/ 韩国，首尔
Architect/ 建筑师：Doojin Hwang (Doojin Hwang Architects)
Photos/ 摄影：Youngchae Park
Area/ 面积：209.83m²
Completion date/ 竣工时间：2011

Key materials: Façade – brick
主要材料：立面——砖

Overview
The West Village building, located in the West Village near Gyeongbok Palace, is a low-rise, high-density, mixed-use building. This area is full of multi-layered beauty of Seoul. To preserve historic and cultural ambience of the area, the architects proposed a typical 'rainbow cake' building, a concept developed by DJHA. This 3-storey building incorporates both residential and commercial functions vertically.

Detail and Materials
To design an "ordinary but not ordinary" building, the architects tried to create a building rooted in its location and tried not to disturb the ambience of the historic West Village area, where the building is located. Large northern window commends a panoramic view towards Mt. Bukak and Mt. Inwang. On the southern façade, a unique brick pattern was used as a visual filter to screen the view of the building in front while allowing sunlight in. This unique pattern of bricks produces various shadow patterns by change of the time and the season, and makes the space rich and alive.

Main material of The West Village is bricks. The architects attempted to emphasise the natural quality of the materials and avoid using too many different materials. Likewise, the interior was finished with paint rather than expensive, unique finish materials. Residential spaces on the first and second floors are partitioned by built-in furniture, which is integrated with lighting. Lightings were installed at the upper part of the furniture; the indirect light illuminates the ceilings.

项目概况
西村楼位于靠近景福宫的西村里，是一座低层高密度混合使用建筑。这一区域到处都呈现出首尔多层次的美。为了保护该地区的历史文化氛围，建筑师提议建造一座典型的"曹洪蛋糕"建筑。这座三层高的建筑在垂直方向中结合了住宅和商业功能。

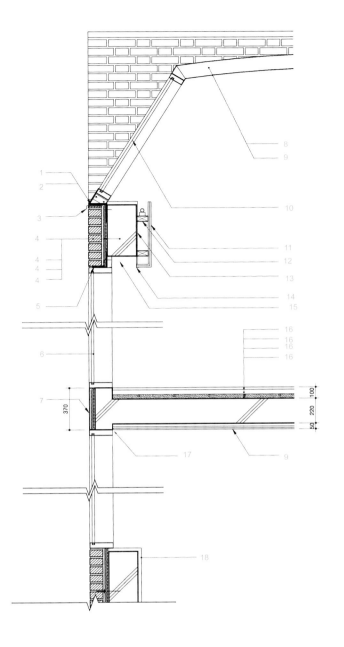
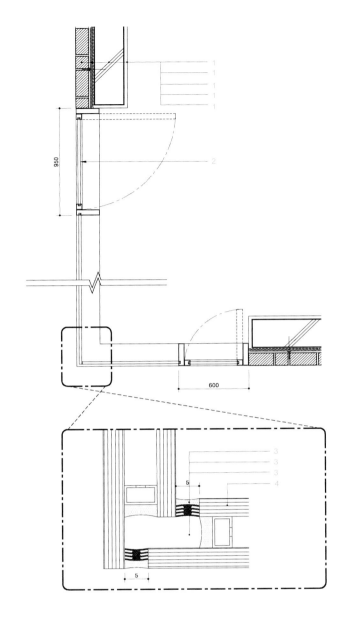
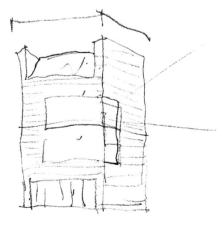

Details of Front Part of Floor Windows
1. L – 5nn SST PL
2. 160x100x10mm SST angle
3. 1.2mm zinc plates
4. 0.5 B brick (joint 15mm)
   13mm thermal insulation
   Hydro isolation
   T250 concrete wall
5. 130x130x10mm SST angle
6. Thk 22 insulating glass
7. SST PL
8. M-125x75x2.3(T) ST pipe
9. Thk 9.5 custom paint painted gypsum board – 2PLY
10. Thk 29 insulating glass
11. Thk 12.5 custom paint painted gypsum board – 2PLY
12. 100x45 stud
13. Custom paint coating
14. 1.2mm SST PL
15. Curtain box
16. 20mm floor finishes
    40mm finishing mortar
    XL-pipe
    40mm lightweight foamed concrete
17. 10x10 reveal
18. T30 custom paint coating finish in mortar

落地窗前半部节点
1. L – 5nn SST PL 不锈钢板
2. 160x100x10mm SST 不锈钢角材
3. 1.2mm 锌板
4. 0.5 B 砖（接缝 15mm）
   13mm 隔热层
   防水层
   T250 混凝土墙
5. 130x130x10mm SST 不锈钢角材
6. 22 厚隔热玻璃
7. SST PL
8. M-125x75x2.3(T) ST 不锈钢管
9. 9.5 厚定制涂层石膏板 – 双层
10. 29 厚隔热玻璃
11. 12.5 厚定制涂层石膏板 – 双层
12. 100x45 螺柱
13. 定制涂料涂层
14. 1.2mm SST PL 不锈钢板
15. 窗帘盒
16. 20mm 地面饰面
    40mm 抹面灰浆
    XL 管
    40mm 定制泡沫混凝土
17. 10x10 窗侧
18. T30 定制涂料涂抹灰浆饰面

Details of Floor Flat Windows
1. 0.5B brick (joint 15mm)
   13mm thermal insulation
   Hydro isolation
   T250 concrete walls
   T30 custom paint coating finish in mortar
2. Thk 22 insulating glass
3. Elastic joint
   Sealing gasket
   Drainage channel
4. Insulating glass

落地窗节点
1. 0.5B 砖（接缝 15mm）
   13mm 隔热层
   防水层
   T250 混凝土墙
   T30 定制涂料涂抹灰浆饰面
2. 22 厚隔热玻璃
3. 弹性接缝
   密封垫片
   排水沟
4. 隔热玻璃

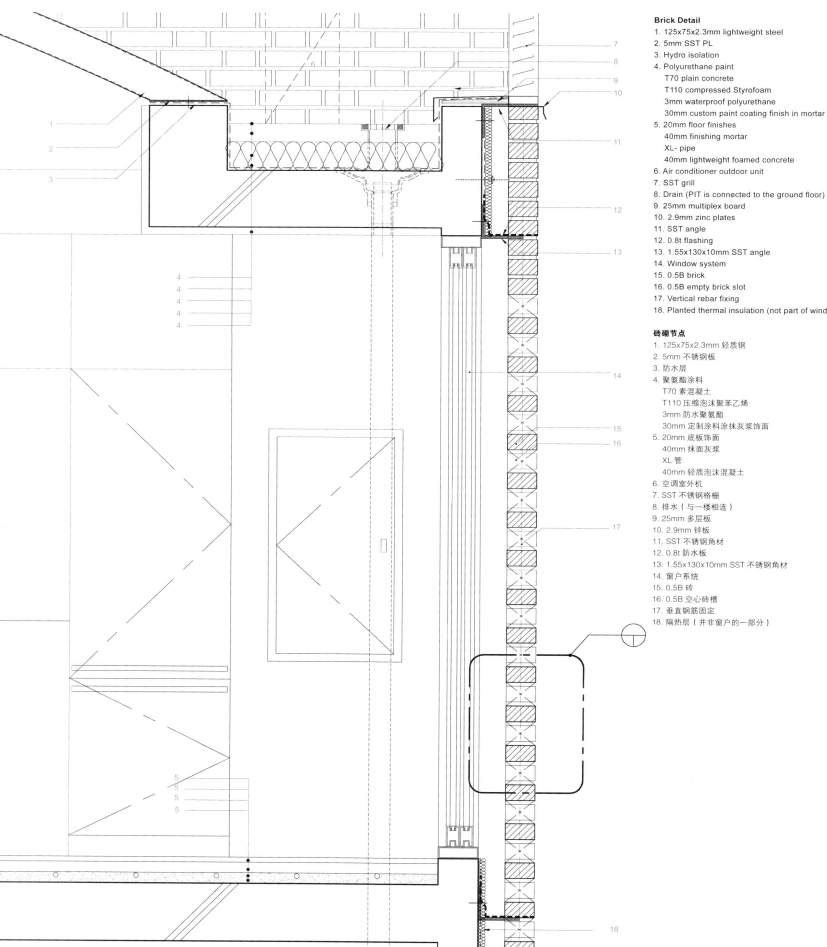

Brick Detail

1. 125x75x2.3mm lightweight steel
2. 5mm SST PL
3. Hydro isolation
4. Polyurethane paint
   T70 plain concrete
   T110 compressed Styrofoam
   3mm waterproof polyurethane
   30mm custom paint coating finish in mortar
5. 20mm floor finishes
   40mm finishing mortar
   XL- pipe
   40mm lightweight foamed concrete
6. Air conditioner outdoor unit
7. SST grill
8. Drain (PIT is connected to the ground floor)
9. 25mm multiplex board
10. 2.9mm zinc plates
11. SST angle
12. 0.8t flashing
13. 1.55x130x10mm SST angle
14. Window system
15. 0.5B brick
16. 0.5B empty brick slot
17. Vertical rebar fixing
18. Planted thermal insulation (not part of windows)

砖砌节点

1. 125x75x2.3mm 轻质钢
2. 5mm 不锈钢板
3. 防水层
4. 聚氨酯涂料
   T70 素混凝土
   T110 压缩泡沫聚苯乙烯
   3mm 防水聚氨酯
   30mm 定制涂料涂抹灰浆饰面
5. 20mm 底板饰面
   40mm 抹面灰浆
   XL 管
   40mm 轻质泡沫混凝土
6. 空调室外机
7. SST 不锈钢格栅
8. 排水（与一楼相连）
9. 25mm 多层板
10. 2.9mm 锌板
11. SST 不锈钢角材
12. 0.8t 防水板
13. 1.55x130x10mm SST 不锈钢角材
14. 窗户系统
15. 0.5B 砖
16. 0.5B 空心砖槽
17. 垂直钢筋固定
18. 隔热层（并非窗户的一部分）

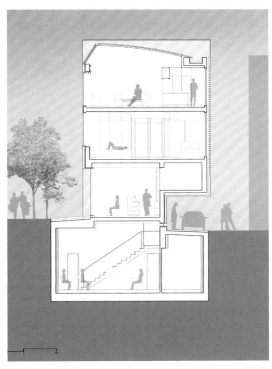
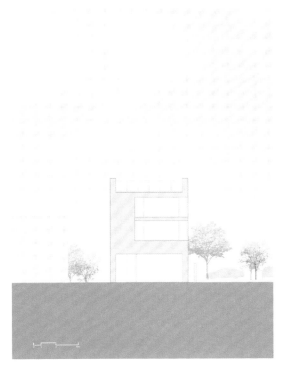
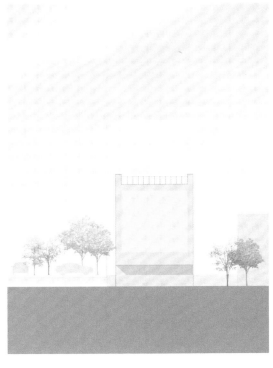
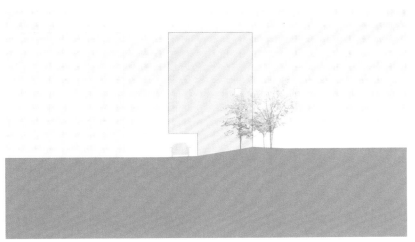
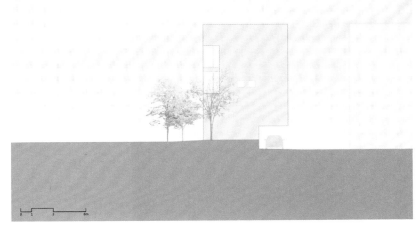

细部与材料

为了设计一座"平凡而又不平凡"的建筑，建筑师力求让建筑一方面立足于自身的地理位置，一方面又不打扰西村地区的历史氛围。宽大的北窗让建筑享有北岳山和仁王山的壮丽景色。在南侧立面，独特的砖砌图案形成了一面屏风，过滤了建筑前方的风景，同时也让阳光照射进来。随着时间和季节的变换，砖砌图案形成了变化万千的光影效果，让整个空间变得丰富而富有生气。西村楼的主要建筑材料是砖。建筑师希望突出材料的自然品质，避免使用过多种类的材料。

同样，室内也仅采用涂料进行涂装，并没有选择昂贵的特征饰面材料。二楼和三楼的住宅空间通过嵌入式家具隔开。家具与照明设施也被整合起来，照明设施被安装在家具的上部，间接光线点亮了天花板。

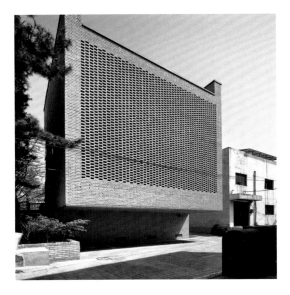

Building for Laboratories and Administrative Offices

实验楼与行政楼

Location/ 地点: Belval, Luxembourg/ 卢森堡，巴伐尔
Architect/ 建筑师: Bruck + Weckerle Architekten

Key materials: Façade – clinker brick, aluminium panel
Structure – concrete
主要材料：立面——烧结砖、铝板
结构——混凝土

Overview
The Building for Laboratories and Administrative Offices is built in Belval-Ouest, on the south east corner of the so-called "Terrasse des Hauts Fourneaux", directly next to the historical blast furnaces. With its 60 metres of height it plays an important role in the urban context.

The foundation block grows out to a tower. The foundation borrows its geometry and height out of the urban context, and draws a line to the neighbouring houses: the former governor building and the concert hall "Rockhal".

The two existing sycamore trees on the plot had to been taken into consideration. The space the trees required around them added up to the sculptural form of the building and, together with the trees, created a generous space at the entrance.

Detail and Materials
A simple and compact geometry is characteristic for the building. The façade is covered in clinker bricks. This material was often used in early industrial areas and stands out very robust and durable. The window frames and the enclosure for the window blinds are made of aluminium. The project combines building tradition with innovation to create a building that is constant and sustainable.

To make the building feel lighter and higher, the percentage of aluminium panels is increasing and the coverage of the bricks decreasing as going higher on the surface of the façade. According to the weather, the upper part of the tower seems to migrate with the sky.

The concept of the façade is that of a traditional façade, with brick layout and opening win-

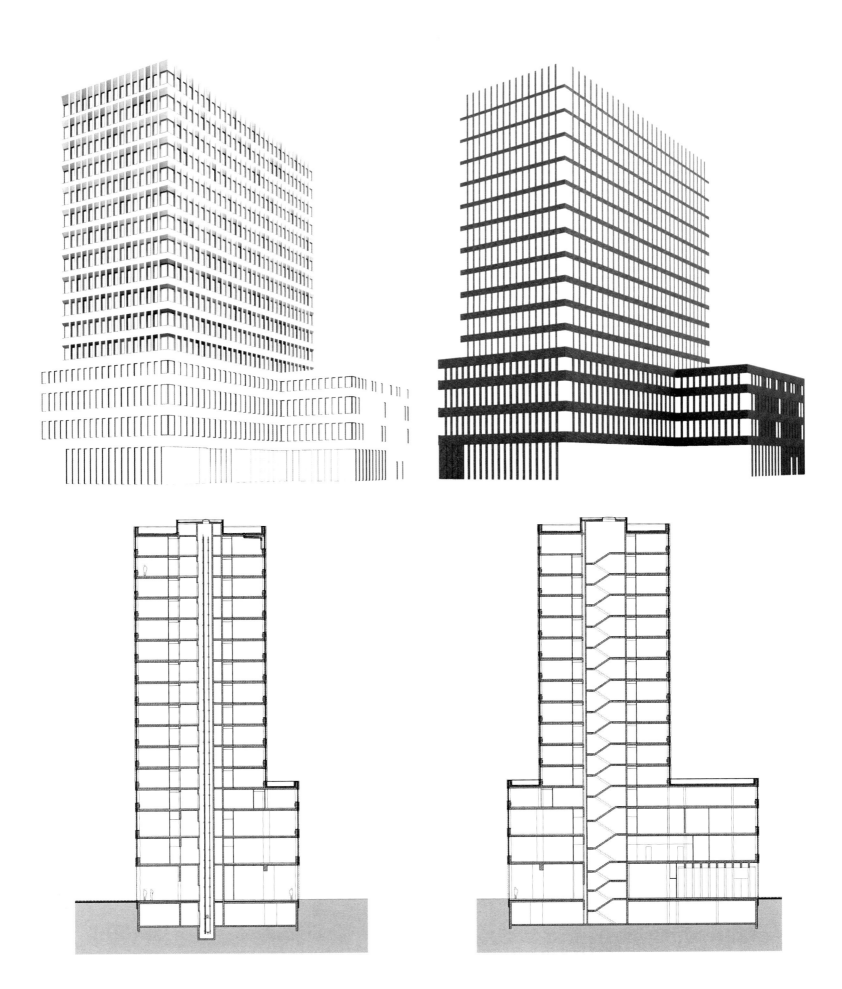

dows. The structure is composed of columns in concrete on each axis of 1.25 m. This concrete structure is covered with insulation and with clinker bricks on the prefabricated lintel. On each axis there is a triple glazed window. The electric window blinds are installed on the outside and integrated in the lintel.

The façade responds to all the demands of the client and the surrounding buildings in a simple way. The percentage of transparent surface on the façade is limited to 42% to ease the maintenance. The chosen massive façade increases the persistence of the building and protects against excessive heat. The balustrades also serve as vertical fire protection.

细部与材料

简单而紧凑的结构是建筑的主要特色。建筑立面由烧结砖覆盖而成。这种材料常见于早期的工业建筑，十分经久耐用。窗框和遮阳帘的外框由铝材制成。项目融合了传统建筑形式和创新技术，打造了一座经典而富有可持续价值的建筑。

为了让建筑更轻、更高，随着建筑越来越高，铝板在结构中所占的比例被提高，而砖块的覆盖面积被缩小。在晴好的天气，塔楼的上半部分几乎与天空融为一体。

立面设计采用了传统设计概念，呈现为砖墙和开放的窗口。建筑结构由轴长1.25米的混凝土柱构成。混凝土结

项目概况

实验楼与行政楼建在卢森堡巴伐尔区，位于"高炉露台"的东南角，紧邻那些历史悠久的高炉。60米高的楼高让建筑在城市环境中扮演了重要的角色。

建筑由底座体块和上方的塔楼组成。底座的造型和高度与周围的城市环境相匹配，基本与周边的建筑（包括政府大楼、岩石音乐厅等）平齐。

场地上已有的两棵美国梧桐树必须被纳入建筑设计的考量之中。树木四周必需的空间为建筑增添了雕塑感，这些空间与树木一起，共同打造了一个宽敞的入口空间。

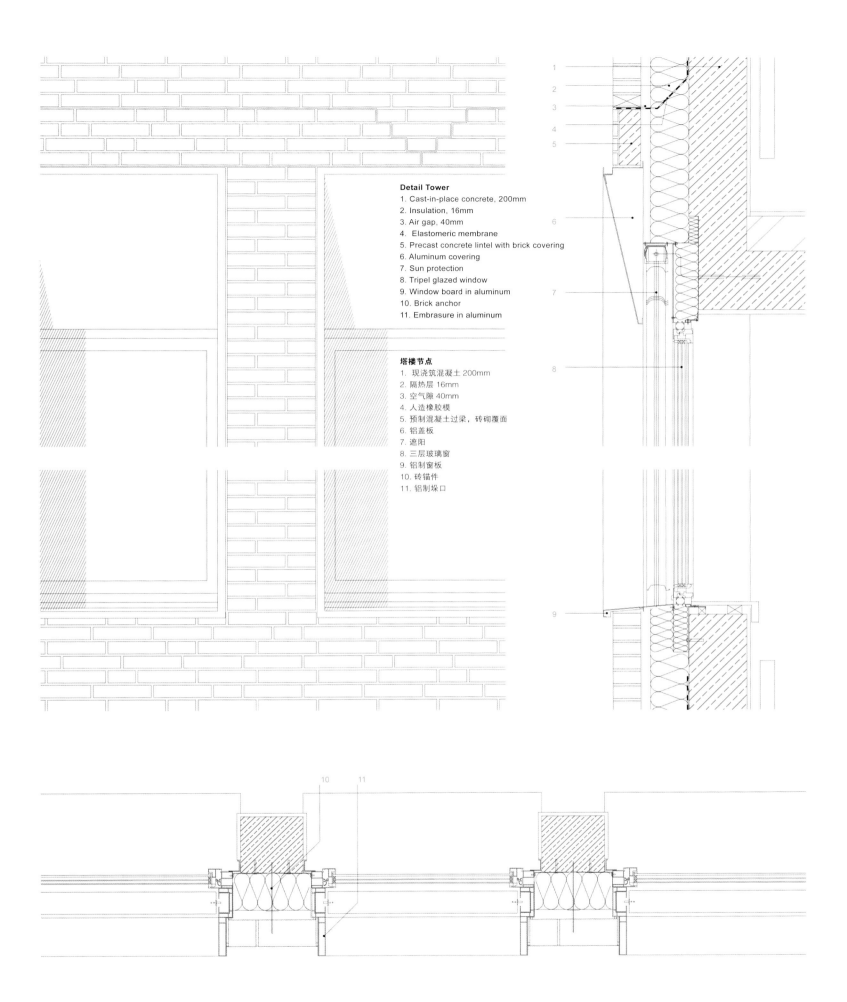

Detail Tower
1. Cast-in-place concrete, 200mm
2. Insulation, 16mm
3. Air gap, 40mm
4. Elastomeric membrane
5. Precast concrete lintel with brick covering
6. Aluminum covering
7. Sun protection
8. Tripel glazed window
9. Window board in aluminum
10. Brick anchor
11. Embrasure in aluminum

塔楼节点
1. 现浇筑混凝土 200mm
2. 隔热层 16mm
3. 空气隙 40mm
4. 人造橡胶模
5. 预制混凝土过梁，砖砌覆面
6. 铝盖板
7. 遮阳
8. 三层玻璃窗
9. 铝制窗板
10. 砖锚件
11. 铝制垛口

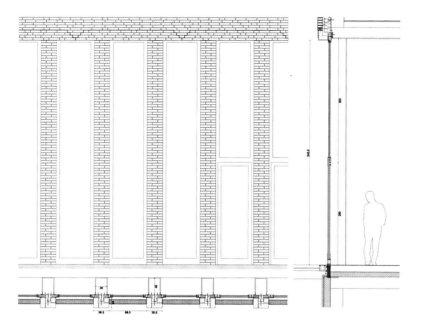
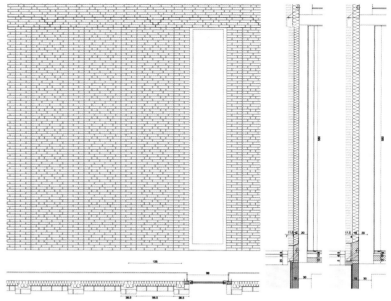

构上覆盖着隔热层，在预制过梁上砌有烧结砖。每根轴上有一个三层玻璃窗。电动遮阳帘安装在外侧，与过梁结合起来。

建筑立面满足了客户的所有要求，以简单的形式应对了周边建筑。为了便于维护，透明表面的百分比被限制在42%。大面积立面的设计增加了建筑的稳定度，同时也能防止楼体过热。栏杆还能起到垂直防火保护的作用。

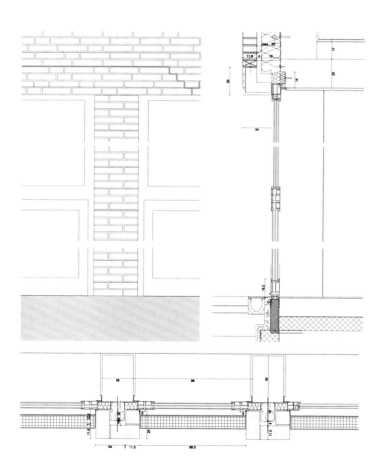
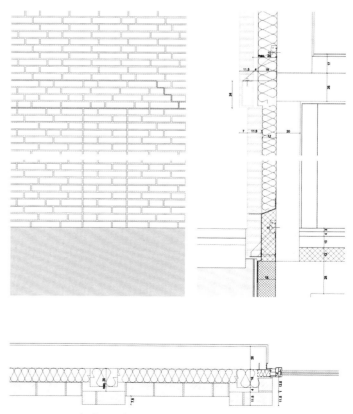
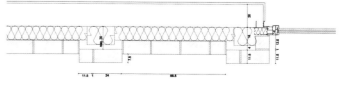

Lyric Theatre
歌词剧院

Location/ 地点: Belfast, UK/ 英国，贝尔法斯特
Architect/ 建筑师: O'Donnell + Tuomey
Photos/ 摄影: Dennis Gilbert
Gross floor area/ 总楼面面积: 5,500m²

Key materials: Façade – brick
Structure – concrete
主要材料：立面——砖
结构——混凝土

Overview

The Lyric is sited between the characteristic grid pattern of the surrounding Belfast brick streetscape and the serpentine parkland setting of the river Lagan. The architectural design concept responds to these conditions by housing each the 3 principal functional elements of the building within its own distinctive brick box, with the public circulation spaces and staircases wrapping around the fixed forms of the theatre, studio and rehearsal, standing on the sloping ground of the site like rocks in a stream. The skyline of the building will display the constituent ingredients of the conceptual design. The solid sculpted brick volumes linked by transparent permeable public spaces are intended to visually connect the street through the Lyric woods to the continuous flowing line of the river through the city.

Detail and Materials

The team has worked closely with Gilbert Ash, the appointed contractor, during this crucial stage of construction to ensure a high quality of construction and craftsmanship. The structure is up, the scale and volume of the building is established on the skyline and can be seen reflected in the river.

Brick, concrete and timber are the fundamental building materials and the visual effect of their combination can be experienced inside the building and glimpsed through the scaffolding. All the building materials are selected to endure and will be crafted to weather with age.

项目概况

歌词剧院坐落在富有特色的贝尔法斯特砖石建筑街景和蜿蜒曲折的拉根河畔公园之间。建筑设计参考了周边环境条件，将三个主要功能元素分别放置在独立的砖砌盒式结构中，而公共通道和楼梯则环绕着剧院、工作室和排练厅展开。三个不同的空间坐落在斜坡场地上，就像是溪流中的岩石。建筑的空中轮廓充分展示了概念设计的构成元素。封闭的雕刻砖砌结构与通透的公共空间通过"歌词树林"与街道实现了视觉联系，并且与穿过城市的河流相互连接。

细部与材料

在施工的重要阶段，设计团队与指定承包商保持了紧密的合作，从而保证了高品质的施工和工艺。当结构完成后，建筑的空中轮廓充分显示了它的规模和体块，建筑的外观还能倒映在水面上。

砖、混凝土和木材是主要的建筑材料，它们所组成的视觉效果在建筑内部和脚手架上都十分明显。所有建筑材料的选择都以经久耐用为原则，材料还能够随着时间的变化而逐步风化。

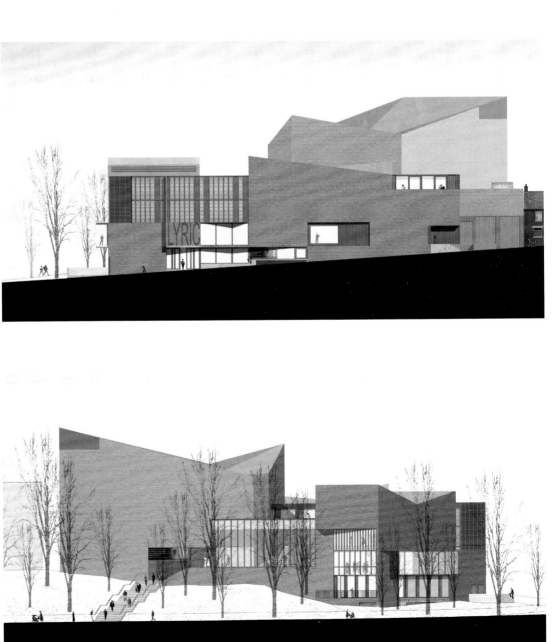

Façade Detail

1. 100mm concrete roof screed with Larsen Fiberflex admixture on 80mm extruded polystyrene insulation on hydrotech roof membrane system on structural precast concrete slabs supported on steel trusses
2. Perimeter in-situ concrete coping with SS structural dowel bar connection to concrete roof screed
3. Continuous roof membrane lapped and sealed into cavity
4. 100mm Ibstock Heritage Red blend brick
5. 28mm BOEN Boflex Stadium flooring with iroko hardwood finish on 18mm plywood substrate on supporting timber joists @ 400 ctrs on in-situ RC slab upstand structure with 75mm mineral wool acoustic insulation between
6. Acon cavity wall tie
7. Full fill cavity wall insulation
8. 215mm reinforced concrete wall
9. Untreated hardwood iroko screens with 3 coats clear Bondex seal to internal face
10. Continuous galvanised steel fixing plate to support coping
11. Folding aluminium coping PPC RAL 9006
12. Roof build-up: 100mm pebbles layer on protection sheet on extruded polystyrene insulation on Hydrotech roof membrane system on 200mm RC slab structure
13. Continuous galvanised steel angle supporting brick
14. Low-e double glazing
15. Solid iroko bar counter supported off external screen
16. Meinertz Finned Tube radiator
17. 5mm PPC interior steel cover plate (colour red oxide)
18. Schoeck Isokorb loadbearing connection providing thermal break within cantilevered concrete structure
19. Floor Build-up: 220x54x54 Wienerberger "Terre De Rose" brick paver with sand cement grout (sealed) on 75mm concrete screed with underfloor heating on 75mm extruded polystyrene insulation on reinforced concrete slab with soffit lightly sandblasted & sealed externally
20. Untreated hardwood iroko bottom rail
21. Galvanised steel insulated exterior fascia
22. Galvanised plate steel exterior cladding
23. Louvre system fixed to mechanical ventilation units

立面节点

1. 100mm 混凝土屋顶砂浆，配 Larsen Fiberflex 掺合剂
   80mm 挤塑聚苯乙烯隔热层
   反渗透屋顶防水膜系统
   结构预制混凝土板，钢桁架支承
2. 边界现浇筑混凝土顶盖，SS 结构传力杆连接
   至混凝土屋面砂浆
3. 连续屋面防水层，与气腔重叠密封
4. 100mm Ibstock Heritage Red 混合砖
5. 28mm BOEN Boflex Stadium 地面，配绿柄桑硬木饰面
   18mm 胶合板底层支承木托梁 @ 400 ctrs 现浇筑钢筋
   混凝土板竖立构件结构，配 75mm 款屋面隔音层
6. Acon 空心墙系铁
7. 满填充空心墙隔热层
8. 215mm 钢筋混凝土墙
9. 未处理绿柄桑木板，三层 Bondex 清胶密封与内面接合
10. 连续镀锌钢固定板，支承顶盖
11. 折叠铝顶盖 PPC RAL 9006
12. 屋顶构成：100mm 碎石层
    保护板
    挤塑聚苯乙烯隔热层
    反渗透屋顶防水膜系统
    200mm 钢筋混凝土板结构
13. 连续的镀锌角钢，支承砖块
14. 低辐射双层玻璃
15. 实心绿柄桑木吧台，与外幕墙相连
16. Meinertz Finned Tube 散热器
17. 5mmPPC 室内钢板（氧化铁红色）
18. Schoeck Isokorb 承重连接，提供断热效果位于悬臂式混凝土结构内部
19. 楼面构成：250x54x54 Wienerberger "Terre De Rose" 砖铺面砂水泥浆
    75mm 混凝土砂浆，配地热供暖装置
    75mm 挤塑聚苯乙烯隔热层
    钢筋混凝土板，地面喷砂密封
20. 未处理绿柄桑木下横栏
21. 镀锌钢隔热外过梁横带
22. 镀锌钢板外包层
23. 遮阳系统，固定于机械通风机组上

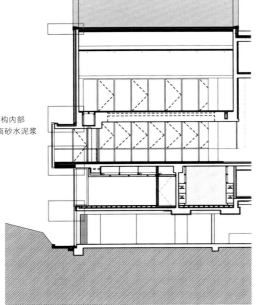

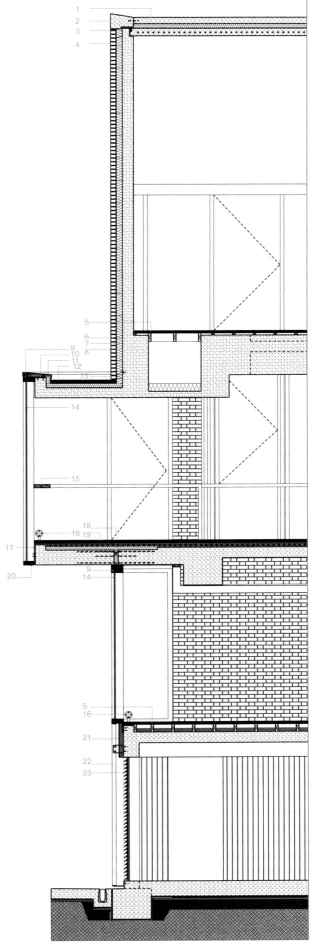

Sport & Cultural Centre NIARA
尼亚拉体育文化中心

Location/ 地点：Valladolid, Spain/ 西班牙，巴利亚多德
Architect/ 建筑师：Amas4arquitectura
Photos/ 摄影：Pablo Guillén Llanos & Fernando Zaparaín
Surface area/ 楼面面积：4,459.60m²
Cost/ 成本：€4,263,675.46/4,263,675.46 欧元

Key materials: Façade – glass, coloured polycarbonate, Palau red clinker brick
主要材料： 立面——玻璃、彩色聚碳酸酯、红色烧结砖

Overview
The Niara Sports Club needed a place of its own for its activities. After finding a suitable site in an easily accessible, outlying residential area with a low population density, the Schola Foundation decided to concentrate a large number of its activities into a single building, thus fulfilling a number of complex requirements that included an educational and recreational area for children at two different stages of development – "small" and "big" children; a separate area for parents' meetings and activities, an area for offices, a residential area, and a block of independent services. All of the above were laid out around the chapel which, together with the sports complex, is the nucleus of the centre.

Primarily occupying the Barcelona Street side of the plot and looking out onto the public park in which it is located, it was decided that the heterogeneity and complexity of the project be formally reflected in the building. Moreover, this was to include a number of open spaces, what could be described as "urban spaces", which the area did not have: namely, a porch connecting the public road and the entrance to the complex, a square which has become the centre's main hall since the building was opened, and a small garden for peace and relaxation.

Detail and Materials
This project is one of a series that explores the liberating massivity of space, which makes what "has not been clone" the protagonist. This "blank space" in the form of porches, extractions or vacuums articulates the entire building. Light reaches the interior through indirect, distinctive openings. The structure is comprised of large superficial elements that take the form of thick enclosures where storage elements, installations, climate conditioning and light openings have been installed. Thanks to this supporting mass storage system, large expenses of façade walls have been freed up to administer shades and transparencies of a unique artistic value. Red re-

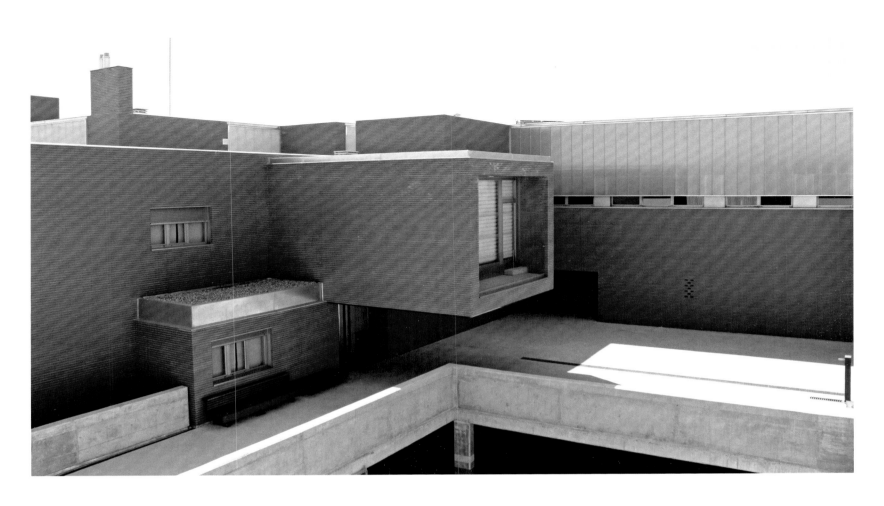

inforced masonry was used to create a continuous plane into which noteworthy openings were inserted. A single ceramic piece was used to enable savings in materials and to address and promote passive energy savings, which was of special importance.

项目概况

尼亚拉体育俱乐部需要一个场所来开展自己的活动。俱乐部选址在一处交通便利的场地，位于住宅区外围。由于圣经学院基金会决定将大量活动集中在一座单一建筑之中，建筑将面临一系列复杂的要求，其中包括大小两个年龄段的儿童教育与娱乐区、家长聚会活动区、办公区、居住区以及独立服务区。以上所有设施都环绕着礼拜堂展开，礼拜堂与体育中心共同组成了项目的核心。

项目主要占据了场地临巴塞罗那街的一侧，远眺公园景观。建筑师决定在建筑外观上反映出项目内部空间的层次感和复杂性。此外，项目还将融入一系列开放空间，填补所在城市区域的空白，其中包括：连接公共道路与建筑入口的门廊、作为主大厅的中央广场以及一个用于休闲放松的小花园。

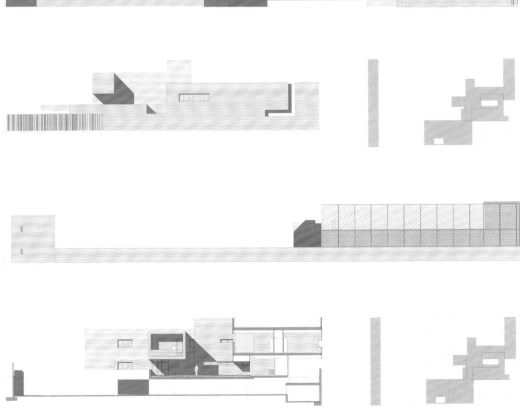

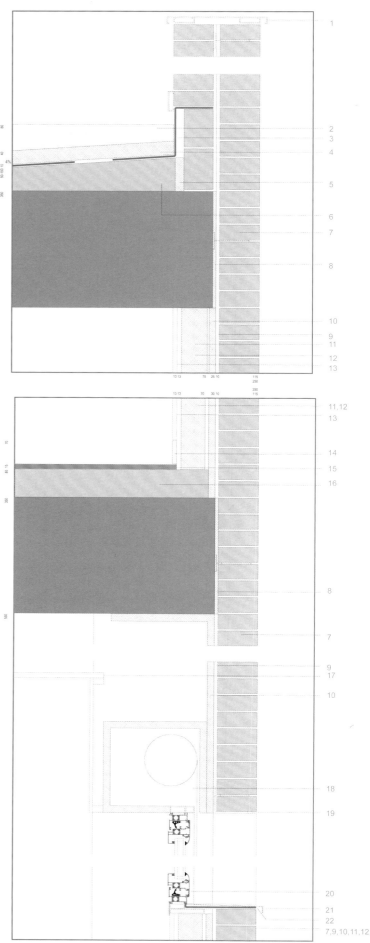

Façade Detail
1. Galvanised steel wall capping fixed with pression strap
2. Gravel bedding
3. Waterproofing system formed by: polypropylene geotextile, double bituminous sheet and polypropylene geotextile
4. Thermal-Isolation: 40mm High Density Extruded Polystyrene board (XPS)
5. Thermal-Isolation: 25mm High Density Expanded Polystyrene board (EPS)
6. 4% slope finished by regulating mortar layer
7. Selfstanding brickwork wall reinforced with Structura G.H.A.S. System every 10 course, secured to slabs and pillars every 50cm. Palau red clinker brick
8. Thermal-Isolation: 15mm cellular glass placed as lost-workform in slab structure
9. Water-repellent mortar
10. Projected Polyurethane in air cavity
11. 60mm Thermal-Isolation: Mineral wool with vapour barrier
12. Selfstanding Gypsum board structure every 40cm
13. Thermal isolated gypsum board coating
14. Laminated skirting board
15. 8mm laminated flooring over acoustic-isolation sheet
16. Regulating mortar layer
17. Suspended ceiling: Gypsum Board 15mm
18. Sun protections: Exterior aluminium blind
19. Steel profile lintel suspended from slab with steel straps
20. Hinged aluminium window with polyamide thermal break: KL-57
21. Aluminium windowsill fixed with polyurethane foam
22. Waterproofing system formed by bituminous sheet

立面节点
1. 镀锌钢墙顶盖，压力带固定
2. 碎石层
3. 防水系统：聚丙烯土工布、双层沥青膜、聚丙烯土工布
4. 隔热层：40mm 高密度挤塑聚苯乙烯板（XPS）
5. 隔热层：25mm 高密度发泡聚苯乙烯板（EPS）
6. 4% 斜坡，找平砂浆层饰面
7. 自承重砖墙，Structura G.H.A.S. 系统加固，每个 50cm 安装板材和立柱保护；Palau 红色烧结砖
8. 隔热层：15mm 泡沫玻璃，安装在墙板结构上
9. 防水灰浆
10. 凸出聚氨酯，位于空气腔内
11. 60mm 隔热层：矿物棉 + 隔热层
12. 自承重石膏板结构，间隔 40cm
13. 隔热石膏板包层
14. 夹层踢脚板
15. 8mm 复合地板，下设隔音板
16. 找平砂浆层
17. 吊顶：15mm 石膏板
18. 遮阳：外部铝百叶
19. 钢型材过梁，通过钢带悬挂于墙板上
20. 铰链式铝窗，带有聚酰胺断热装置：KL-57
21. 铝窗台，通过聚氨酯泡沫固定
22. 防水系统，沥青层构成

细部与材料

项目力求解放空间的整体感，将不同的功能空间分散开。各种门廊形式的"空白空间"将整个建筑连接起来。光线通过富有特色的间接门窗开口进入室内。建筑结构由大型表面组件构成，厚厚的外墙内设置了存储元件、安装设施、空调系统以及照明接口。这个支承存储系统让大面积的立面墙面得以独立，形成了富有独特的审美价值的光影效果。红色配筋砌筑体的应用形成了连续的平面，建筑师在上面添加了引人注目的开口。单体陶瓷块的使用实现了材料的节约，并且提升了建筑的被动节能性能。

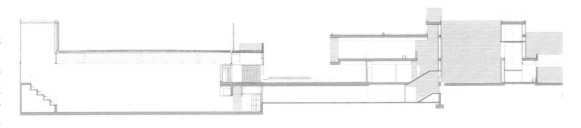

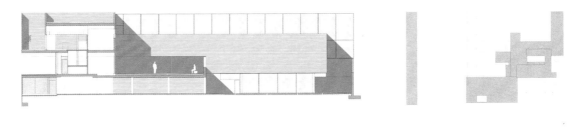

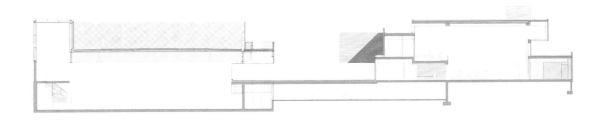

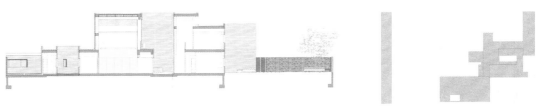

Primary School ZV Zavelput
ZV小学

Location/ 地点：Brussel, Belgium/ 比利时，布鲁塞尔
Architect/ 建筑师：dmvA
Photos/ 摄影：Serge Brison
Built area/ 建筑面积：1,350m²
Completion date/ 竣工时间：2011

Key materials: Façade – brick
主要材料：立面——砖

Overview
Urban Concept
The building is interpreted as a fusion of two typologies. Along the street side it is, because of its three levels, a continuation as well as a closing of the existing ribbon development. At the backside the elongated building consists of two levels and is partly sunk into the terrain. This way it refers to the existing pavilions on the plot.

Together with the blurring of boundaries between the two building-typologies, the line between city and nature vanishes. Between the new school building and the existing pavilion, an internal street penetrates the school area. A big entrance gate closes this street, meanwhile the main building maintains accessible after school hours.

Architectural Concept
Duality between restrictive urbanisation and frivolity of the spread pavilions is one of discipline and chaos, hard and soft, knowledge and creativity, strain and relaxation, studying and playing. Teaching children to coop with this duality is an important training instruction. This duality is reflected in the design and materiality of the building.

Detail and Materials
Skin of Patchwork in Red Masonry
Referring to the brick façades of the neighbouring apartment blocks dmvA opted for red masonry as the shell of the new school.

The masonry is built up with different red bricks applied in different bonds. This tactile patchwork has a downscaling effect on the edifice. The new skin could be a piece of art made by and for children. It invites children to touch it.

Primary school ZV is a Flemish school in a French speaking neighbourhood. The new skin also symbolises the multicultural character of the quarter.

项目概况

城市概念

建筑被视为两种类型的结合体。在临街一侧,因为有三层楼高,它是已有带状开发的延续和终点。后面的细长建筑由两层楼构成,并且还部分沉入地势之中。这让建筑与场地上原有的建筑结构联系起来。

在两种建筑类型的融合之下,城市与自然的界限变得模糊乃至消失。在新的教学楼和已有的建筑之间是一个贯穿学校的内部街道。入口大门将这条街道封闭起来,而主楼在放学之后仍可对外开放。

建筑概念

严格的城市规划与分散的原有建筑形成了鲜明对比:一个标准,一个混乱;一个坚硬,一个柔软;一个以知识为主,一个以创意为主;一个紧张,一个放松;一个致力于学习,一个致力于游戏。让学生正确处理这种双重性是一项重要的教学任务。这种双重性反映在建筑的设计和材料中。

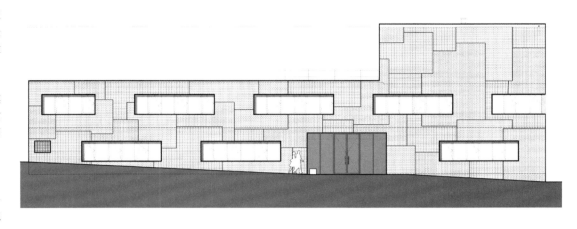

细部与材料

红色砖砌表皮

为了呼应附近公寓楼的砖砌墙面,建筑师选择以红砖作为新教学楼的外壳。

墙面采用不同的红砖以不同的砌筑方式构建而成。这种纹理拼接有一种在视觉上削减建筑规模的效果。新的表皮就像是专为学生制作的艺术品,让孩子们忍不住去触摸。

ZV 小学是一所位于法语社区中的佛兰德语(比利时)学校。全新的建筑表皮同时也象征着这个区域的多文化特征。

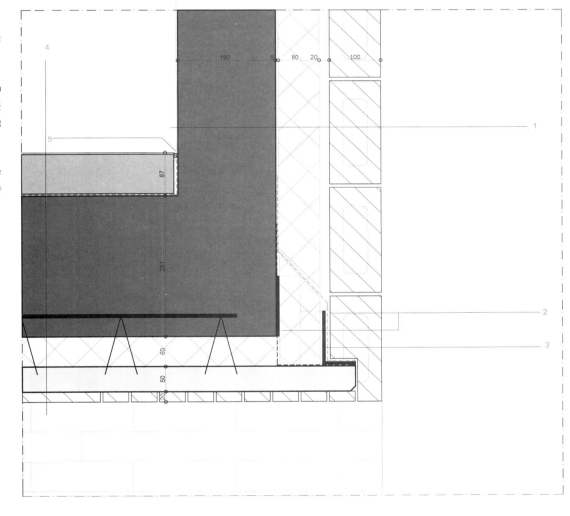

Façade Detail
1. Façade brick
   Air cavity
   Insulation
   Concrete beam
   Plaster
2. EPDM foil
3. L-shaped beam
4. PU pouring floor
   Screed
   PE foil
   Concrete slab
   Insulation
   Precast concrete panel
   Thin façade brick panels glued on concrete
5. Flexible sealant + insulation

立面节点
1. 墙面砖
   空气腔
   隔热层
   混凝土梁
   石膏
2. 三元乙丙橡胶膜
3. L 形梁
4. 聚氨酯浇注地面
   砂浆层
   聚乙烯膜
   混凝土板
   隔热层
   预制混凝土板
   薄墙面砖板,胶合于混凝土上
5. 软密封剂 + 隔热层

Section | **剖面**
1. Entrance | 1. 入口
2. Toilet | 2. 卫生间
3. Waiting room | 3. 等候室
4. Administration offices | 4. 管理办公区
5. Polyvalent room | 5. 多功能区
6. Kitchen | 6. 厨房
7. Classroom | 7. 教室
8. Hall | 8. 大厅
9. Storage | 9. 储物区

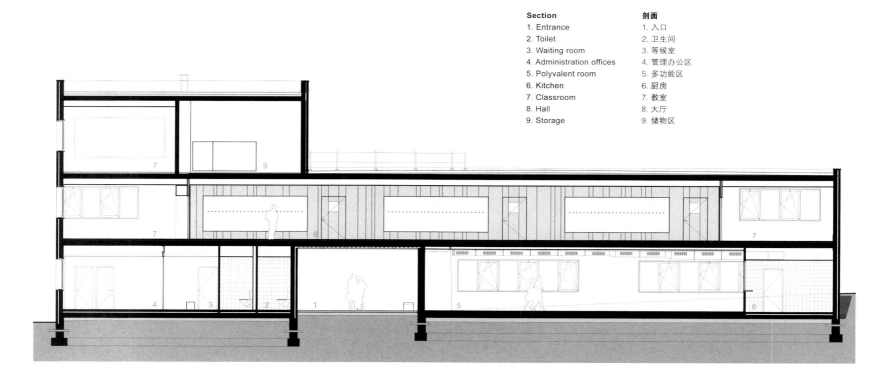

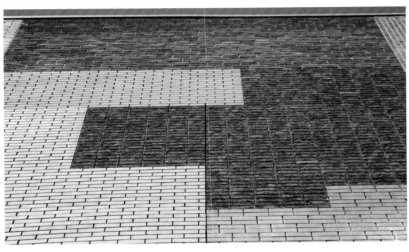

Fort Cortina
科尔蒂纳堡垒——公司办公楼

Location/ 地点：Amsterdam, The Netherlands/ 荷兰，阿姆斯特丹
Architect/ 建筑师：Karelse & den Besten, Rotterdam
Photos/ 摄影：Sjaak Henselmans, Marcel van der Burg; Jan Derwig
Area/ 面积：1,000m²
Building budget/ 建造预算：2,700,000 euro/2,700,000 欧元

Key materials: Façade – clay brick
主要材料：立面——黏土砖

Overview

As a graphic design agency Karelse & den Besten already had a long cooperation with Cortina, a wholesale in gift items, before being asked to design their new headquarters. It was their first architectural assignment.

Fort Cortina is built on the premises of the NDSM-wharf, a former shipyard located on the banks of the river IJ in Amsterdam. In this harsh environment the architects designed an office and warehouse building that looks like a Morrocan fort.

It is a commercial building that almost fits like a home. The patios at the heart of the building are inspired on a monastery tour. It is a space for reflection and a way to give light to the interior and to create outdoor recreation areas with footpaths on the roof, which is planted with sedum for an optimal indoor climate control.

Detail and Materials

Clay bricks infused with metal shavings make up the orthogonal exterior of the building. The building is monolithic structure, rough on the outside with metallised brick walls and smooth on the inside patios are lined with cedar wood. The dynamic lay-out of the façades is a reflection of the different rooms that vary in height and size, a result of the intention to make an exciting interior with small and big views over the river.

The type of brick used is produced by: CRH Clay Solutions in The Netherlands. The type of brick is called: Kardinaal Rood Metallic (BU1150). The bricks are tailor-made metallised on the flat side. The format of the bricks in millimeter is: Length: 210 x Width: 100 x Height: 65. The bricks are built with the flat side to the front. The wrong side, so to say. By doing so one can see the notch making a strong, graphic image, full of contrast and expression. By using a dark mortar and carefully hidden seams, the architects managed to create a monolithic building.

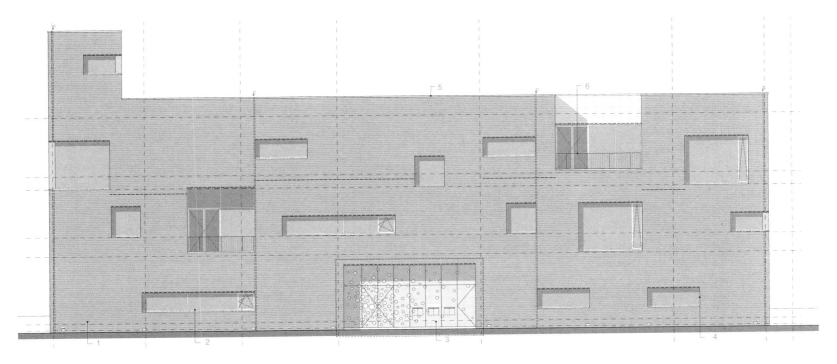

Fragment brickwork (scale 1:10)
Specifications masonry brick
- hand shape with "frog"
- dimensions: 205×98×65mm
- size class: T2
- stretch size: 213mm (butt joint 8mm)
- low size: 106mm (horizontal joint 8mm)

片段砖砌结构（比例 1:10）
砖块规格
– 正面为磨砂面
– 尺寸 205×98×65mm
– 尺寸级：T2
– 延伸尺寸：213mm（对接接缝 8mm）
– 下方尺寸：106mm（水平接缝 8mm）

Fragment fencing (scale 1:20)
Steel slat fencing
- baluster 70×10mm
- top and bottom line 50×20mm
- laminate 40×10mm

片段式围栏（比例 1:20）
钢条围栏
– 栏杆 70x10mm
– 顶部及底部衬线 50x20mm
– 层压板 40x10mm

Front façade
1. Brickwork
2. Aluminum frame
3. Aluminum frame with perforated zinc plate
4. Steel finishing
5. Aluminium roof
6. Steel rail

正面
1. 砖砌结构
2. 铝框
3. 带穿孔锌板的铝框
4. 钢饰面
5. 铝屋檐
6. 钢栏杆

It was the intention to make a building that did not needed much maintenance. Therefore bricks were chosen. They can last a long time and are used for centuries in The Netherlands.

项目概况

作为一家平面设计公司，Karelse & den Besten 与科尔蒂纳公司（专营礼品批发）已经合作了很长时间。此次科尔蒂纳公司委托他们为其建造新的总部办公楼，这是 Karelse & den Besten 的第一个建筑项目。

科尔蒂纳堡垒建在 NDSM 码头，位于阿姆斯特丹 IJ 河岸边。在这个严峻的环境中，建筑师将办公兼仓储楼设计成了类似于摩洛哥堡垒的造型。

这是一座像家一样的商业楼。建筑中央的天井的设计灵感来自于修道院的设计。它是一个沉思的场所，还能为室内提供充足的光线。天井与屋檐下的小径共同组成了户外休闲区。屋顶上种植着景天属植物，起到了优化室内气候的作用。

细部与材料

注入了金属碎屑的黏土砖构成了建筑的外立面。建筑采用单体结构，朝外的金属砖墙显得粗糙奔放，而朝内的天井墙面则镶有雪松木板，光滑细腻。立面的动感布局反映了内部房间多样化的高度和尺寸。各种不同的空间都享有或多或少的河景。

项目所使用的砖块由荷兰 CRH Clay Solutions 公司制造，具体类型名称为 Kardinaal Rood Metallic (BU1150)。砖块特别在平坦的一面注入了金属碎屑。砖块的尺寸为（以毫米为单位）：长 210，宽 100，高 65。砖块的平面朝前，从而营造出强烈的图形效果，充满了冲击感和表现力。深色灰浆和隐式接缝的运用让建筑更富有整体感。

建筑师特别设计了一座无需过多养护的建筑，因此选择了砖作为主要材料。它们可以使用很长时间，并且在荷兰已有几百年的使用历史。

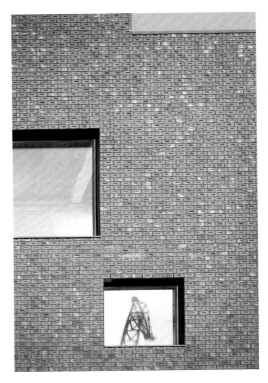

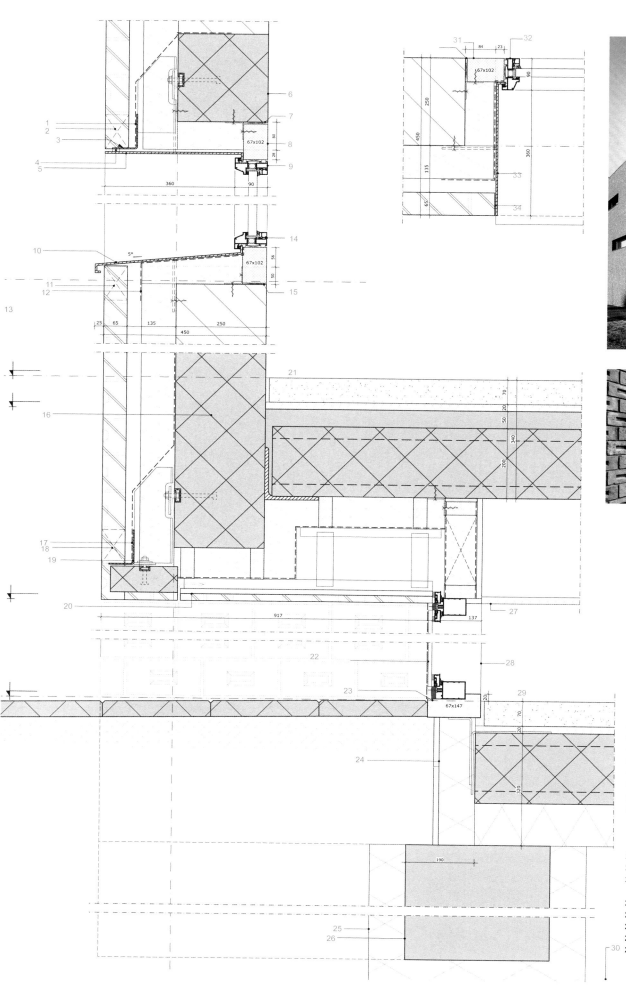

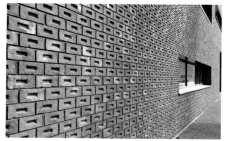

Detail
1. Waterproof membrane
2. Open joints
3. Steel façade carrier
4. Cell bolt
5. Steel framework, with welded strips
6. Precast concrete lintel, with collapsed anchor rail
7. Cell bolt
8. Hardwood mounting frame
9. Aluminum frame, Reynaers CS 38-SL
10. Steel framework, with welded strips
11. Open joints
12. Waterproof membrane
13. Façade construction:
    - brickwork 65mm
    - air space 40mm
    - insulation 95mm
    - limestone 250mm
14. Aluminum frame, Reynaers CS 38-SL
15. Cell bolt
16. Precast concrete lintel with collapsed anchor rail
17. Waterproof membrane
18. Open joints
19. Steel bracket anchors with brick-concrete lintel
20. Masonry strips glued on fermacell HD 15mm
21. Screed 70mm
    Insulation 20mm
    Low pressure 50mm
    Hollow core slab 200mm
22. Perforated zinc plate, laminated glass
23. Stone sill
24. Insulated side shelf
    Frame adjuster bracket
25. PS formwork
26. Foundation
27. Plaster ceiling with sprayer
28. Aluminum curtain wall, Reynaers CW 50-SC
29. Screed 70mm
    Insulation 20mm
    Insulated hollow core slab

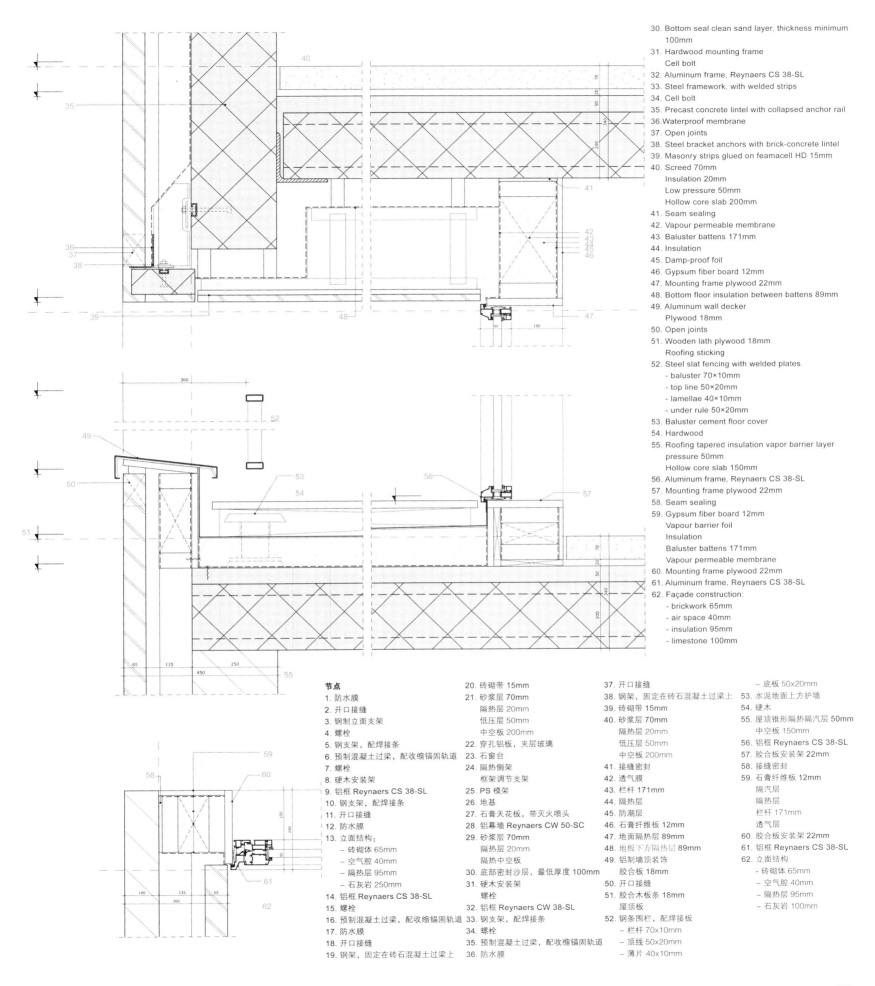

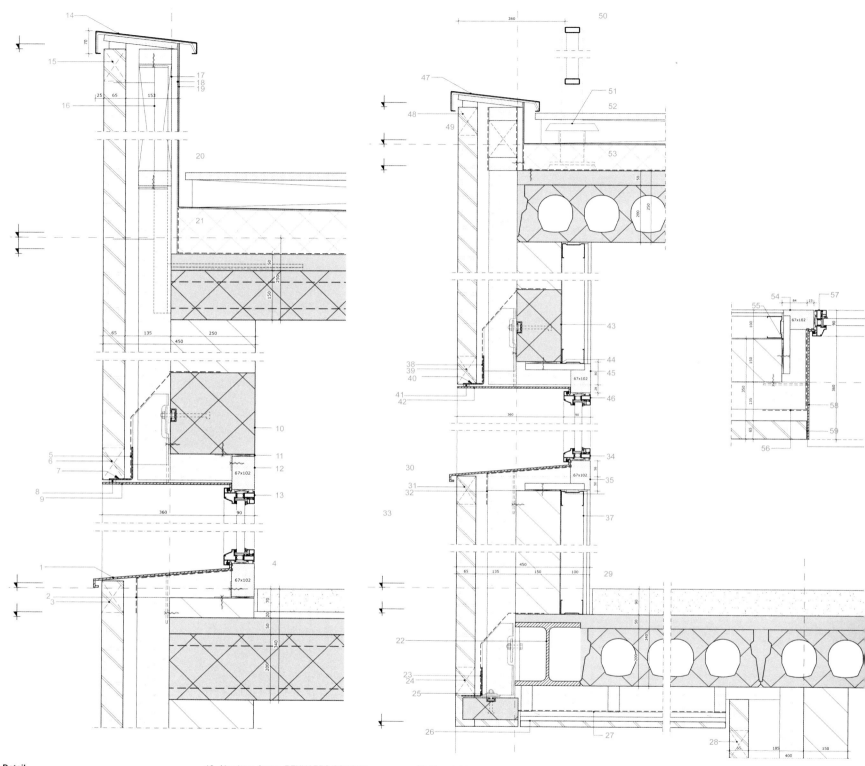

Detail
1. Steel framework with welded strips
2. Waterproof membrane
3. Open joints
4. Screed 70mm
   Insulation 20mm
   Low pressure 50mm
   Hollow core slab 200mm
5. Waterproof membrane
6. Open joints
7. Steel façade carrier
8. Cell bolt
9. Steel framework with welded strips
10. Precast concrete lintel with collapsed anchor rail
11. Cell bolt
12. Hardwood mounting frame
13. Aluminum frame, REYNAERS CS 38-SL
14. Aluminum
    Plywood 18mm
15. Open joints
16. Steel parapet support, with welded wall ties
17. Wooden battens
18. Plywood 18mm
19. Roofing slabs
20. Hardwood parts with anti-skid on hardwood joists
21. Roofing tapered
    Insulation
    Vapour barrier layer
    Low pressure 50mm
    Hollow core slab 150mm
22. Steel girder
23. Waterproof membrane
24. Open joints
25. Steel bracket anchors with brick-concrete lintel
26. Masonry strips glued on fermacell HD 15mm
27. Bottom floor insulation between battens 89mm
28. Open joints
29. Screed 90mm
    Low pressure 50mm
    Hollow core slab 200mm
30. Steel frame with attached loads stripping
31. Open joints
32. Waterproof membrane
33. Technique:
    - brickwork 65mm
    - air space 40mm
    - insulation 95mm
    - limestone 150mm
34. Aluminum frame, REYNAERS CS 38-SL
35. Hardwood plywood 22mm mounting frame
36. Cell bolt
37. Plasterboard 10mm
    OSB plate 12mm on steel frame
38. Waterproof membrane
39. Open joints
40. Steel façade carrier
41. Cell bolt
42. Steel frame with welded strips
43. Precast concrete lintel with collapsed anchor rail
44. Cell bolt
45. Hardwood plywood 22mm mounting frame
46. Aluminum frame, REYNAERS CS 38-SL
47. Aluminum
    Plywood 18mm

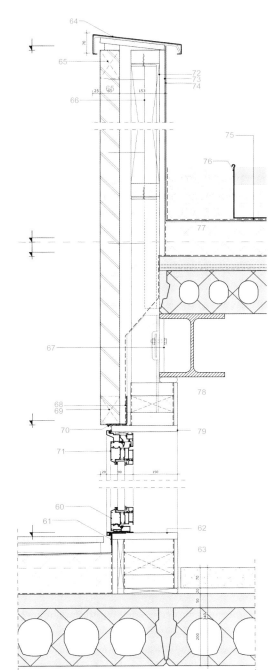

48. Open joints
49. Wooden battens 18mm
    Plywood
    Roofing slabs
50. Steel slat fencing with welded plates:
    - baluster 70x10mm
    - top line 50x20mm
    - lamellae 40x10mm
    - under rule 50x20mm
51. Baluster cement floor cover
52. Hardwood share with slip-on hardwood joists
53. Roofing tapered
    Insulation
    Vapour barrier layer
    Pressure 50mm
    Hollow core slab 200mm
54. Hardwood plywood 22mm mounting frame
55. Cell bolt
56. Waterproof membrane
57. Aluminum frame, REYNAERS CS 38-SL
58. Steel frame with welded strips
59. Cell bolt
60. Aluminum frame, REYNAERS CS 38-SL
61. Lead flashing
62. Mounting frame 22mm plywood
63. Screed 70mm
    Insulation 20mm
    Low pressure 50mm
    Hollow core slab 200mm
64. Aluminum
    Plywood 18mm
65. Open joints
66. Steel parapet support, with welded wall ties
67. Steel beam
68. Waterproof membrane
69. Open joints
70. Steel façade carrier
71. Aluminum frame, REYNAERS CS 38-SL
72. Wooden battens
73. 18mm plywood
74. Roofing sticking
75. Building sedum package:
    - substrate
    - water retention layer
    - with drainage filter layer
    - sliding layer
76. Edge profile
77. Roofing tapered
    Insulation
    Vapour barrier layer
    Pressure 50mm
    Hollow core slab 150mm
78. Vapour permeable membrane
    Baluster battens 171mm
    Insulation
    Vapour barrier foil
    Gypsum fibreboard 12mm
79. Mounting frame 22mm plywood

节点
1. 钢支架，配焊接条
2. 防水膜
3. 开口接缝
4. 砂浆层 70mm
   隔热层 20mm
   低压层 50mm
   中空板 200mm
5. 防水膜
6. 开口接缝
7. 钢制立面支架
8. 螺栓
9. 钢支架，配焊接条
10. 预制混凝土过梁，配收缩锚固轨道
11. 螺栓
12. 硬木安装架
13. 铝框 REYNAERS CS 38-SL
14. 铝材
    胶合板 18mm
15. 开口接缝
16. 钢护栏，配焊接系墙铁
17. 木板条
18. 胶合板 18mm
19. 屋顶板
20. 硬木配件，配防腐硬木托梁
21. 锥形屋顶
    隔热层
    隔汽层
    低压层 50mm
    中空板 150mm
22. 钢梁
23. 防水膜
24. 开口接缝
25. 钢架，固定在砖石混凝土过梁上
26. 砖砌带 15mm
27. 底层地面隔热层 89mm
28. 开口接缝
29. 砂浆层 90mm
    低压层 50mm
    中空板 200mm
30. 钢架，配承重条
31. 开口接缝
32. 防水膜
33. 技术：
    – 砖砌体 65mm
    – 空气腔 40mm
    – 隔热层 95mm
    – 石灰岩 150mm
34. 铝框 REYNAERS CS 38-SL
35. 硬木胶合板安装架 22mm
36. 螺栓
37. 石膏板 10mm
    定向刨花板 12mm，安装在钢架上
38. 防水膜
39. 开口接缝
40. 钢立面支架
41. 螺栓
42. 钢架，配焊接条
43. 预制混凝土过梁，配收缩锚固轨道
44. 螺栓
45. 硬木胶合板安装架 22mm
46. 铝框 REYNAERS CS 38-SL
47. 铝材
    胶合板 18mm
48. 开口接缝
49. 木板条 18mm
    胶合板
    屋顶板
50. 钢条栏杆，配焊接板：
    – 栏杆 70x10mm
    – 顶线 50x20mm
    – 薄片 40x10mm
    – 底板 50x20mm
51. 水泥地面上方护墙
52. 硬木配件，配硬木托梁
53. 锥形屋顶
    隔热层
    隔汽层
    压力层 50mm
    中空板 200mm
54. 硬木胶合板安装架 22mm
55. 螺栓
56. 防水膜
57. 铝框 REYNAERS CS 38-SL
58. 钢架，配焊接条
59. 螺栓
60. 铝框 REYNAERS CS 38-SL
61. 铅防水板
62. 胶合板安装架 22mm
63. 砂浆层 70mm
    隔热层 20mm
    低压层 50mm
    中空板 200mm
64. 铝材
    胶合板 18mm
65. 开口接缝
66. 钢护栏，配焊接系墙铁
67. 钢梁
68. 防水膜
69. 开口接缝
70. 钢制立面支架
71. 铝框 REYNAERS CS 38-SL
72. 木板条
73. 胶合板 18mm
74. 屋顶板
75. 景天属植物层：
    – 基质
    – 保水层
    – 排水过滤层
    – 滑动层
76. 边缘型材
77. 锥形屋顶
    隔热层
    隔汽层
    压力层 50mm
    中空板 150mm
78. 透气膜
    栏杆 170mm
    隔热层
    隔汽层
    石膏纤维板 12mm
79. 胶合板安装架 22mm

Piazza Céramique
陶瓷广场

Location/ 地点 : Maastricht, The Netherlands/
荷兰，马斯特里赫特
Architect/ 建筑师 : Jo Janssen Architecten -
Prof. ir. Wim van den Bergh architect
Photos/ 摄影 : Atelier Kim ZWarts

Key materials: Façade – brick
主要材料：立面——砖

Overview
The programme to be housed within Piazza Céramique was that of an integrated form of dwelling and working, which was partly integrated within the living units and partly separated from them. The basic architectural concept of the project, in terms of structure and infra-structure, allowed a large degree of flexibility while remaining within the limits of the buildings envelop. This flexibility the client exploited to the maximum. During design and construction, the client changed his mind several times about the numbers of apartments, work units, home offices and commercial spaces, which resulted in the present project with its 92 apartments, 27 with home offices, a commercial space and 7 individual work units.

Detail and Materials
The choice of construction was an uncommon one for a housing project. Normally the floors span from apartment to apartment, however, to keep the subdivision of the integrated apartments as flexible as possible, the floors in this project span from the outer façade to an inner ring of service spaces, which concentrates all infra-structure. As earlier explained this allowed a large degree of flexibility.

The Céramique quarter in Maastricht was named after the Société Céramique established on the site since 1859. A century later they merged with the "Royal Sphinx" industry that produced porcelain and earthenware. As a continuation of the site's tradition, the masterplan demanded the use of red brick on a plinth of traditional blue stone. The architects further opted for a low tech building with natural ventilation and traditional folding shutters to filter the sunlight through their perforations and provide privacy for the inhabitants.

In terms of materialisation, a deliberate attempt has been made to harmonise with the surrounding buildings so that individualisation within the architectural collective can express itself, exactly in the concept of "turning inside out". In the extension of this concept, an integral solution to the problem of the outdoor space was sought, consisting of the usual individual interpretation of balconies and terraces with sun blinds and privacy protection measures. This was found, on the one hand, in the application of private outdoor spaces behind the façade and, on the other, in the folding shutters, by means of which every individual can filter his privacy and sunlight to various degrees in each room. Despite the strict rhythmic structure of the façade, this possibility ensures a dynamic expressive picture of the façade, as a consequence of the fact that these folding shutters are manually operated.

项目概况

陶瓷广场将集住宅与办公于一身，其中一部分办公空间与住宅单元相结合，另一部分与其独立。从结构和基础设施来讲，项目的基本建筑理念是在有限的建筑外壳设计中实现最大限度的空间灵活性，以便委托人可以最大限度的使用空间。在设计和施工阶段，委托人几次修改公寓、办公、家庭办公和商业空间的数量配置。最终，项目容纳了92套公寓、27套家庭办公室、一个商业空间和7个独立办公单元。

细部与材料

对于住宅项目来说，本项目所选用的施工方式是极为罕见的。通常来说，楼面跨度一般是从公寓到公寓；然而，为保证综合公寓细分的灵活性，本项目的楼面从外立面跨越到内环服务空间（集中了所有基础设施）。这种设计保证了项目的高度灵活性。

陶瓷广场的命名来自于1859年成立的陶瓷公司。百年之后，公司被并入专门生产陶瓷和陶器的"皇家斯芬克斯"产业。作为场地传统的延续，项目的总体规划要求在建筑底座使用红砖。建筑师进一步选择建造一座低技术建筑，配有自然通风和传统的折叠式百叶窗，后者既能过滤阳光又能保证居民的隐私。

在材料方面，建筑师特意让建筑先与周边的建筑和谐统一，然后再表现出建筑自身的个性，完美地体现"内外翻转"的概念。为了进一步实践这一概念，建筑师为室外空间实施了整合措施：阳台和露台都安装了遮阳装置和隐私保护措施。一方面，私人室外空间被设在了立面后面；另一方面，折叠式百叶窗让个人可以随心调节私密度和日照度。虽然建筑立面呈现出规则的韵律感，但是折叠式百叶窗的开合为其增添了活力，形成了富有表现力的多样化外观。

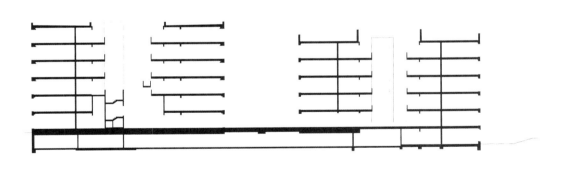

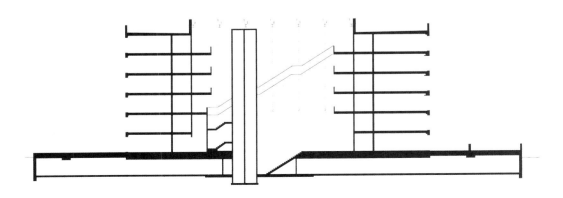

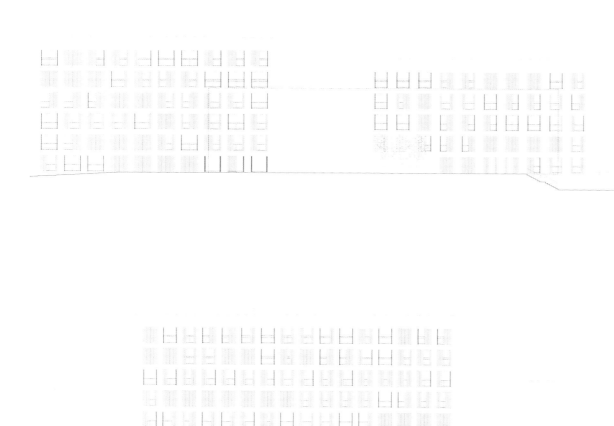

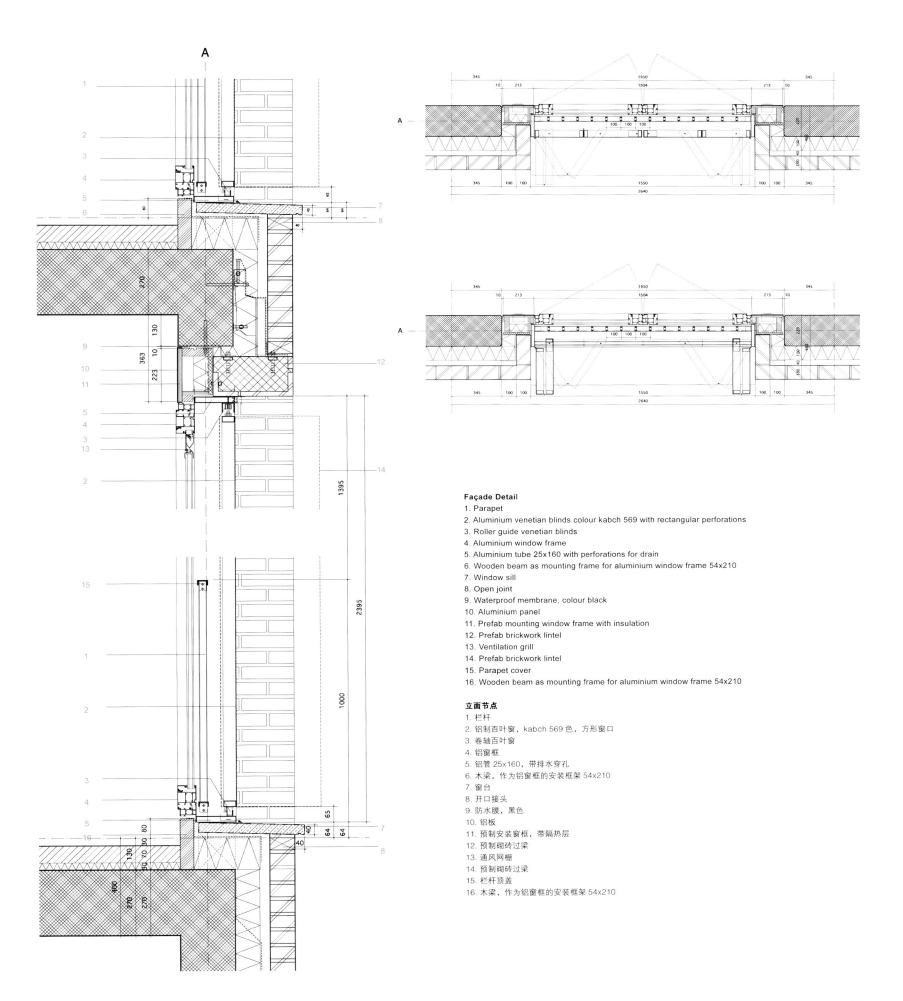

The Curving House
曲线住宅

Location/ 地点：Gyeonggi-do, Korea/ 韩国，京畿道
Architect/ 建筑师：JOHO Architecture
Photos/ 摄影：Sun Namgoong
Built area/ 建筑面积：140.57m²
Completion date/ 竣工时间：2012

Key materials: Façade – brick
主要材料：立面——砖

Overview

It was a rare residential lot with an open view to the south at the dead end of a small path beneath Mt. Gwanggyo. What was unique about this lot was that it was very hard to turn the car to come out of the path after more than 2 cars parked, because it was a small path only 4 m in width. Ironically, the fundamental challenge was not only solving the parking problem but also creating a space for both parking and gardening to coexist. It was closely related to the lifestyle of home owner to decide whether to create a garden directly accessible from the living room or to emphasise a visual garden. To resolve this issue, the overall shape was formed to encase the lot with more curves and lifted about 2 m from the ground using pilotis for more efficient parking. The shape of the mass resembling a concave lens was created by the parking needs and the topographical condition of the lot.

The mountains penetrate the sky and the sky contains the mountains as nature. Here, the mountains form lines and the lines remember the mountains in the land. The terrains of Mt. Gwanggyo flow low above the lot and the lot displays the entire view as if it responds to the graceful flow. At this site, the land is the proof of space and everything about the substance. The shape created here contains the sky as an earthenware jar and displays the potentiality of land as a spatial substance. It draws a shape, but creates a space that shows the sky outside the shape to hide itself in nature. Should the line be hidden in nature or should the nature be displayed in the hidden line? This was the essential challenge of this land and the sincere response to the background. This is directly related to how the topographies should be interpreted in Korean traditional spaces. Korean traditional spaces have pursued the shape that is not completely hidden in nature yet beautifully harmonised with surrounding nature. It is based on the post-dualistic beauty of harmony that proves its existence while hiding in nature rather than dominating

Contrast and Harmony of Texture

The rough texture of the traditional bricks interprets the lot in a different way in combination with the property of highly reflective stainless steel. The skies and nature reflected on the stainless steel surface distort what the true substance is to break the boundaries between shapes and texture. Unlike the rough texture of ceramic bricks, the stainless steel used on the front and on the side reflects the surrounding landscapes to make itself disappear. If the bricks reveal themselves by the change of light and shadow, the stainless steel de-materializes itself by making itself disappear in nature. Such contrasting textures have different properties and confront each other in a single mass, but they ultimately establish balance through the extinction and reflection of light.

纹理的对比与统一

传统砖块的粗糙表面与高光反射的不锈钢从另一个角度诠释了项目场地。天空大自然扭曲地倒映在不锈钢表面上，打破了造型与纹理的界限。与瓷砖的粗糙纹理不同，建筑正面和侧面所使用的不锈钢能倒映出周边的景观，从而将自身隐藏起来。如果说砖块通过光影变化中显露出来，那么不锈钢就能虚化自己，消失在自然中。这种对比纹理拥有不同的属性，它们在同一座建筑中相互对立，又通过吸光和反光实现了最终的平衡。

nature with its shape and lines.

Detail and Materials

The ash-coloured bricks (traditional bricks) embrace the concrete surface as fish scale while slightly altering the angles. The traditional bricks used for this project have silver water-repellent coating on the surface and show sentimentality different from the rough surfaces of their tops and bottoms. The bricks with two different surfaces were piled to form a certain pattern from angles 1° through 25°. In other words, the variation of angle is another way how the outer skin in the shape of a concave lens facing south defines its existence. The shadow of the brick wall caste as the sun moves converts the flow of lines into the subtle change of the outer skin. The variation of the brick surface is intended to read the entire mass differently according to the perspective of incomer and the perspective of viewing the images from the mountains.

项目概况

项目场地十分稀有，南面朝向一条通往光桥山脚的小路的尽头。该场地最大问题是：当停车数超过两辆时，很难从小路向外掉头，因为这条小路只有4米宽。然而，项目最根本的挑战不是解决停车问题，而是打造一个集停车位与花园于一身的空间。房屋的业主希望打造一个能与客厅直接相连的花园或是突出设计一个视觉花园。为了解决这一问题，建筑的整体造型通过更多的曲线将场地包围起来，并且高出地面两米，用底层架空柱来实现高效的停车。建筑的造型参考了凹透镜的形式，由停车需求和场地的地形条件共同决定。

山峰融入天空，天空包容山峰。山峰的轮廓线让人将其与地面联系起来。光桥山低低地略过场地，呈现出优雅而流畅的曲线。在场地上，土地是空间存在的证据，也是一切存在的基础。建筑的造型向陶器一样将天空容纳起来，同时也展示了土地作为空间实质的潜力。建筑的造型将天空展示在外，把自身隐藏在自然之中。究竟是应该把线条隐藏在自然中还是将自然展示在隐藏的线条中呢？这是这片土地所面临的主要挑战，也让建筑与环境真实地联系起来。这与韩国传统空间构造中的地形学有着直接的联系。韩国传统空间追求将建筑造型半隐在自然中，从而形成和谐美观的统一环境。它以和谐之美的二元性为基础，显示了自身的存在感。它隐藏在自然之中，而不是用自身的造型和线条来统治自然。

细部与材料

浅灰色传统砖块像鱼鳞一样将混凝土表面包裹起来，砖块的角度呈现出微妙的变化。项目所使用的传统砖块表面带有银色防水涂层，与建筑顶部和底部的粗糙表面形成了质感上的对比。两种不同表面的砖块堆积起来形成了特殊的纹理，角度在1°到25°之间。换言之，角度的变化从另一方面突出了南向凹透镜造型表皮的存在感。砖墙的阴影随着日照变化而变化，将线条转化成外表皮上的微妙变化。砖面的变化让整座建筑从不同的角度（无论是逐渐走进建筑，还是从山顶欣赏建筑）看起来变化多端。

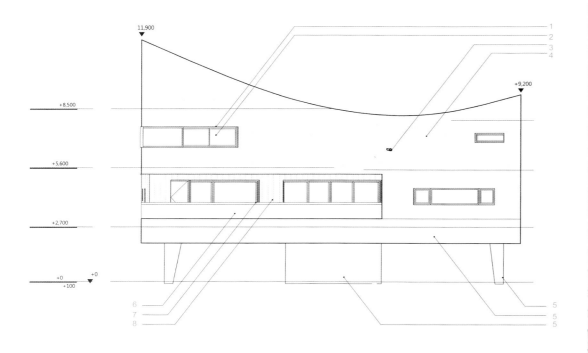

Front Elevation
1. Aluminium flashing
2. Thk 24 tinted thermopane
3. Gutter
4. Brick finish
5. Exposed concrete finish_specified pattern
6. Thk 10 tempered glass (round)
7. Aluminium cap finish
8. Thk 1.5_STS panel mirror type

正面
1. 铝槽
2. 24 厚有色隔热玻璃
3. 排水沟
4. 砖饰面
5. 裸露混凝土装饰（特殊图案）
6. 10 厚钢化玻璃（圆角）
7. 铝盖板
8. 1.5 厚 STS 板（镜面型）

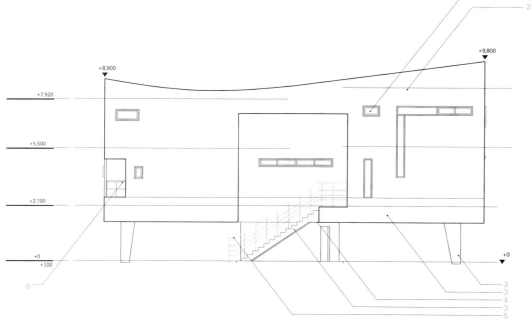

Back Elevation
1. Thk 24 tinted thermopane
2. Brick
3. Exposed concrete finish_specified pattern
4. Steel plate Thk 10 + powder coating
5. Thk 10 steel plate handrail
6. Steel plate handrail

背面
1. 24 厚有色隔热玻璃
2. 砖
3. 裸露混凝土装饰（特殊图案）
4. 10 厚钢板（带有喷粉涂层）
5. 10 厚钢扶手
6. 钢扶手

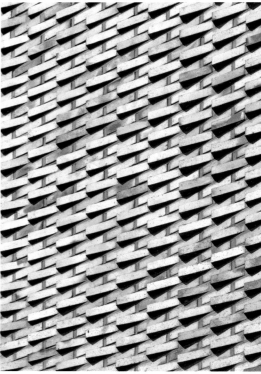
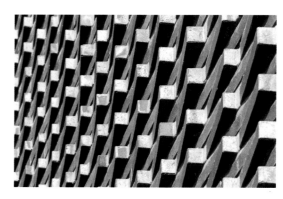

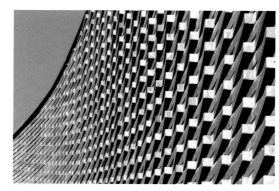

Brick Façade Variation
砖立面的变化

Type A. plain pattern / typical brick stacking
A 型：普通样式 / 标准堆砌法

Type B. striped pattern / typical brick stacking
B 型：条纹样式 / 标准堆砌法

Type C. plain and striped pattern mixed / typical brick stacking
C 型：普通 + 条纹样式混合 / 标准堆砌法

Type D. plain pattern / one row turned
D 型：普通样式 / 单排翻转

Type E. striped pattern / one row turned
E 型：条纹样式 / 单排翻转

Type F. plain and striped pattern mixed / plain row turned
F 型：普通 + 条纹样式混合 / 普通样式单排翻转

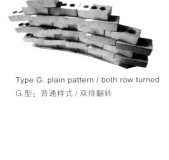
Type G. plain pattern / both row turned
G 型：普通样式 / 双排翻转

Type H. striped pattern / both row turned
H 型：条纹样式 / 双排翻转

Type I. plain and striped pattern mixed / striped row turned
I 型：普通 + 条纹样式混合 / 条纹样式单排翻转

Type A. one row tilted with same angle
A 型：单排呈相同角度倾斜

Type B. one row turned according to the end of the lower one with various angle
B 型：单排根据下层端头呈不同的角度翻转

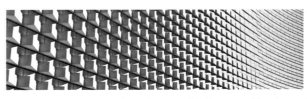
Type C. one row turned according to the centre of the lower one with various angle
C 型：单排根据下层中央呈不同的角度翻转

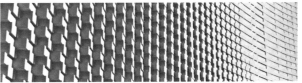
Type D. both rows turned with various angle changes
D 型：两排呈不同角度翻转

Brick Main Façade
砖砌主立面

Guidelines arranged for the angle variation
角度变化的参考线

Bricks following the guideline from the rooftop
砖块从屋顶沿着参考线向下

Each brick have a different angle value to check
每块砖都有一个不同的角度值用于检查

Bricks wrap the corner of the façade and continue to be turned
砖块包裹住里面的转角，然后继续翻转

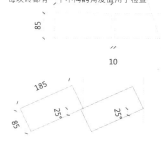

Odd-row horizontally arranged
单数排水平放置

Even-row turned with angle variation
双数排呈变化角度翻转

Typical brick turning plan (top view)
- Based on the centre of lower brick, gradually increase the turning angle value starting from 0°
标准砖块翻转方式（顶部视角）
– 以下排砖中心为基础逐渐增大角度（从 0° 开始）进行翻转

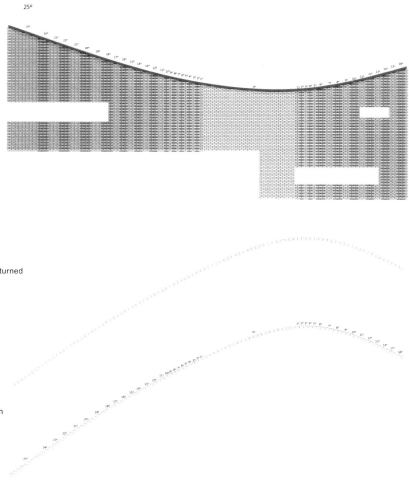

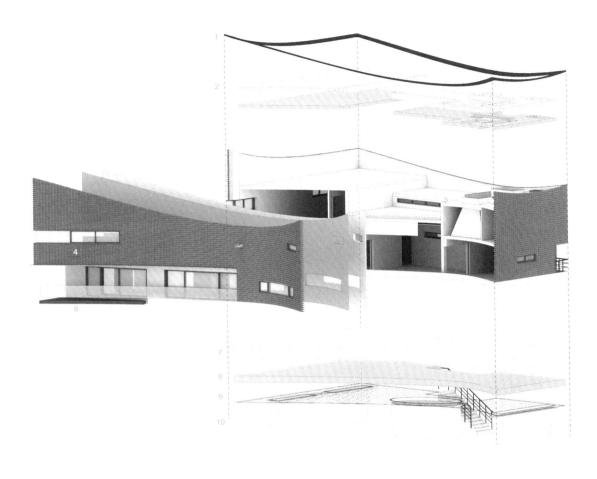

Detail
1. Al.cap (Al.cap on top of the brick façade to highlight the curve)
2. Roof (concrete + insulation + concrete + waterproof paint)
3. Rooftop window (different sizes of rooftop windows bring warmth and light)
4. Ash-coloured brick (brick reflects the light, carefully arranged scales shine)
5. STS panel (mirror type) (The STS holds the sky and reflects the four season changes)
6. Outer deck (the balcony can work as the extension of the living room)
7. Patterned exposed concrete (pine board pattern gives a warm feeling to cold grey concrete)
8. Slab insulation (super insulation at the bottom slab)
9. Mesh (mesh applied on top of the parking place)
10. Base structure (base structure to create pilotis and parking space)

节点
1. 铝顶盖（位于砖砌立面的最上方，突出曲线）
2. 屋顶（混凝土＋隔热层＋混凝土＋防水漆）
3. 屋顶窗（不同尺寸的屋顶窗为室内带来温暖和阳光）
4. 浅灰色砖（砖块反射光线，呈鱼鳞状巧妙排列）
5. STS（特殊处理钢）板（镜面型）（钢板倒映出天空，反映四季的变化）
6. 外平台（阳台可作为客厅的延伸）
7. 纹理混凝土（松木板纹理为冰冷的灰色混凝土带来了温暖的感觉）
8. 楼板隔热（底部楼板安装了超级隔热层）
9. 网状铺装（停车场地面顶层覆盖着网状铺装）
10. 底部结构（底部结构打造了底层架空柱和停车空间）

3D axonometric view
三维轴测图

Assembling drawing
装配图

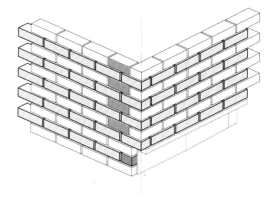

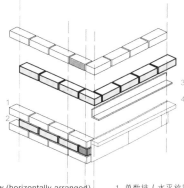

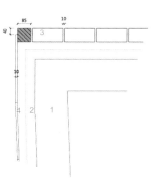

1. Odd-row (horizontally arranged)
2. Even-row (turned w/angle variation)
3. 'L' shaped steel frame
4. STS panel mirror type

1. 单数排（水平放置）
2. 双数排（变化角度翻转）
3. 角型钢框
4. 特殊处理钢板（镜面型）

Even-row top view
1. Concrete wall thk.200
2. Outside insulation thk.50
3. Even-row (turned w/angle variation)
4. STS panel mirror type

双数排俯视图
1. 200厚混凝土墙
2. 50厚外部绝缘结构
3. 双数排（变化角度翻转）
4. 特殊处理钢板（镜面型）

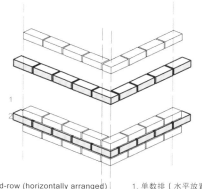

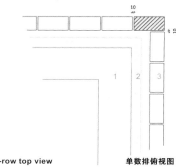

1. Odd-row (horizontally arranged)
2. Even-row (turned w/angle variation)

1. 单数排（水平放置）
2. 双数排（变化角度翻转）

3Odd-row top view
1. Concrete wall thk.200
2. Outside insulation thk.50
3. Odd-row (horizontally arranged)

单数排俯视图
1. 200厚混凝土墙
2. 50厚外部绝缘结构
3. 单数排（水平放置）

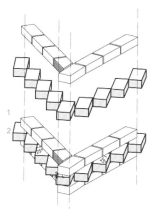

1. Odd-row (horizontally arranged)
2. Even-row (turned w/angle variation)

1. 单数排（水平放置）
2. 双数排（变化角度翻转）

3Odd-row top view
1. Concrete wall thk.200
2. Outside insulation thk.50
3. Odd-row (horizontally arranged)

单数排俯视图
1. 200厚混凝土墙
2. 50厚外部绝缘结构
3. 单数排（水平放置）

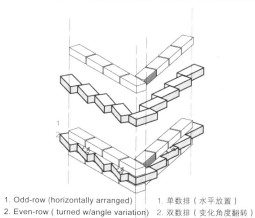

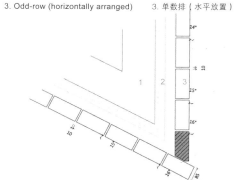

Brick corner detail
砖墙拐角节点

1. Odd-row (horizontally arranged)
2. Even-row (turned w/angle variation)

1. 单数排（水平放置）
2. 双数排（变化角度翻转）

3Odd-row top view
1. Concrete wall thk.200
2. Outside insulation thk.50
3. Odd-row (horizontally arranged)

单数排俯视图
1. 200厚混凝土墙
2. 50厚外部绝缘结构
3. 单数排（水平放置）

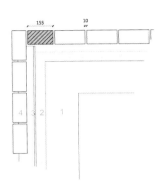

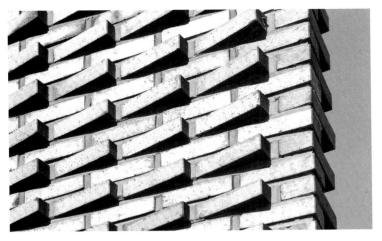

Odd-row top view
1. Concrete wall thk.200
2. Outside insulation thk.50
3. 'L' shaped steel frame
4. Odd-row (horizontally arranged)

单数排俯视图
1. 200 厚混凝土墙
2. 50 厚外部绝缘结构
3. 角型钢框
4. 单数排（水平放置）

Even-row top view
1. Concrete wall thk.200
2. Outside insulation thk.50
3. Even-row (turned w/angle variation)

双数排俯视图
1. 200 厚混凝土墙
2. 50 厚外部绝缘结构
3. 双数排（变化角度翻转）

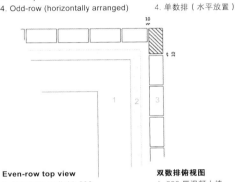

Even-row top view
1. Concrete wall thk.200
2. Outside insulation thk.50
3. Even-row (turned w/angle variation)

双数排俯视图
1. 200 厚混凝土墙
2. 50 厚外部绝缘结构
3. 双数排（变化角度翻转）

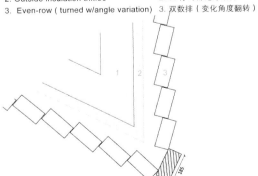

Even-row top view
1. Concrete wall thk.200
2. Outside insulation thk.50
3. Even-row (turned w/angle variation)

双数排俯视图
1. 200 厚混凝土墙
2. 50 厚外部绝缘结构
3. 双数排（变化角度翻转）

Chengdu Skycourts
成都空中庭院

Location/ 地点: Chengdu, China/ 中国，成都
Architect/ 建筑师: Höweler and Yoon Architecture, LLP
Photos/ 摄影: Yihuai Hu
Gross area/ 总面积: 6,225m²

Key materials: Façade – brick
主要材料: 立面——砖

Overview

The Skycourt building in the Intangible Culture Park takes its inspiration from the traditional Chinese courtyard house and the relationship between the architecture and the natural context. The multiple courtyards in the compound frame views of the sky above. The courtyards also frame small gardens that bring the natural world inside the house.

The proposal interpreted the courtyard house type as an inward focused building with an irregular perimeter. The entire footprint of the site was organised around courtyards, with each courtyard having specific characteristics that correspond to a poem by the Tang poet, Li Bai. The poem, "Tianmu Mountain Ascended as in a Dream" was an original inspiration for the sequence of courtyards. A visitor to the building would experience the spaces, each one referring to a line in the poem. For example the visitor enters through the "Hidden Garden" and passes through the "Garden of Perception," glimpses the "Heavenly Terrace" before entering the "Moonlit Shadow Garden." Each courtyard has its own characteristics, making the experience of the building analogous to "inhabiting" the poem.

The emergence of a contemporary Chinese architecture negotiates the desire for a site specific architecture of the "local" with an awareness of the "global." The design of the Skycourt building balances the narratives and construction logics of Chengdu with contemporary building practices to blend Tradition and Modernity in a particularly Chinese expression.

Detail and Materials

The perimeter wall varies in height from 11 to 15 metres, and is faced with a traditional grey brick. The bricks for the perimeter wall are always oriented in the same grain, regardless of the oblique angle of the perimeter walls. The result is of an overall project "grain" or oriented texture. The main entrance is located in a wall

that is smooth, while around the corner the perimeter wall takes on a subtle texture as bricks maintain the grain against the angle of the wall. From the exterior the building will look smooth or serrated, depending on the orientation. Local construction materials and techniques are enlisted to produce nontraditional articulations and effects.

The exterior wall is punctuated by a series of tapered openings that express the 600 mm thickness of the wall. The incised frame and tapered faces are made of a local wood panel that adds a rich accent colour to the windows: a smooth cut to contrast with the varied texture of the brick wall. The irregular pattern of openings is produced by combining a set of window proportions with a set of opening proportions.

项目概况

国际非物质文化遗产博览园中的空中庭院从中国传统庭院住宅以及建筑与自然环境的关系中获得了设计灵感。建筑中的多重庭院不仅享有天空的美景，还通过小花园将自然世界带进了建筑内部。

设计重新诠释了庭院住宅，将其演绎成一座拥有不规则边界的内向型建筑。场地的整个占地都围绕着庭院展开。庭院以唐代诗人的名作《梦游天姥吟留别》为主题串联起来。建筑中的每个空间都代表着一个诗句。游客从"微茫园"进入，穿过"感知园"，一瞥"天台"，然后进入"月影园"。每个庭院都独具特色，让建筑宛如栖身于诗歌之中。

现代中国建筑的出现为"本地"特色建筑注入了"全球化"的感觉。空中庭院的设计以独特的中国式表达调和了成都本地建筑在传统和现代之间的矛盾。

细部与材料

建筑的外墙高度在11至15米之间，采用传统的青砖覆盖。无论外墙的倾斜角度如何，外墙的砖块统一朝向同一个纹理。主入口所在的墙面是光滑的，而在转角处，外墙又呈现出砖块细微的纹理，保持墙面的纹理与墙壁的角度相对。感觉不同的朝向，建筑的外墙时而光滑，时而粗糙。本地建筑材料和建造技术的应用产生了非传统的表达和效果。

外墙上设有一系列锥形开口，显露出墙壁600毫米的厚度。切入的框架和锥面均由本地木板构成，为窗口添加了丰富的色彩。光滑的切口与纹理不一的砖墙也形成了奇妙的对比。窗口以一定的比例组合起来，形成了各种各样的木板开口。

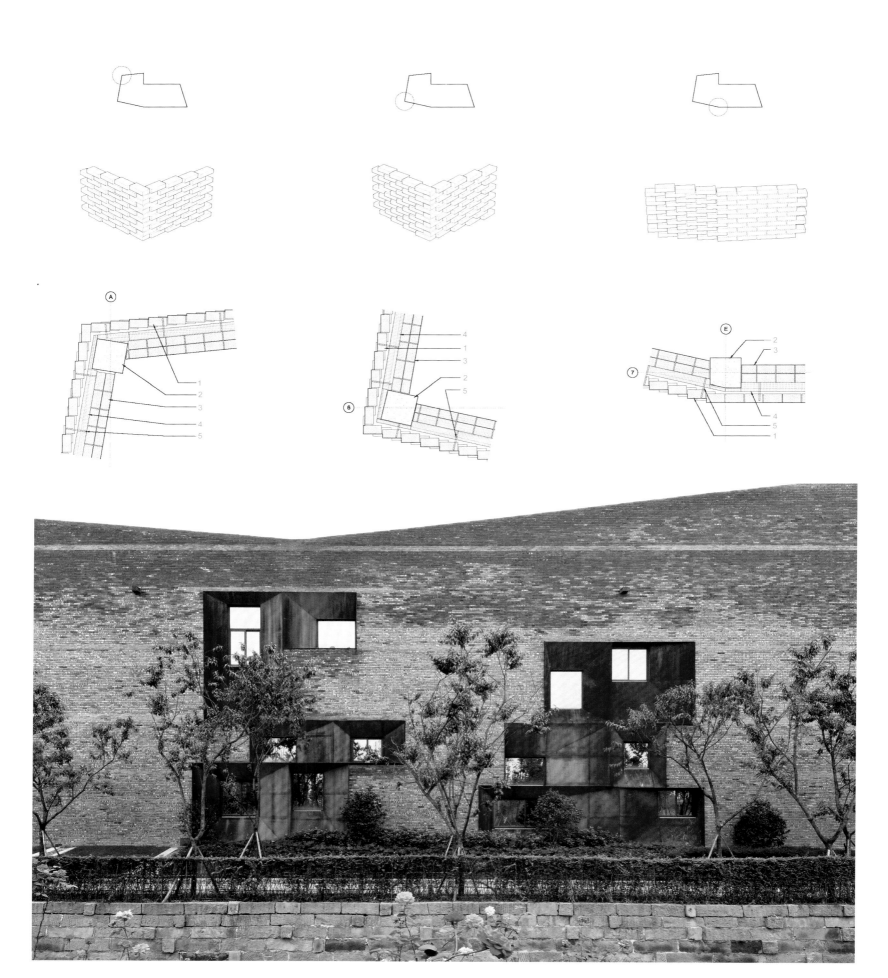

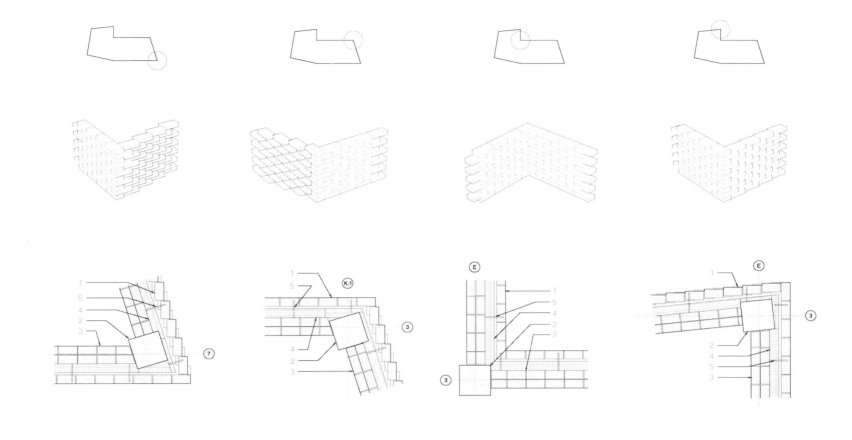

1. Exterior brick　　　　1. 外墙砖
2. Concrete column　　　2. 混凝土柱
3. Double wythe brick interior　3. 双层内墙转
4. Rigid insulation　　　4. 刚性隔热层
5. Relieving angle　　　5. 外露角材

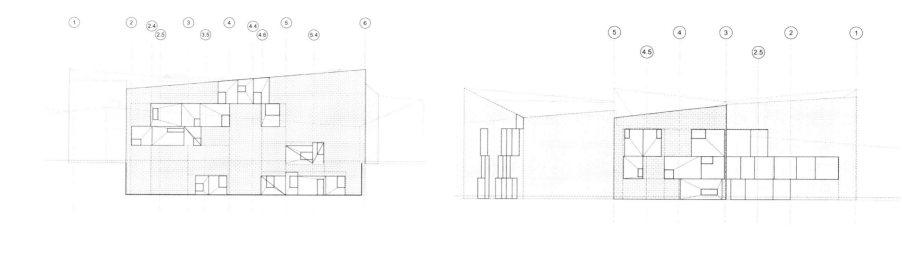
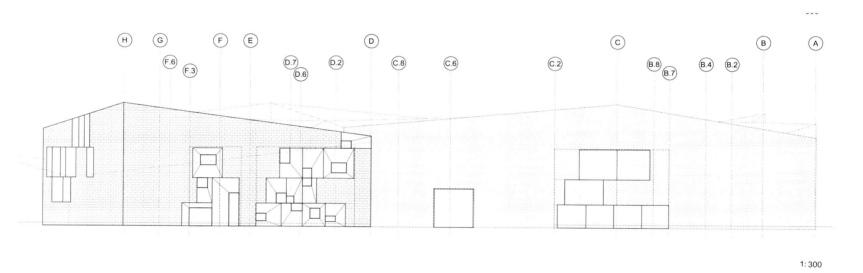
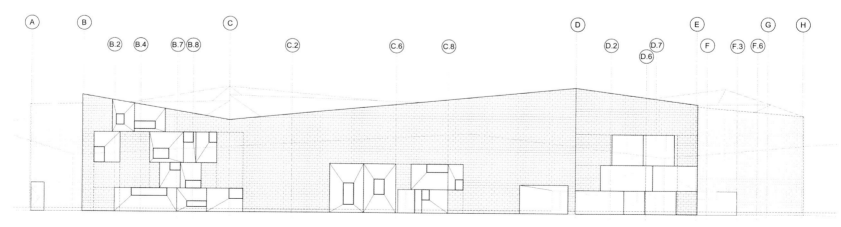

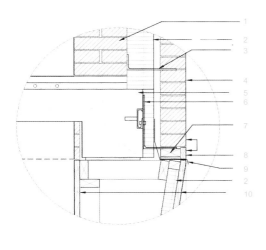

1:10

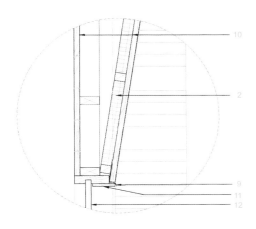

1:10

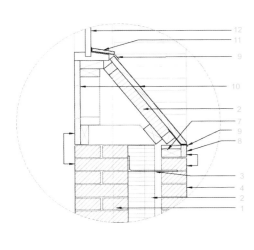

1:10

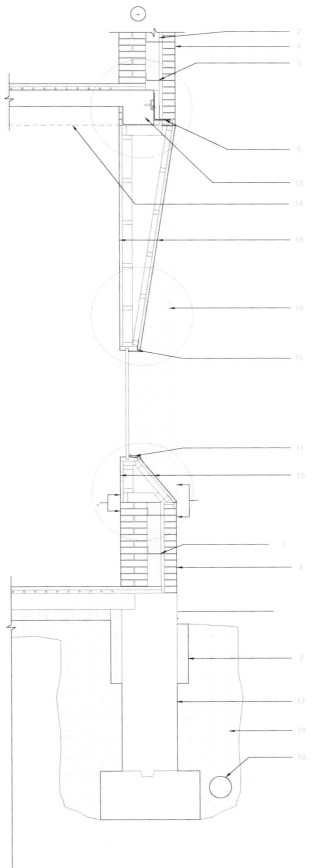

1:20

Façade Detail
1. Double wythe brick interior
2. Rigid insulation
3. Relieving angle
4. Brick façade
5. Concrete slab/beam
6. STL angle to support brick façade
7. WD blocking
8. Wood trim
9. MTL flashing
10. Wood infill panel tongue and groove
11. Wood window frame
12. Glazing
13. Concrete beam
14. Line of finished ceiling
15. Wood infill panel w/rigid insulation
16. Wood panel beyond
17. Concrete footing
18. Drainage gravel
19. Drainage pipe

立面节点
1. 双层内墙砖
2. 刚性隔热层
3. 外露角材
4. 砖立面
5. 混凝土板 / 梁
6. STL 角材，支撑砖立面
7. WD 砌块
8. MTL 防水板
10. 木填充板，凹凸缝
11. 木窗框
12. 玻璃
13. 混凝土梁
14. 天花板线
15. 木填充板，配刚性隔热层
16. 木板后面
17. 混凝土底脚
18. 碎石排水层
19. 排水管

23 Housing Blvd de Hollande Béthune
奥朗德贝蒂纳23户住宅楼

Location/ 地点: Béthune, France/ 法国，贝蒂纳
Architect/ 建筑师: FRES architectes
Associated Architect/ 合作设计: KENK architecten
Photos/ 摄影: Philippe Ruault
Cost/ 成本: 2,800,000 €/2,800,000 欧元
Design/completion date/ 设计 / 竣工时间: 2006.2/2012.7

Key materials: Façade – brick (Hagemeister)
Structure – concrete
主要材料: 立面——砖 (Hagemeister)
结构——混凝土

Façade material producer:
外墙立面材料生产商:
Hagemeister

Overview

The project situated in the outskirts of Béthune, groups together 23 "semi-collective" social housing units. The inspiration for the design came from research into new qualities developed at the interface between the individual and the collective. The building is established in the angle of two urban axes. It respects the alignments on street and marks the angle by an extra height. It presents cuts which give rhythm to the façade and create private and shared terraces. Five "running through" duplex houses in the ground floor, and four duplex on a passageway in the first floor take place as "individual houses" along the boulevard de Hollande. Along the boulevard de Varsovie take place "collective dwellings", multi-directed simplex up to the first floor and five triplex apartments with double-heights in the highest floors.

Detail and Materials

The façades are isolated double walls, concrete-isolation-brick. The sliding joineries are made of movingui. They are maintained in overhang of the façades by massive wooden preframes fixed to the concrete wall. Every joinery constitutes a space in itself. Their projection towards the outside allows to enlarge the space of the dwellings and to animate façades with the movement of shadows.

The wide proportion of glass surface and the use of light railings allow to propose particularly bright and qualitative apartments.

Brick products are very present in Pas-de-Calais. By opting for this material, the architects aimed at a good integration of the building in its context, while reinterpreting the traditional approach with a contemporary touch brought among other things by the unusual colour and the implementation. The imperfections of the brick give live to the façade with infinity of light variations. The sky reflects on the façade, which subtly changes its colour according to time and weather.

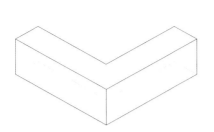
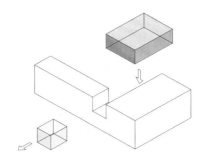
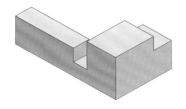
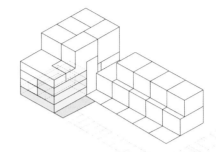
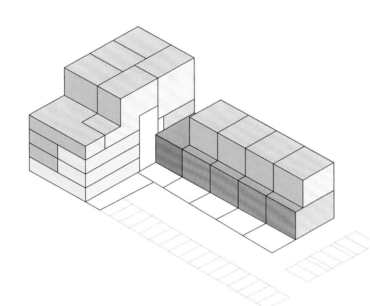
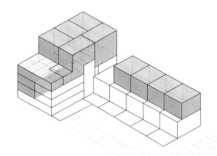
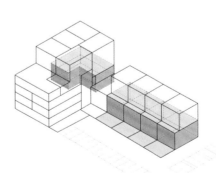

The ground-floor dwellings are slightly above the level of the street to protect a certain intimacy. It is used as a pretext to create a small landscape project in the courtyard. The front doors of individual houses integrate bells and mailbox while assuming the protective function of canopies. Elements are subtlety graded from the public to the private parking, staircases, joineries,... and the hall receive architectural specific treatments as many qualitative common spaces to be crossed to return "at home".

项目概况

项目坐落在贝蒂纳郊外，汇集了23户"半集体式"社会福利住宅单元。设计的灵感来自于对独立住宅与集体住宅之间交界的新品质研究。建筑位于两条城市轴线的夹角处，遵循了街道建筑的排列方式，并且用高度来标志出转角。建筑体块的切口赋予了外墙节奏感，并且形成了私人露台和共享露台。5套双联式住宅位于一楼；4套复式公寓沿着二楼的走廊一字排开，作为沿着奥朗德大道的"独立住宅"。"集体住宅"沿着华沙大道展开，多朝向的单层公寓设在二楼，5套三室双层公寓则设在最上面。

细部与材料

建筑立面是分离的双层墙壁，依次是混凝土、隔热层、砖。门窗框架选用双蕊苏木作为主要材料，它们通过预制木框架固定在混凝土墙上，呈现为立面突出的部分。每个框架都自成一个独立的空间。它们向外突出，既扩展了居住空间，又为外墙增加了运动感。

大面积的玻璃装配和轻质围栏的运用让公寓空间更加明亮和有品位。

砖制品在加莱海峡十分流行。建筑师通过这种材料让建筑很好地融入了所在的环境，同时也以现代风格重新诠释了传统设计方式，还添加了独特的色彩和应用设施。砖块的不完美感让墙面变得鲜活，充满了无尽的光影变化。天空倒映在墙面上，让墙面随着时间和气候的变化而不断变换色彩。

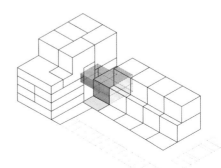

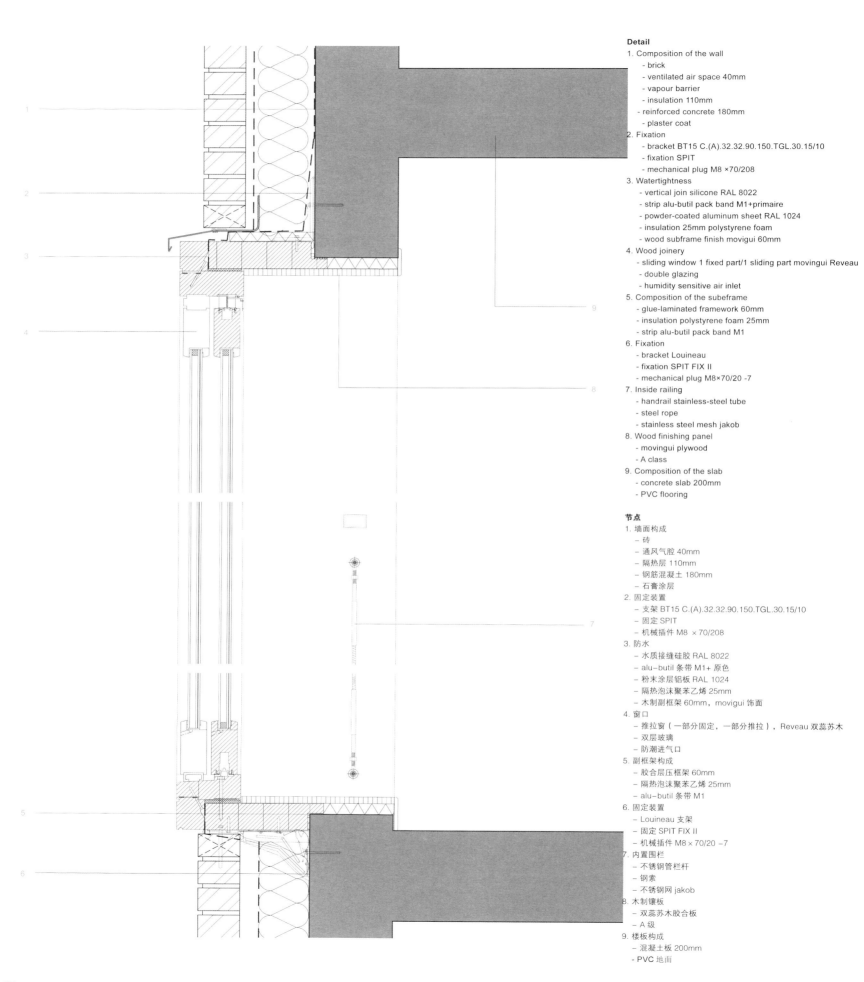

Detail

1. Composition of the wall
   - brick
   - ventilated air space 40mm
   - vapour barrier
   - insulation 110mm
   - reinforced concrete 180mm
   - plaster coat
2. Fixation
   - bracket BT15 C.(A).32.32.90.150.TGL.30.15/10
   - fixation SPIT
   - mechanical plug M8 ×70/208
3. Watertightness
   - vertical join silicone RAL 8022
   - strip alu-butil pack band M1+primaire
   - powder-coated aluminum sheet RAL 1024
   - insulation 25mm polystyrene foam
   - wood subframe finish movigui 60mm
4. Wood joinery
   - sliding window 1 fixed part/1 sliding part movingui Reveau
   - double glazing
   - humidity sensitive air inlet
5. Composition of the subeframe
   - glue-laminated framework 60mm
   - insulation polystyrene foam 25mm
   - strip alu-butil pack band M1
6. Fixation
   - bracket Louineau
   - fixation SPIT FIX II
   - mechanical plug M8×70/20 -7
7. Inside railing
   - handrail stainless-steel tube
   - steel rope
   - stainless steel mesh jakob
8. Wood finishing panel
   - movingui plywood
   - A class
9. Composition of the slab
   - concrete slab 200mm
   - PVC flooring

节点

1. 墙面构成
   - 砖
   - 通风气腔 40mm
   - 隔热层 110mm
   - 钢筋混凝土 180mm
   - 石膏涂层
2. 固定装置
   - 支架 BT15 C.(A).32.32.90.150.TGL.30.15/10
   - 固定 SPIT
   - 机械插件 M8 ×70/208
3. 防水
   - 水质接缝硅胶 RAL 8022
   - alu-butil 条带 M1+ 原色
   - 粉末涂层铝板 RAL 1024
   - 隔热泡沫聚苯乙烯 25mm
   - 木制副框架 60mm，movigui 饰面
4. 窗口
   - 推拉窗（一部分固定，一部分推拉），Reveau 双蕊苏木
   - 双层玻璃
   - 防潮进气口
5. 副框架构成
   - 胶合层压框架 60mm
   - 隔热泡沫聚苯乙烯 25mm
   - alu-butil 条带 M1
6. 固定装置
   - Louineau 支架
   - 固定 SPIT FIX II
   - 机械插件 M8 × 70/20 -7
7. 内置围栏
   - 不锈钢管栏杆
   - 钢索
   - 不锈钢网 jakob
8. 木制镶板
   - 双蕊苏木胶合板
   - A 级
9. 楼板构成
   - 混凝土板 200mm
   - PVC 地面

一楼的住宅略微高于街面，起到了保护隐私的作用。这种设计也为庭院中的小片景观设施提供了空间。独立住宅的前门不仅包含了门铃和邮箱，还有雨篷保护设施。建筑元素被精心划分为公共元素、私人停车位、楼梯间、门窗框等。通过精心的设计处理，大厅像其他高品质公共空间一样，为人们提供了居家感。

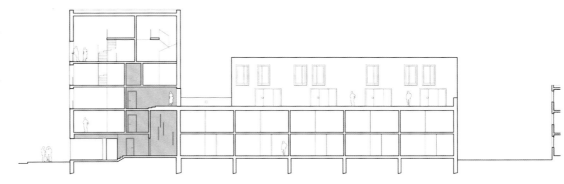

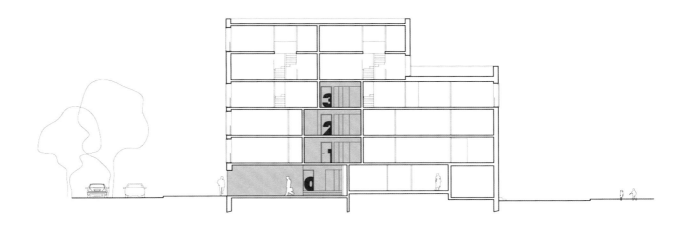

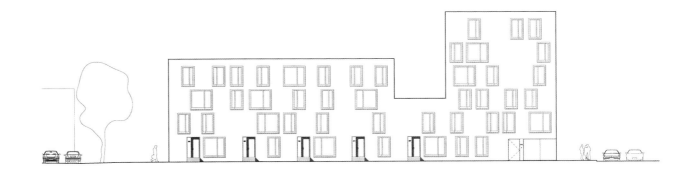

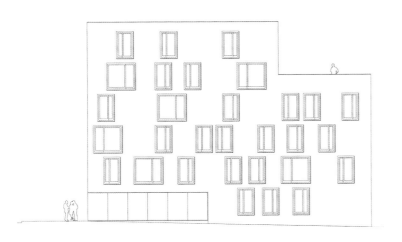

Amstelveen College
阿姆斯特尔芬学院

Location/地点: Amstelveen, The Netherlands/荷兰，阿姆斯特尔芬
Architect/建筑师: DMV architects
Gross floor area/总楼面面积: 13,200m²
Completion date/竣工时间: 2012
Construction costs/建造成本: €9,900,000/9,900,000 欧元

Key materials: Façade – brick
主要材料: 立面——砖

Overview

The building of four floors is designed as a "gallery" school. Translated into a house it means that if you enter through the front door you first find the wardrobe to hang your jacket. Through the hallway, where the toilet is, you walk into the living room. This idea is reflected in the Amstelveen College. Each of the seven education departments, with about two hundred pupils, has its own entrance and a matching outdoor space in the form of south-facing terrace. To emphasise the individuality of the departments, they each have their own colour. Common areas have a natural green colour.

Detail and materials

The brickwork contributes to the sculptural character of the building. The building stands in his environment like a shining black pearl as the silver glow on the black stone allows sunlight to be reflected.

Large masonry surfaces seem freely balancing along the ascending stairs, they are held up by a steel structure in the cavity which is out of sight.

In order to make the masonry look like a shell, the brick outer leaf is not captured by visible steel lintels, but a detail is designed with the bottom layer of bricks hung on special anchors.

项目概况

这座四层高的建筑被设计成一座画廊式学校。学校的设计类似一座住宅，走进正门，你首先会看到放置外套的衣柜。然后沿着走廊（侧面设有洗手间），缓缓步入起居室。阿苏斯特芬学院的设计充分展示了这一概念。学院共有200多名学生，七个独立教学部门均配有独立入口和露天空间（呈现为南向露台的形式）。为了突出各个部门的独立性，每个部门都有其特有的色彩；公共区域则是天然的绿色。

细部与材料

砖砌结构赋予了建筑雕塑特色。黑色石材上通过太阳反射出闪闪的银光，让建筑宛如一颗闪耀的黑珍珠。

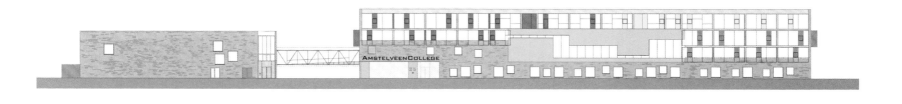

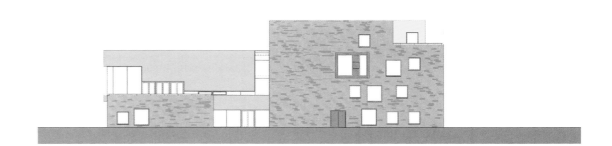

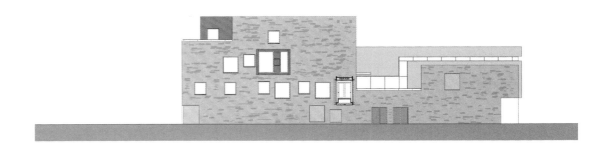
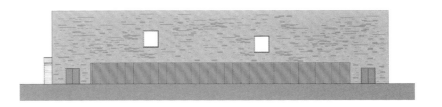

大面积的砖砌表面看起来与上升的楼梯相互独立，实际上它们通过隐藏在空心墙中的钢铁结构支撑。

为了让砖砌墙面看起来更像一个外壳，砖块的外部并没有采用显露出来的钢过梁串联，但是最底层的砖块被悬挂在特殊的锚固装置上。

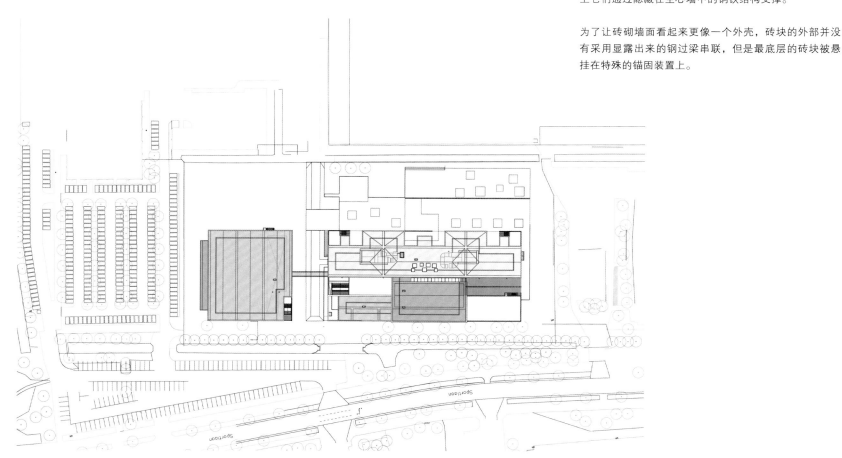

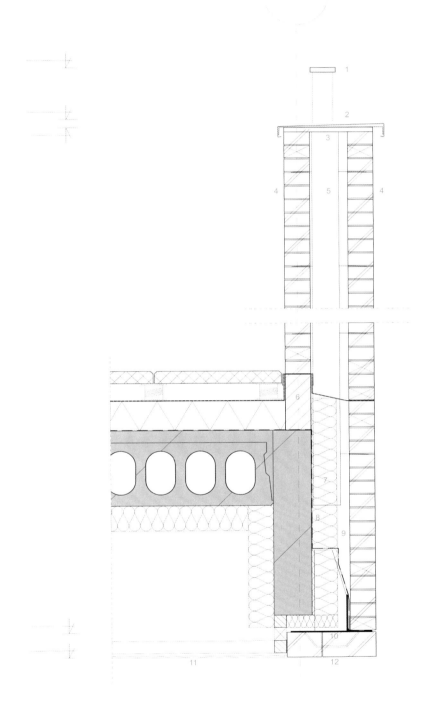

Detail
1. Steel railing coated RAL 9006
2. Aluminium cover coated RAL 9006
3. Underlayment 18mm
4. Brickwork format 210x100x50mm
5. Steel construction with masonry anchors
6. Ytong 100mm
7. Insulation Rd>=4.0m²K/W
8. Waterproof foil
9. Air cavity
10. Steel construction with lintel anchors
11. Coloured trespa 6500+P
12. Brickwork on lintel anchors

节点
1. 钢轨，RAL 9006 色喷漆
2. 铝盖，RAL 9006 色喷漆
3. 衬垫材料 18mm
4. 砌砖模块 210x100x50mm
5. 钢结构，带砖砌锚固
6. 轻质混凝土 100mm
7. 隔热层 Rd>=4.0m²K/W
8. 防水膜
9. 气腔
10. 钢结构，带过梁锚固
11. 彩色千思板 6500+P
12. 砌砖，与过梁锚固相连

Low-Care and Senior Housing
监护病房及老年公寓

Location/ 地点: 'S-hertogenbosch, The Netherlands/ 荷兰，斯 – 海尔托根博斯
Architect/ 建筑师: HILBERINKBOSCH architects
Photos/ 摄影: René de Wit, Breda, The Netherlands (http://www.architectuurfotografie.com)
Gross floor area/ 总楼面面积: 7,400m²
Building cost/ 建造成本: € 6,500,000/6,500,000 欧元

Key materials: Façade – brick
主要材料：细部——砖

Overview

In the sixties, the city of 's-Hertogenbosch was expanded with a new suburb called "Zuid", located on the south side of the city and composed around a large park: "het Zuiderpark". A spacious, green villa area is situated on the southern edge of this park. In this quiet suburb the demolition of a primary school made space for a new development. Being in the proximity of the park, a small shopping mall and several medical services, the new development consists of 22 senior apartments, 10 patio dwellings for the elderly and 16 low-care units for people with a schizophrenic illness.

This residential programme is divided in three different volumes, which are loosely placed on the plot. The open space between the three volumes becomes a public park.

On the initiative of a group of parents, a foundation was established for the construction and management of a modest and secure housing complex for schizophrenia patients living in the community. These 16 two-room units are grouped around a green courtyard which can only be reached through a narrow passageway on the north side. The courtyard side of the dwellings has also been kept fairly closed in order to reduce direct stimuli from outside to a minimum.

Detail and Materials

The architectural composition of spaces and façades in the three separate buildings is based on Dom van de Laan's dimensional and proportional principles and uses his Plastic Number. (The Plastic Number works essentially on a perceptual bases and goes beyond the two-dimensionality of other systems such as the golden ratio. It brings into play the space and the material.)

Together with the restrained colour palette (yellow brick, pale grey concrete elements and dark grey frames) and the meticulously

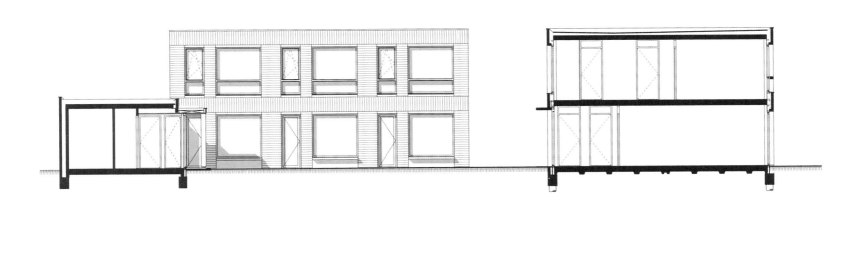
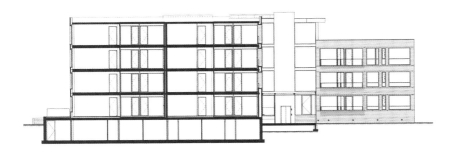
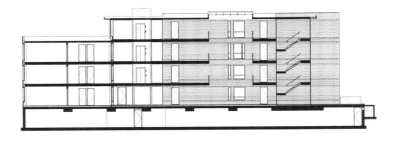

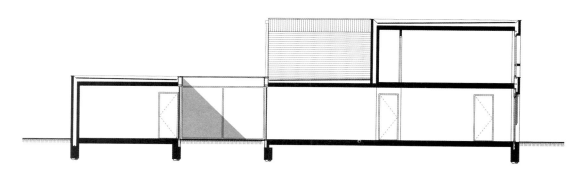

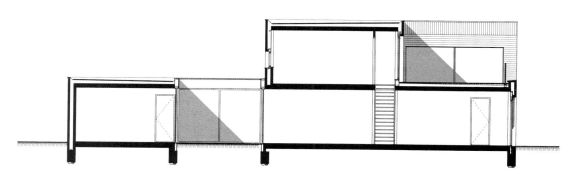

designed garden and park, the complex offers all residents a tranquil and very high-quality living environment.

项目概况

在建市 60 余年之际，斯－海尔斯登波斯市新扩张了一个郊区——"祖德区"，该区位于城市南侧，环绕着一个大型公园展开。公园的南端是一片绿树成荫的大型别墅区。在这宁静的郊区，一座刚刚拆除的小学原址为新开发提供了建造空间。紧邻公园、小型购物中心和若干个医疗服务设施，新开发的项目包含 22 套老年公寓、10 套老年庭院住宅以及 16 套精神分裂症患者监护病房。这个住宅项目被划分为三个不同的空间结构，分散地布置在场地各处。三者之间的开放空间形成了一个公园。

在一群家长的倡议下，社区为精神分裂症患者成立了一个基金会，旨在为他们建造一个低调而安全的住宅环境。这 16 套双室病房围绕着一个绿地庭院展开，仅与场地北侧的一条狭窄的通道相连。庭院一侧的病房保证了良好的密封性，尽量减少对内部人员造成直接刺激的机会。

细部与材料

三座独立建筑的空间及立面构成全部以当姆·范德朗的空间及比例原则为基础，采用了他所提出的"可塑系数"。（可塑系数以感知为基础，超越了黄金分割率等其他系统的二维性，它适用于空间与材料）

建筑在色彩的应用上十分克制：黄色砖块、浅灰色混凝土构件和深灰色框架。它们与精心设计的花园和公园共同为居住者提供了一个宁静的高品质居住环境。

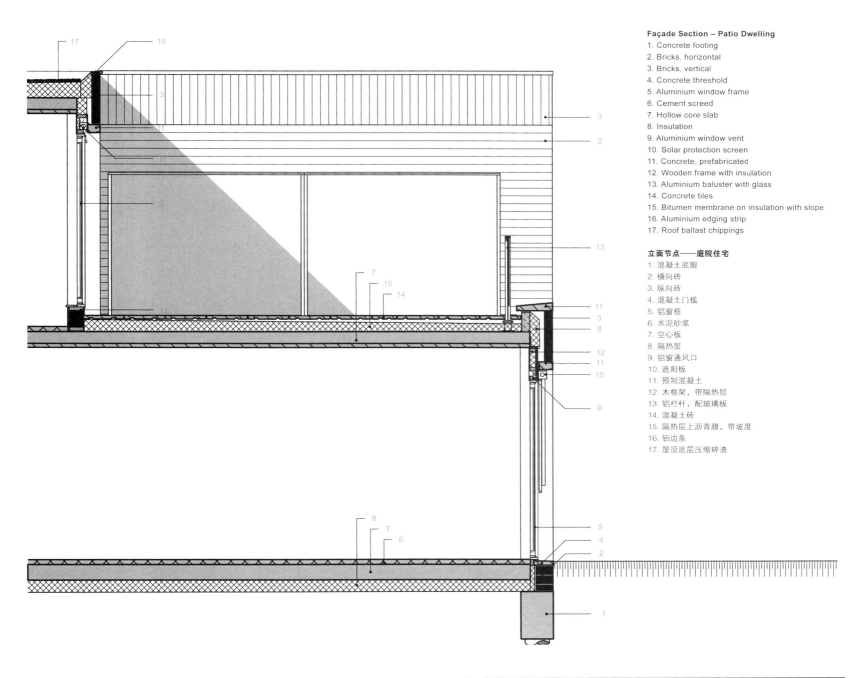

Façade Section – Patio Dwelling
1. Concrete footing
2. Bricks, horizontal
3. Bricks, vertical
4. Concrete threshold
5. Aluminium window frame
6. Cement screed
7. Hollow core slab
8. Insulation
9. Aluminium window vent
10. Solar protection screen
11. Concrete, prefabricated
12. Wooden frame with insulation
13. Aluminium baluster with glass
14. Concrete tiles
15. Bitumen membrane on insulation with slope
16. Aluminium edging strip
17. Roof ballast chippings

立面节点——庭院住宅
1. 混凝土底脚
2. 横向砖
3. 纵向砖
4. 混凝土门槛
5. 铝窗框
6. 水泥砂浆
7. 空心板
8. 隔热层
9. 铝窗通风口
10. 遮阳板
11. 预制混凝土
12. 木框架，带隔热层
13. 铝栏杆，配玻璃板
14. 混凝土砖
15. 隔热层上沥青膜，带坡度
16. 铝边条
17. 屋顶底层压缩碎渣

56 Dwellings and Office Space at the Bloemsingel
布罗伊姆辛格尔住宅与办公空间

Location/ 地点: Rotterdam, The Netherlands/ 荷兰，鹿特丹
Architect/ 建筑师: Marlies Rohmer Architects and Planners
Gross floor area/ 总楼面面积: 9,312m²
Building cost/ 建造成本: €6,290,500/6,290,500 欧元

Key materials: Façade – brick
主要材料：立面——砖

Overview
The project includes 56 dwellings for starters and 1,450 m² (GFA) office space at the Bloemsingel, the former water corporation site at Groningen. The houses are situated around a public courtyard with meandering galleries creating enough space for a dining table. The meandering galleries generate sunny outdoor spaces for the housing, each of which includes a French balcony. The homes are alternately covered with light gray and light green panels allowing the individual house to remain recognisable. With the vegetation roof and the trees the courtyard gets a green character and takes the shape of an inner garden.

Detail and Materials
In this project, sustainability is realised by application of the principle of "richness and abundance". In the past, manpower was cheap and materials were relatively expensive; building were built to last "forever" and were splendidly decorated and detailed with high craftsmanship, often using very durable materials such as hardwood. Now the reverse is true. Using industrial and less labour-intensive techniques, the architects try to reinstate that rich feeling of the past in a contemporary way. The façade-filling brickwork elements with patterned bonds and relief were therefore prefabricated for this project. This procedure yielded a "rich" façade for a relatively low budget. The industrial prefabrication of custom building components also has advantages for energy efficiency. Prefabrication is less energy-intensive and the transportation of ready assembled components to the building site means less building waste will be produced there. This cuts the energy consumed for the processing and transportation of waste materials.

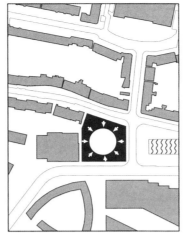 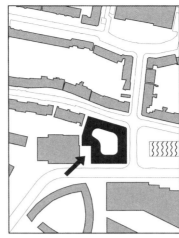 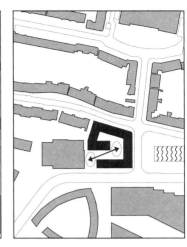

项目概况

项目包含 56 套住宅和 1,450 平方米的办公空间，位于布罗伊姆辛格尔，罗格宁根供水公司的原址。住宅围绕着一个公共庭院而建，曲折的长廊为餐桌提供了充足的空间。蜿蜒的长廊为住宅提供了阳光明媚的户外空间，每套住宅都配有一个法式阳台。住宅单元通过浅灰色和浅绿色墙板的相互交替进行区分。植物屋顶和庭院中的树木为项目带来了绿色风格，形成了一个内部花园。

细部与材料

"丰富而充盈"的设计原则实现了项目的可持续性。在过去，人力成本低廉而材料成本相对昂贵；建筑的建造目标是"永久性"使用，通常都通过高超的手工艺进行华丽而精致的装饰，并且采用硬木等十分耐用的材料。现在，情况正好相反。建筑师利用工业技术和非劳动密集型技术用现代的方式重现了过去的丰富感。因此，他们为项目特别定制了带有花纹和浮雕的立面填充砖砌元素。这种设计用相对较低的预算实现了"丰富"的立面设计。定制建筑构件的工业预制作同样有利于提高能效。预制作过程耗能更低，而在建筑场地现场装配也将产生更少的建筑垃圾。这不仅大大削减了制作能耗，也较少了建筑废料的运输成本。

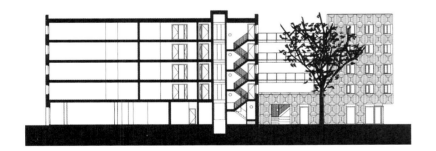 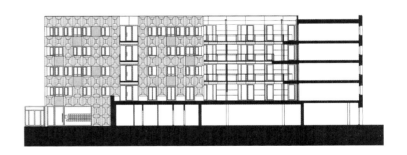

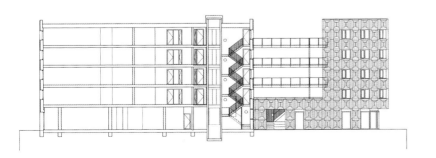 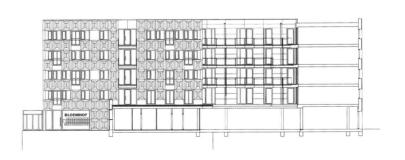

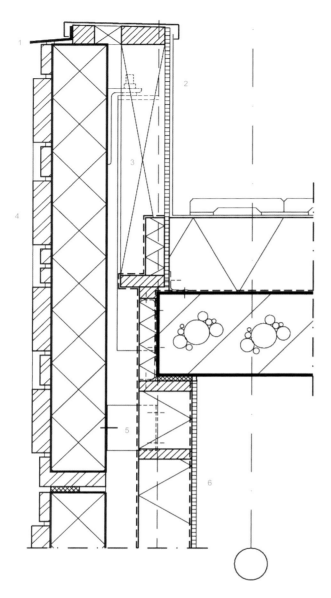
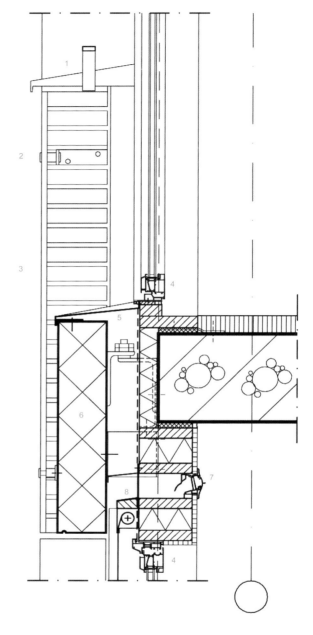
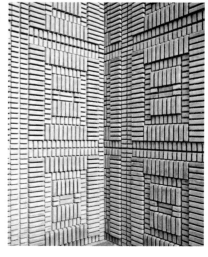

Detail 1
1. Steel profile 5mm
2. Cempanel 10mm
3. Steel baluster façade unit
4. Prefabricated concrete unit provided with brick: Kooy 146.P size 210x98x49mm
5. Steel console for the promotion of hinge prefabricated façade unit
6. Wooden inner leave: Rc > cf EPN RW-foil, Tyvek o.g.
    Storey high timber frame unit
    Mineral wool
    Damp-proof course
    Fermacell 15mm o.g.

节点 1
1. 钢型材 5mm
2. 水泥板 10mm
3. 钢栏杆立面元件
4. 预制混凝土砖，Kooy 146.P 提供，尺寸 210x98x49mm
5. 钢支架，用于提升铰链式预制立面元件
6. 木制内叶：Rc > cf EPN RW-foil, Tyvek o.g.
    与楼层等高木的框架
    矿物棉
    防潮层
    Fermacell 纤维板 15mm o.g.

Detail 2
1. Steel fall prevention strip 15x40 mm powder coated
2. Mounted on glass spacer steel corner promotion of glass panel mounting
3. Glass panel screen print
4. Aluminum window frame Reynaers CS38SL stoved in colour
5. Walkable aluminum water hammer provided with powder coated colour, felt bottom
6. Continuous prefabricated concrete, painted in color
7. Air grate buva topstream 14
8. Shading Harsol 5060 in color RAL 9007, side guide with tensioning cables

节点 2
1. 涂层钢防跌保护带，15x40mm
2. 玻璃板钢角垫片，支撑安装玻璃板
3. 丝印玻璃板
4. 彩色铝窗框 Reynaers CS38SL
5. 可行走式铝制防水板，彩色粉末涂层，毛毡基层
6. 连续预制混凝土，彩色涂料
7. 空气格栅 buva topstream 14
8. 遮阳，Harsol 5060，RAL 9007 色，侧面拉绳

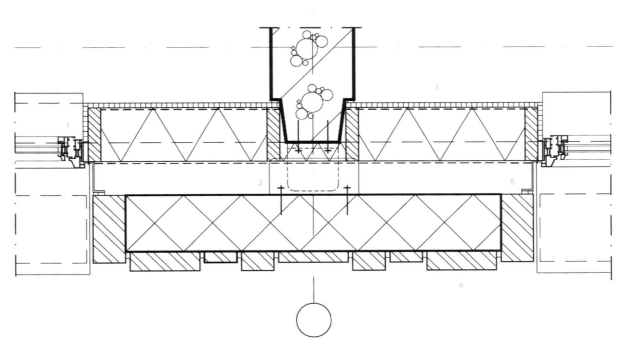

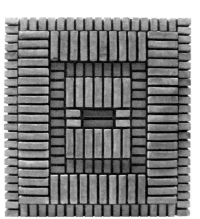 

Detail 3

1. Aluminum window frame Reynaers CS38SL stoved in colour
2. Façade holder
3. Wooden inner leave: Rc > cf EPN RW-foil, Tyvek o.g.
   Storey high timber frame unit
   Mineral wool
   Damp-proof course
   Fermacell 15mm o.g.
4. Prefabricated concrete unit provided with brick: Kooy 146.P size 210x98x49mm
5. Aluminum side panel with powder coating

节点 3

1. 彩色铝窗框 Reynaers CS38SL
2. 立面支架
3. 木制内叶：Rc > cf EPN RW-foil, Tyvek o.g.
   与楼层等高的木框架
   矿物棉
   防潮层
   Fermacell 纤维板 15mm o.g.
4. 预制混凝土砖，Kooy 146.P 提供，尺寸 210x98x49mm
5. 粉末涂层铝侧板

Split View
错层住宅

Location/ 地点 : Helmond, The Netherlands/ 荷兰, 海尔蒙德
Architect/ 建筑师 : UArchitects
Photos/ 摄影 : Daan Dijkmeijer

Key materials: Façade – brick
主要材料：立面——砖
Masonry: with two special colour mad bricks of clay: CRH Clay Solutions (www.crhbricks.com)
砖砌结构：两种特殊色彩的黏土砖：CRH Clay Solutions (www.crhbricks.com)

Façade material producer:
外墙立面材料生产商：
masonry（砖）
- CRH Clay Solutions (www.crhbricks.com);
wood（木材）
- Plato Wood (www.platowood.nl);
glass（玻璃）
- Scheuten (www.scheuten.nl)

Overview

The spatial concept of the architect, the reaction to the surrounding and the living programme wishes of the client for a private house resulted into a free standing split-level house.

The house has an interesting spatial concept, material and detailing. Every side of this building has its own architectural language and on the inside of the house all the 5 different levels are in open spatial connection with each other. The open terrace on the backside of the house and the sun terrace are part of these levels. It is based on the concept of open living level and is organised like the leaves of a tree. The wooden staircase connects the different open levels of the house.

Detail and Materials

The house stands at the cross point of two roads. The architects proposed to lift the house up from their surrounding and to mark the different floors levels with steel beams which are also a reference to the road protection beams at the side to the roads in the Netherlands. The other materials are the two colours bricks (LxBxH= 240x90x40 mm), strong horizontal deep lying cement and vertical wood elements with different size (Plato wood: special modified wood), and the steel beams are galvanised and marks the beginning of the split in the house. The bricks are specially made in the colour with certain amount clay. The architects try to find a balance in the façades and they try to express the split view theme at the outside.

项目概况

建筑的空间理念、建筑对周边环境的反应以及客户对私人住宅的居住功能要求共同决定了项目的结果，形成一座独立的错层式住宅。

住宅的空间概念、材料选择和细节设计都十分有趣。建筑的每一面都有独特的建筑语言，住宅内的五层独立空间以开放的方式相互连接。建筑后面的开放露台和日光浴平台也是这

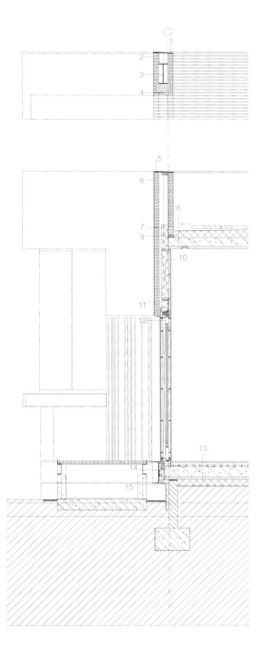 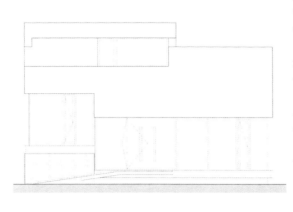

些层级的一部分。建筑以开放式居住层的概念为基础，像树叶一样展开布局。各个不同的开放空间通过木制楼梯与主住宅相连。

细部与材料

住宅坐落在两条路的交叉口处。建筑师提议将住宅抬高，用钢梁来区分不同的楼层。钢梁的设计还参考了荷兰公路两侧常见的道路防护栏。其他主要建筑材料包括双色砖（240x90x40 mm）、强力横向深埋水泥以及不同尺寸的纵向木制元件（帕拉图木材：一种新型改性木材）。钢梁经过镀锌处理，标志出住宅错层的起点。特定的黏土含量实现了特殊的砖块色彩。建筑师试图在立面之间找到一个平衡点，并且力求在建筑外观上表现出错层感。

Vertical Detail
1. Zinc-plate bended on plywood
2. Masonry horizontal joint deep 15mm and 15mm high joint and vertical joint open of 1mm till 3mm
3. Profile IPN 360 galvanised
4. Steel loops for masonry
5. Zinc-plate bended on plywoo
6. Masonry horizontal joint deep 15mm and 15mm high joint and vertical joint open of 1mm till 3 mm
7. Rain drain by too much water on the roof
8. Concrete tiles 50x50mm anthracite without
   Facet on rubber spacers and
   Darkgray gravel width of around 30cm from front terrace
   EPDM roofing
   Flameproof high pressure resistant insulation sloped 15mm
   200mm concrete slab
9. Thermal bridge break hard insulation foam glass
10. Plato wood element straight parts with varying widths treated twice with sansinenviro stain sdf
11. Masonry horizontal joint deep15 mm and 15mm high joint and vertical joint open of 1mm till 3 mm
12. Mahogany door on both sides with straight parts with varying widths as the outside Plato wood fraké, both treated twice with sansinenviro stain sdf
13. Floor level 700 + above street level, 80mm cement floor with heating
    PIR insulation 120mm
14. Profile UNP 400 galvanised
15. Stucco with a fine of 1-2mm grain on an insulation board suitable for exterior stucco

垂直节点
1. 弯曲锌板，贴于胶合板上
2. 砖砌结构，水平接缝，深 15mm；15mm 高缝，垂直接缝 1~3mm
3. IPN 360 型材，镀锌
4. 砖砌结构钢环
5. 弯曲锌板，贴于胶合板上
6. 砖砌结构，水平接缝，深 15mm；15mm 高缝，垂直接缝 1~3mm
7. 屋顶雨水排水
8. 混凝土砖 50x50mm
   橡胶垫层面
   深灰色碎石，宽度约 30cm
   EPDM 屋面
   防水高压隔热层，坡度 15mm/m1
9. 断热桥硬泡沫玻璃隔热层
10. 帕拉托木材元件，宽度不同，sansinenviro stain sdf 二次处理
11. 砖砌结构，水平接缝，深 15mm；15mm 高缝，垂直接缝 1~3mm
12. 红木门，两侧帕拉托木板条宽度不同，均经过 sansinenviro stain sdf 二次处理
13. 楼面高于街面层 700，80mm 水泥地板，带地热设施
    PIR 隔热层 120mm
14. UNP 400 型材，镀锌
15. 1~2mm 细颗粒灰泥，涂于隔热板上，与外层灰泥相适

School Restaurant of Paul Bert and Léon Blum Scholar Group, Lille-Lomme

保罗·伯特与莱昂·布鲁姆学校集团餐厅

Location/ 地点: Lille-Lomme, France/ 法国，里尔
Architect/ 建筑师: D'HOUNDT+BAJART Architects & Associates, Architects (Tourcoing)
Surface area/ 面积: 417m²
Completion date/ 竣工时间: 2014
Cost/ 成本: 1M €HT/100 万欧元

Key materials: Façade – brick
主要材料: 立面——砖

Overview
The urban challenge was to create a frame edge structuring this corner plot whose boundaries are based on low hosted terraced buildings, at the Marx Dormoy and Léon Blum streets. At the corner, the school restaurant's parapet ascends gently, creating a bow that hides the restaurant's technical equipment (central air handling and heat pump) while freeing the ground floor.

Detail and Materials
The architectural treatment is contextual in its urban form and its materiality; the use of the brick, which dominates the landscape, has received a contemporary and qualitative review showcasing the polychrome emblematic brick bonding of the regional architecture. The layout of the masonry, with an almost random pattern, evokes the snake skin and activates the imaginary potential, architecture and young children dream-like: in a world where you have to respect the rules of the urban game, the building delivers a message of imagination, and seems, with its "snake skin" and "scale-like windows" to have the power to move.

The façades of the restaurant, monolithic and protective, form a boundary, almost an organic membrane, and isolate from the street this place where calm and privacy are needed without locking the user; with its set of windows on two levels, small and big, can, at a glance, kiss the street without being visible from the outside. If the walls form a filter, the inside of the restaurant expands horizontally into two very bright areas open onto an interior garden.

项目概况
项目要求在两条街道的转角处添加一座新建的餐厅结构。在

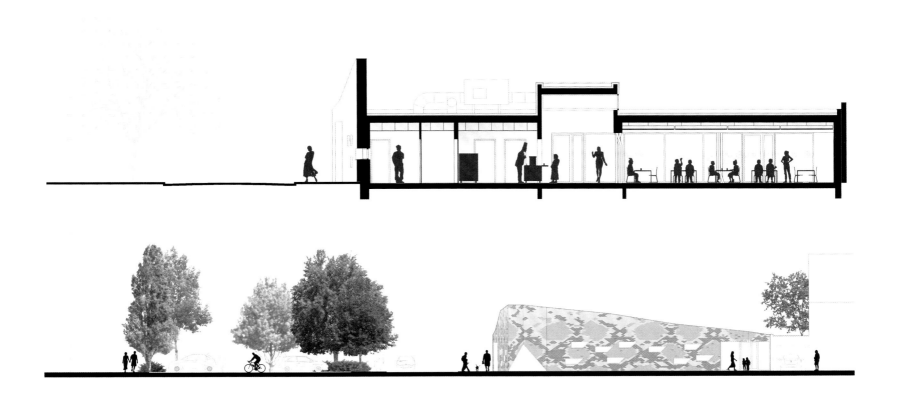

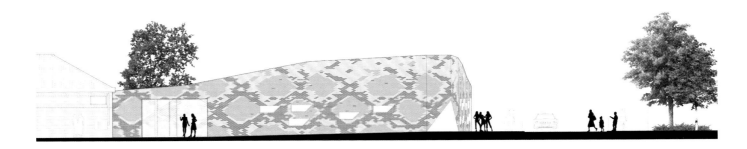

街角，学校餐厅的护墙缓缓上升，既隐藏了餐厅的技术设备（中央换气系统和热泵），又解放了一楼地面。

细部与材料

建筑的处理与它的城市形态和材料选择息息相关。作为周边环境中的主要材料，砖块以现代的方式被运用在建筑上，展示了多彩的砌砖带。砌砖采用近乎随机的图案，有点像蛇纹，激发了建筑和孩子们的想象力：在这个必须遵循诚实游戏规则的世界，建筑传递出一条想象的信息，"蛇纹"和"鳞状窗口"让餐厅看起来好像能向前行走。

餐厅的外墙看起来巨大而具有保护性，形成了一个边界，将街道空间与餐厅隔开，保护着内部用户。两排窗户让餐厅内的人们可以一瞥街道上的风景。餐厅内部呈水平状延伸，两个明亮的就餐空间面向内部花园开放。

Construction Detail
1. White coated steel covering
2. Brickwork/4shades
3. Stainless steel ties
4. 30mm cavity
5. Reinforced concrete
6. Flashing
7. 40mm layer of gravels
8. Two layers sealing membrane
9. 90mm insulation
10. Vapour barrier
11. Suspended ceiling slab
12. 130mm insulation
13. Plasterboard
14. White coated steel sheet
15. Metal structure
16. Steel profile anchored in concrete wall
17. Screed
18. Reinforced concrete
19. 100mm thermal insulation

结构节点
1. 白色涂漆钢顶盖
2. 砖砌结构（4色）
3. 不锈钢筋
4. 30mm 气腔
5. 钢筋混凝土
6. 防水板
7. 40mm 碎石层
8. 双层密封膜
9. 90mm 隔热层
10. 隔汽层
11. 吊顶板
12. 130mm 隔热层
13. 石膏板
14. 白色涂漆钢板
15. 金属结构
16. 钢型材，固定在混凝土墙上
17. 砂浆
18. 钢筋混凝土
19. 100mm 隔热层

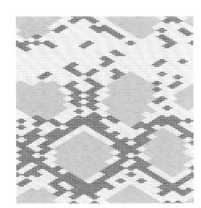

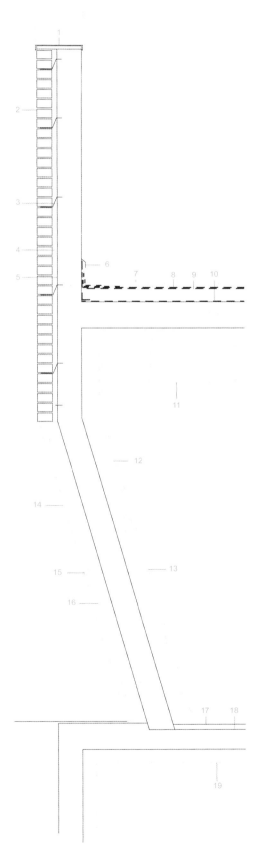

O|2 Laboratory Building of VU Amsterdam University

阿姆斯特丹自由大学O|2实验楼

Location/ 地点: Amsterdam, The Netherlands/ 荷兰，阿姆斯特丹
Architect/ 建筑师: EGM architecten
Photo/ 摄影: Courtesy of Ronald Schlundt Bodien/EGM architecten
Area/ 面积: 33,000m²
Completion date/ 竣工时间: 2015

Key materials: Façade – click bricks
主要材料：立面——拼装砖

Façade material producer:
外墙立面材料生产商：
Daas ClickBricks

Overview

The O|2 Laboratory building is the new research and education building for the Free University of Amsterdam. Strategically located in the Zuidas district, O|2 functions as the intermediary between the renewed campus of the VU and the VU Medical Centre. The building is intended for biomedical and biochemical education and research. Various research groups at the VU and VU MC that work together in the area of Human Health and Life Sciences (H2LS) will be housed there. A total gross floor surface of 33,000 m² will be realised, distributed over 14 floors.

The large floor fields of the floors are pre-eminently fit for flexible use as laboratories, open plan offices, study centres and/or meeting areas. Each floor has a west wing and east wing, separated by the atrium. Both wings are divided into zones for laboratories, offices and services. The O|2 building has biomedical labs up to Bio-safety level 3 (BSL3).

Meeting areas are situated along a network of horizontal and vertical connection routes, with the building facilitating what it was designed for: optimal interaction between users to enable maximum encounters, knowledge exchange and innovation. Flexible, universal laboratory modules in a fixed template, which can be switched if desired, have been designed to accommodate ever-changing research programmes.

Detail and Materials

O|2 has been designed as a compact cube with cut-outs in strategic areas. This results in a fascinating spatial structure, with spectacular lines of sight and daylight that penetrates

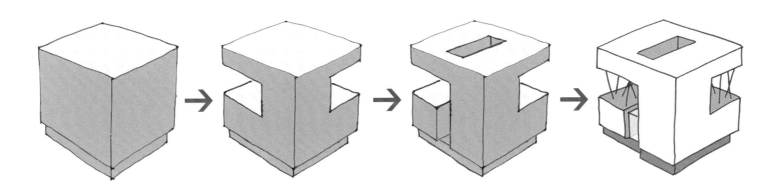

deep into the building. The cut-outs come together in the heart of the building in the central atrium: the new VU meeting point. The staggered terraces in the atrium are furnished as accommodation areas: for study, work and relaxation. The ground floor has an open layout with shops, catering establishments and classrooms, so that the building is specifically focused on the campus.

Daas ClickBricks are clay bricks that are dry stacked together by means of clips to create "wallinging. The bricks are manufactured in such way that when assembled into the wall, they fit into the modular gauge of the structure. The bricks are available in several standard sizes. On the upper and lower side of each brick is a groove. This groove is provided for the fastening of the clips. Over the height and on the header side of each brick there is a recess. This recess is provided for easy placing of wall ties. In addition to the whole brick units, there are also standard half-bricks, corner-bricks, lintel-bricks and window-edge bricks.

The application of the ClickBrick-system is suitable for buildings in which the Façade s have been designed for the modular dimensions of the ClickBrick system. It is important that the dimensions of the building are accurately set-out taking into consideration the modular size of the ClickBrick system. Click-Bricks have very high and tight tolerances, namely 0.1 mm for the height, 1.5 mm for the length and 1.5 mm for the width of each brick. The system allows for some adjustments to make

the header joints somewhat larger or smaller. Adjustment of the position of the outer leaf of the wall while placing wall ties is also possible. There is, however no allowance in the system, other than in the kicker layer, to adjust the height.

项目概况

O|2 实验楼是阿姆斯特丹自由大学新建的研究教学楼。实验楼位于泽伊达斯区，在重建的校区与医疗中心之间形成了过渡。建筑主要供生物医学和生物化学的教学和研究使用。自由大学和大学医疗中心从事人类健康与生命科学领域研究的实验团队将在这里进行科研。建筑的总楼面面积为 33,000 平方米，共有 14 层。

大跨度的楼面设计有利于灵活地设置实验室、开放式办公室、研究中心和会议区。每层楼都有东西两翼，由中庭隔开。两侧空间分别划分为实验室区、办公区和服务区。O|2 实验楼的生物医疗实验室生物安全等级为 3 级。

会议区沿着水平和垂直交通路线而设，能优化用户之间的互动，实现高见面率，有助于信息交换和创新。灵活统一的实验室模块可根据需求进行改造，适用于不断变化的研究课题。

细部与材料

O|2 实验楼被设计成了一个紧凑的立方体，这个立方体在特定区域被切割开，形成了引人注目的空间结构，享有壮观的视野和丰富的自然采光。建筑中心的切割空间形成了中庭，是自由大学全新的聚会场所。中庭错列的平台可被改造成学习、工作、休闲等任何空间。一楼采用开放式布局，设有商店、餐厅和教室。这座实验楼是校园里的一大亮点。

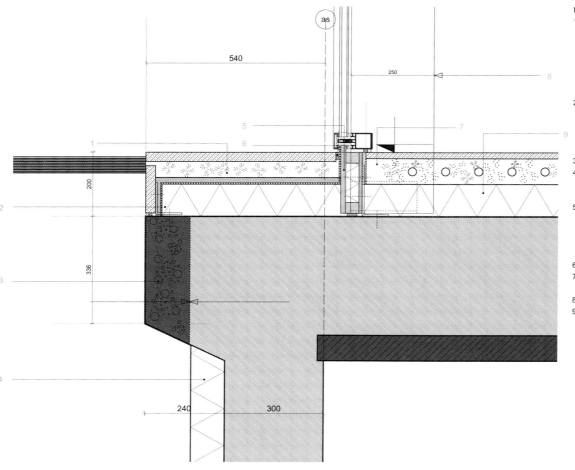

Detail 1

1. 30mm natural stone sloped
   Cement layer
   Plastic drainage mat 16mm thick
   2 layered roofing
   100mm XPS insulation
   Vapor barrier
   Existing reinforced concrete floor
2. 30mm natural stone
   Stainless steel brackets
   Plastic drainage mat
   2 layered roofing
   Sealant on backing material
3. Single layered roofing upon grouted ridge
4. Single layered roofing
   Existing insulation
   Existing reinforced concrete wall
5. Roofing put up under façade
   15mm fiber cement siding
   Battens 34/40mm
   40mm XPS insulation
   Galvanised steel brackets
6. Sealant on backing material
7. Sealant on backing material
   10mm EPS strip
8. Front of curtain wall profiles
9. 20mm natural stone
   Reinforced cement layer
   Floor heating
   Plastic foil
   100mm XPS insulation
   Vapour barrier
   Existing reinforced concrete floor

节点 1

1. 30mm 天然石材坡面
   水泥层
   塑料排水垫 16mm 厚
   双层屋面
   100mmXPS 隔热层
   隔汽层
   原有的钢筋混凝土地面
2. 30mm 天然石材
   不锈钢支架
   塑料排水垫
   双层屋面
   基底材料密封
3. 单层屋面，水泥浆屋脊上方
4. 单层屋面
   原有的隔热层
   原有的钢筋混凝土墙
5. 屋面，立面下方
   15mm 纤维水泥壁板
   板条 34/40mm
   40mmXPS 隔热层
   镀锌钢支架
6. 基底材料密封
7. 基底材料密封
   10mm EPS 条
8. 幕墙型材正面
9. 20mm 天然石材
   加固水泥层
   地面供暖
   塑料膜
   100mmXPS 隔热层
   隔汽层
   原有的钢筋混凝土地面

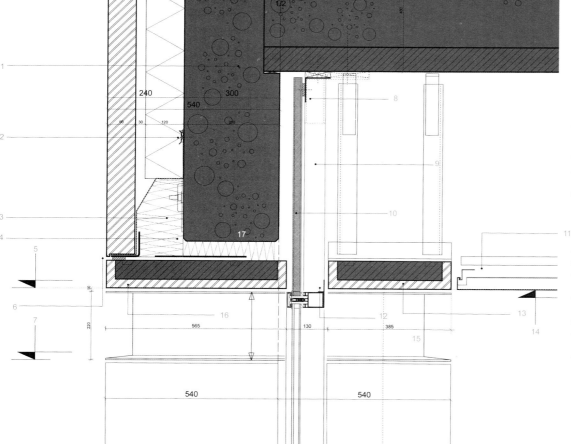

Detail 2

1. Structure façade in situ concrete:
   300mm concrete
   120mm mineral wool insulation
   30mm air cavity
   90mm click-brick masonry
   RC>=3.0m^2.K/W
2. Clamping profile; top sealed
3. Custom made galvanised steel lintel for bearing and suspension of masonry elements
   Bearing felt, siding foil and powder coating
4. Rigid insulation
5. 4,900+bottom masonry
6. Waterproof sealant
7. 4,670+bottom U profile
8. Steel L-profile
   Draft resistant joints top with PIR-foaming
9. Curtain wall narrowed and extended to bottom of concrete floor
10. Sandwich element
    Double galvanised steel sheet
    PIR filling
11. For correct execution and joining
12. Aluminium L.40.20
13. Prefab masonry element suspended to a steel construction a
14. 4,900+floor level
    Bottom of ceiling
15. Steel U-profile around wall
16. Prefab masonry element appearance and division alike façade
17. Mineral wool

节点 2

1. 现浇混凝土结构立面：
   300mm 混凝土
   120mm 矿棉隔热层
   30mm 气腔
   90mm 拼装砖砌体
   RC>=3.0m^2.K/W
2. 钳夹型材，顶部密封
3. 定制镀锌钢过梁，用于支撑和悬吊砌体
   轴承毡圈、侧薄板和粉末涂层
4. 刚性隔热层
5. 4,900+ 底部砌体
6. 防水密封
7. 4,670+ 底部 U 型材
8. L 形钢
   防风接缝，顶部为 PIR 泡沫
9. 幕墙，变窄延伸至混凝土地面底部
10. 夹层元件
    双层镀锌钢板
    PIR 填充
11. 用于正确执行和连接
12. 铝 L.40.20
13. 预制砖砌组件，悬吊在钢结构上
14. 4,900+ 楼面
    天花板底面
15. 环墙 U 形钢
16. 预制砖砌组件，外观和分区与立面相似
17. 矿棉

Detail 3

1. Construction of windows:
   Sill 67/427mm
   Glazing alike façades
   Aluminium glazing bars and visible auxiliary profiles in colour
2. Shaped aluminium sandwich sheet in colour
   Bottom cavity mildly ventilated
3. Aluminium water sill colour alike window frame
   Clamps and end caps
   Noise reducing tape
   Waterproof plastic foil
   120mm EPS bevelled
4. 400+floor level
   Top of masonry
5. Façade structure:
   300mm prefab concrete
   120mm mineral wool insulation
   30mm air cavity
   90mm click brick masonry
   $RC>=3.0m^2.K/W$
6. 500+floor level
   Top of concrete
7. Battens 95/100mm
   Double draft resistant sealing

节点 3

1. 窗结构
   窗台 67/427mm
   玻璃装配与立面类似
   铝制窗压条和外露辅型材，彩色
2. 塑形铝夹层板，彩色
   底部中空通风
3. 铝防水台，色彩与窗框相同
   钳夹与端盖
   减噪胶带
   防水塑料膜
   120mm EPS 斜面
4. 400+ 地面层
   砌体顶部
5. 立面结构：
   300mm 预制混凝土
   120mm 矿棉隔热层
   30mm 气腔
   90mm 拼接砖砌体
   $RC>=3.0m^2.K/W$
6. 500+ 地面层
   混凝土顶部
7. 板条 95/100mm
   双层防风密封

Detail 4

1. Structure façade in situ concrete:
   300mm concrete
   120mm mineral wool insulation
   30mm air cavity
   90mm click-brick masonry
   $RC>=3.0m2.K/W$
2. Clamping profile; top sealed
3. Waterproof plastic foil
   120mm EPS beveled
4. Bottom of masonry slotted and put upon powder coated steel lintel with sliding foil
5. Shaped aluminium L-profile
   Waterproof sealant
   Shaped aluminium water sill
6. Construction of windows:
   Sill 67/427mm
   Glazing alike Façade s
   Aluminium glazing bars and visible auxiliary profiles in color
7. 2,800+floor level
   Bottom of ceiling
8. 8mm fibre cement siding battens
   60mm XPS insulation fixed to concrete mold
9. Mounting frame 56/90mm
   Double draft resistant sealant
   Shaped aluminium L-profile
10. Sealant with backing material
    Wooden batten fixed on brackets
11. Transparent/obscurant shades aluminum alcove in color alike window
    Ventilation through local openings fixation
    Screens individually controllable

节点 4

1. 现浇混凝土结构立面：
   300mm 混凝土
   120mm 矿棉隔热层
   30mm 气腔
   90mm 拼装砖砌体
   $RC>=3.0m^2.K/W$
2. 钳夹型材，顶部密封
3. 防水塑料膜
   120mm EPS 斜面
4. 砌体底部开槽，放置在粉末涂层钢过梁上，配滑动膜
5. 塑形 L 形铝材
   防水密封
   塑形铝防水台
6. 窗结构
   窗台 67/427mm
   玻璃装配与立面类似
   铝制窗压条和外露辅型材，彩色
7. 2,800+ 楼面
   天花板底部
8. 8mm 纤维水泥壁板条
   60mmXPS 隔热，固定在混凝土模架上
9. 安装框架 56/90mm
   双层防风密封
   塑形 L 形铝材
10. 基底材料密封
    木板条，固定在支架上
11. 透明 / 半透明遮阳铝帘，色彩与窗户相同
    窗口通风
    独立控制遮阳帘

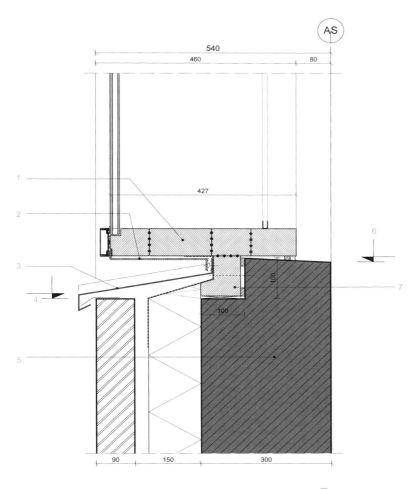

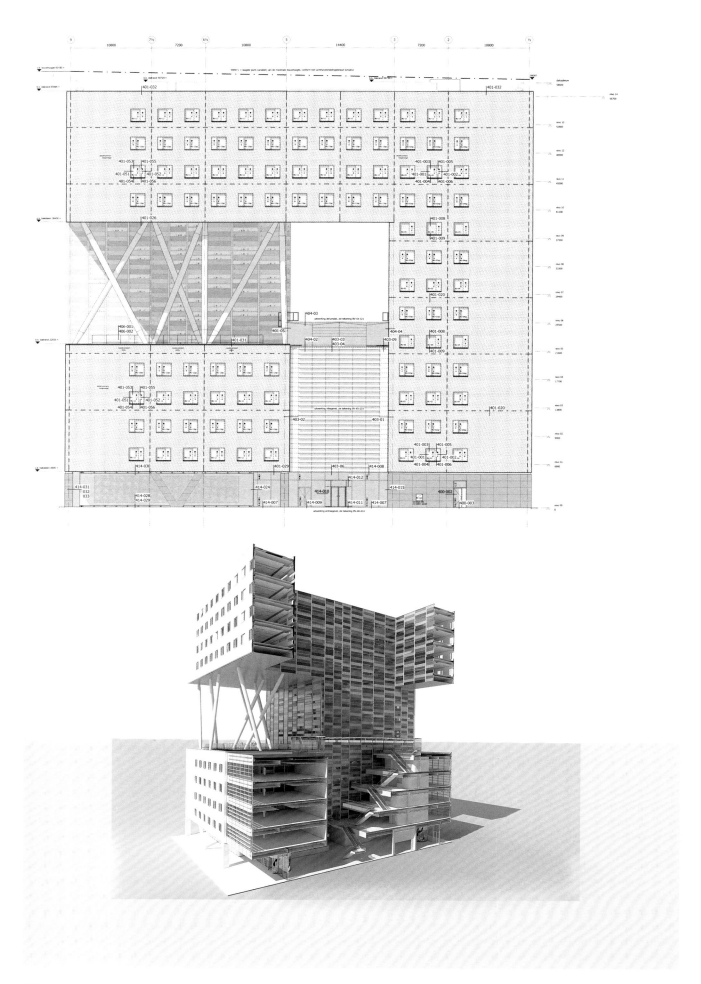

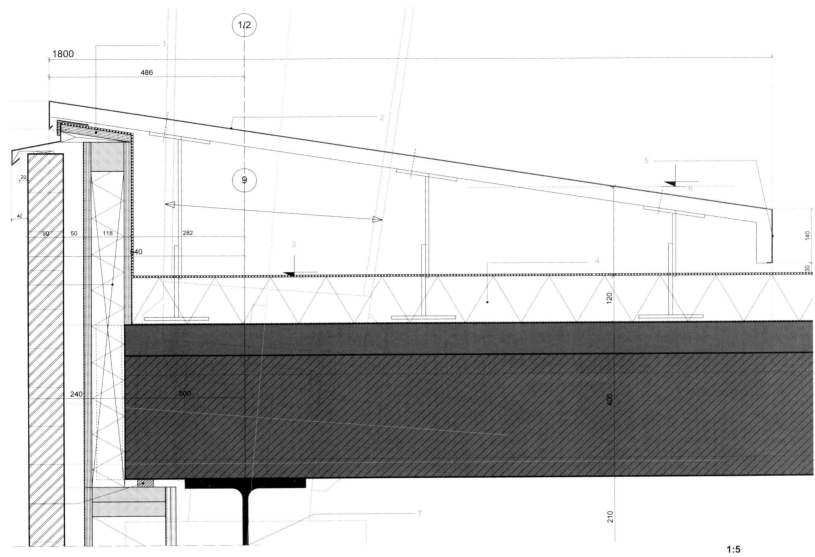

1:5

Detail 5

1. Shaped aluminium profile
   Roofing
   18mm fibre cement siding
2. Shaped demountable aluminium sandwich sheeting with ribs, walkable, screwed to and supported by galvanised steel consoles waterproof sealed into roofing
3. Lowest point of insulation
4. Single layered EPDM roofing
   Sloping roofmate insulation 120mm
   Vapour barrier
   80mm load dividing concrete layer
   Hollow core slates
5. Perforated front
6. Highest point of insulation
7. Steel structure 60 min. fireproof

节点 5

1. 塑形铝型材
   屋面
   18mm 纤维水泥壁板
2. 塑形可拆卸带筋铝夹层板，可行走，由镀锌钢支柱固定并支撑，防水密封于屋面上
3. 隔热层最低点
4. 单层 EPDM 屋面
   坡形屋面隔热层 120mm
   隔汽层
   80mm 承重分段混凝土层
   中空板
5. 穿孔面
6. 隔热层最高点
7. 钢结构，可防火 60min

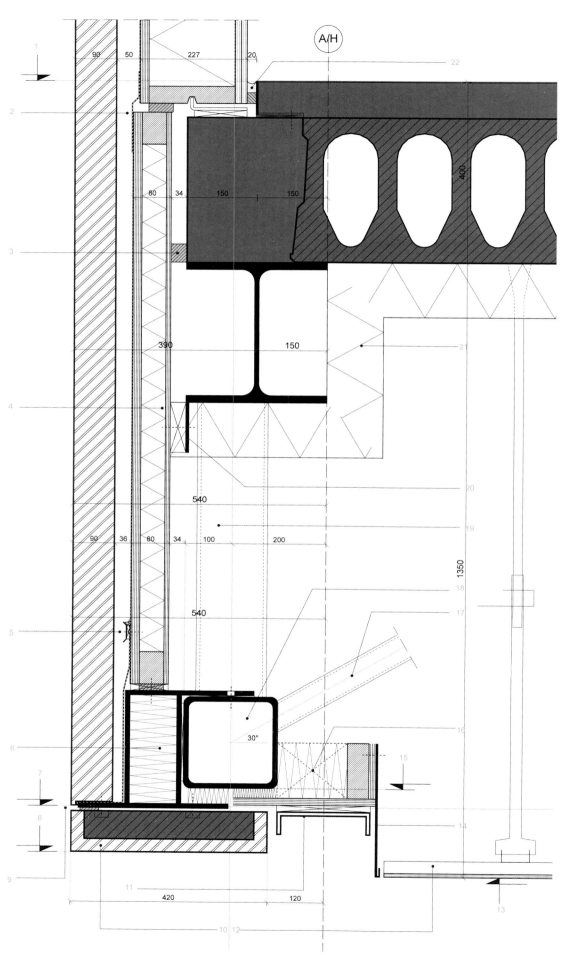

Detail 6

1. 41,000+top finished floor
2. Draft resistant sealant joints between panels to be sealed with waterproof foil
3. Draft resistant sealant
4. Timber panels
   Waterproof and vapor-permeable plastic foil
   18mm finish spruce plywood battens
   50mm high quality insulation
   10mm fibre cement siding
5. Clamping profile; top sealed
6. Custom made galvanised steel lintel for bearing and suspension of masonry and timber frames
   Bearing felt, sliding foil and powder coating hollow spaces filled with insulation
7. 39,500+bottom of lintel
8. 39,400+bottom of masonry element
9. Waterproof sealant
10. Prefab masonry element appearance and division alike façade
11. Extruded aluminium U-profile glued to backing
12. Ceiling
    Stormproof fixation, supported to construction
13. 39,350+bottom of ceiling
14. Shaped metal profile
15. 39,540+bottom box-profile
16. Welded steel strip battens and blocks 46/121mm
    Mineral wool insulation
    15mm fiber resistant sheeting
17. Steel box-profile 60.60.5
    Supports modular 2000mm
18. Steel box-profile 200.200.10
19. Steel box-profile 150.150.8
    Vertically modular 2000mm including supports
20. Steel girder
    Welded steel strip
21. 120mm mineral wool insulation
22. Sealant with backing material
    Galvanised steel strip with brackets

节点 6

1. 41,000+ 顶部整修地面
2. 板材间防风密封接缝，防水膜密封
3. 防风密封
4. 木板
   防水透气塑料膜
   18mm 云杉胶合板
   50mm 高品质隔热层
   10mm 纤维水泥壁板
5. 钳夹型材，顶部密封
6. 定制镀锌钢过梁，用于支撑和悬吊砌体
   轴承毡圈、侧薄板和粉末涂层
7. 39,500+ 过梁底部
8. 39,400+ 砌体底部
9. 防水密封
10. 预制砖砌组件，外观和分区与立面相似
11. 挤制 U 形铝材，黏在衬底上
12. 天花板
    防风暴固定，结构支撑
13. 39,350+ 天花板底部
14. 塑形金属型材
15. 39,540+ 箱式型材底部
16. 焊接钢条板和砌块 46/121mm
17. 钢箱 60.60.5
    支撑模块 2000mm
18. 钢箱 200.200.10
19. 钢箱 150.150.8
    垂直模块 2000mm，包含支架
20. 钢梁
    焊接钢条
21. 120mm 矿棉隔热层
22. 基底材料密封
    镀锌钢条 + 支架

Detail 7
1. Composed aluminium box profile 80/380mm sealed into roofing
   Demountable perforated aluminium coversheet
2. 18mm fibre cement sidings
   Battens 38/121mm
   Mineral wool insulation
   18mm finish spruce plywood
   Waterproof and vapour-permeable plastic foil
3. Composed aluminium box profile
   Demountable perforated aluminium coversheet
   Sealed to masonry
4. 57,400+top of masonry
5. 37,500+floor level
6. Top steel girder, lowest point of beam
7. Anchoring/fixation of timber panels
   Joints sealing are to be draft resistant

节点 7
1. 组合铝箱 80/380，密封于屋面
   可拆卸穿孔铝盖板
2. 18mm 纤维水泥壁板
   板条 38/121mm
   矿棉隔热层
   防水透气塑料膜
3. 组合铝箱
   可拆卸穿孔铝盖板
   密封于砌体结构
4. 57,400+ 砌体顶部
5. 37,500+ 楼面
6. 顶部钢梁，最低点
7. 木板固定
   防风接缝密封

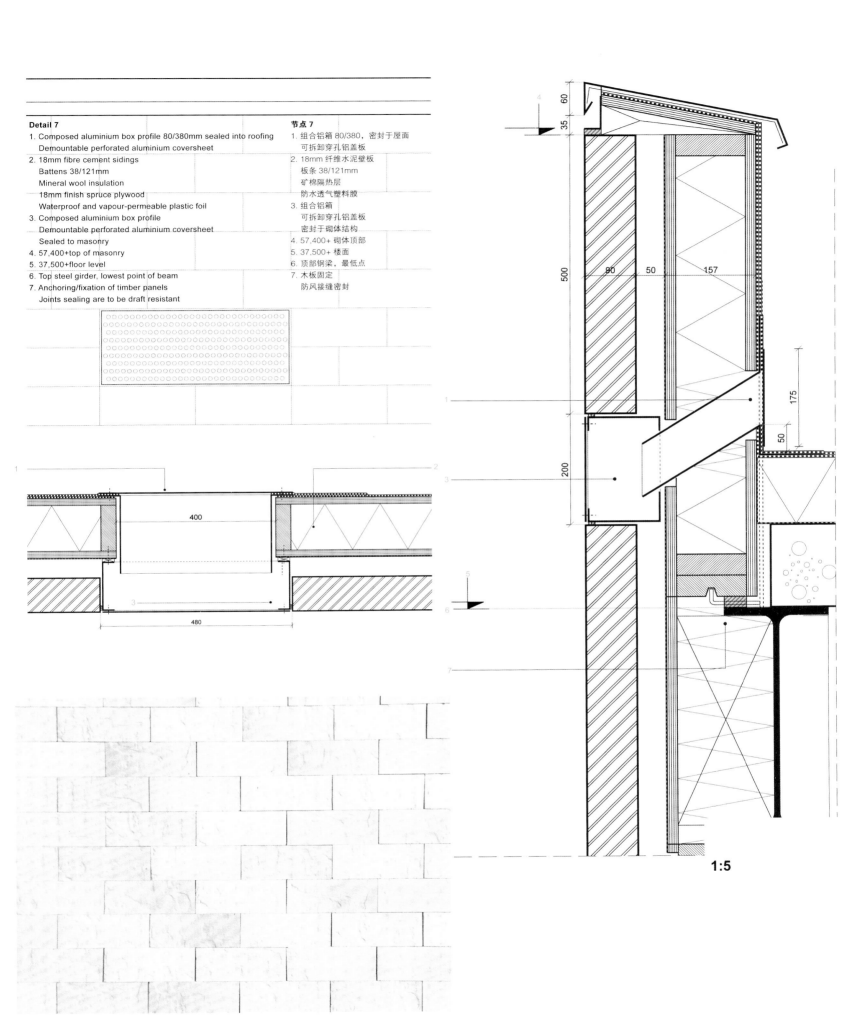

1:5

PARQUE Kindergarten
公园幼儿园

Location/ 地点: Lisbon, Portugal/ 葡萄牙，里斯本
Architect/ 建筑师: PROMONTORIO
Gross floor area/ 总楼面面积: 1,400m²
Budget/ 预算: 1,800,000 Euros/1,800,000 欧元

Key materials: Façade – brick, stucco
主要材料：立面——砖、灰泥

Overview
Since the very beginning on its practice, PROMONTORIO has been involved on educational projects. From schools and cultural facilities to museums and art galleries, a wide range of learning approaches has been tested and implemented with considerable success and accrued experience.

Working in close collaboration with the teachers and pedagogues, PROMONTORIO has attempted to challenge the preconceived classroom space planning principles, into a more fluid and dynamic spatial interaction that meets the increasing demand for flexibility in the learning methods, both at kindergarten and primary school levels.

After research and mock-up testing with the Parque team, the pentagonal classroom module gained consistency as a unifying environment able to respond to a dynamic multitask educational project. Ranging in scale and height; the classrooms, the library, the art room and the canteen flow in a sequence of generous and surprising spaces both from the outside and the inside.

Detail and Materials
Whitewashed in burnt lime, the tectonic articulation between the portent walls of bond brickwork and concrete cast in-situ, contrasts with large and transparent glass panes. The latter, when open, blur the boundaries between outside and inside. In addition, the occasionally punctured brickwork lets air and light into a series of secluded patios.

项目概况
PROMONTORIO 自从业以来一直积极参与教育项目的设计与规划。从学校、文化设施到博物馆、艺术馆，他们获得了大量宝贵的设计经验，也取得了相当大的成功。

PROMONTORIO 与教职员工紧密合作，尝试挑战传统的教室空间规划原则，以更流畅、更动感的空间互动来满足教学

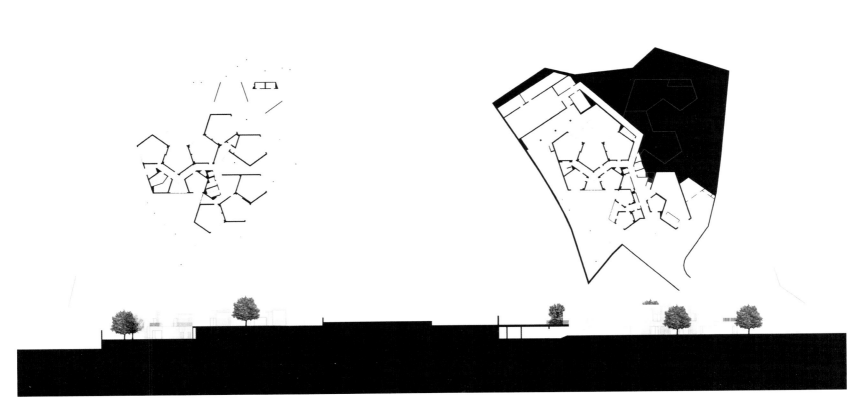

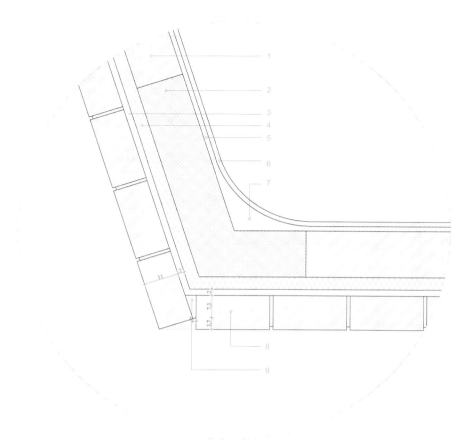

Horizontal Detail
1. 150mm clay brick masonry
2. 150mm reinforced concrete pillar
3. Air space 20mm
4. Thermal insulation (wallmate) 40mm
5. Regularisation plastering 15mm
6. 15mm stucco white painted
7. Reinforcement of plaster and stucco
8. Brick facings uncut 110mm with white exterior painting
9. Elastic mortar 10 to 15mm

水平节点
1. 150mm 黏土砖砌结构
2. 150mm 钢筋混凝土柱
3. 空气层 20mm
4. 隔热层（wallmate）40mm
5. 抹平石膏 15mm
6. 15mm 白色灰泥
7. 石膏与灰泥加固层
8. 毛边砖砌面层 110mm，白色外墙涂料
9. 弹性砂浆 10~15mm

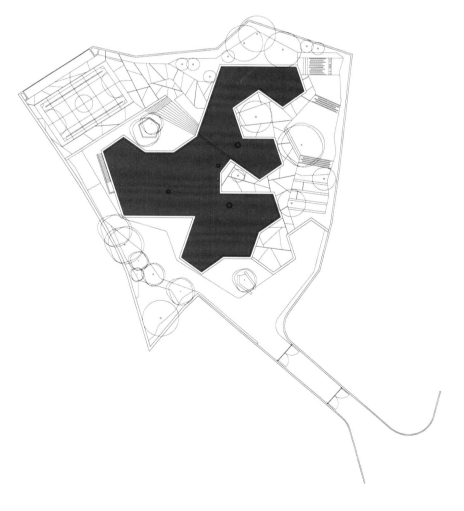

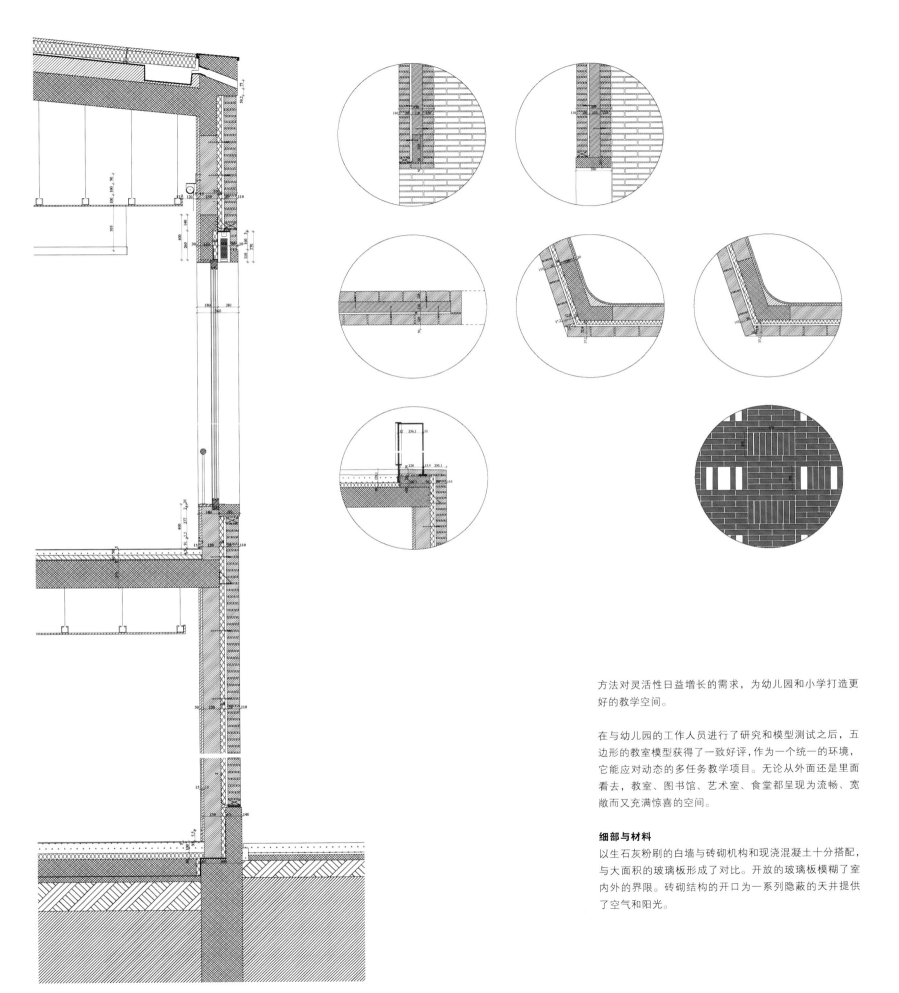

方法对灵活性日益增长的需求，为幼儿园和小学打造更好的教学空间。

在与幼儿园的工作人员进行了研究和模型测试之后，五边形的教室模型获得了一致好评，作为一个统一的环境，它能应对动态的多任务教学项目。无论从外面还是里面看去，教室、图书馆、艺术室、食堂都呈现为流畅、宽敞而又充满惊喜的空间。

细部与材料
以生石灰粉刷的白墙与砖砌机构和现浇混凝土十分搭配，与大面积的玻璃板形成了对比。开放的玻璃板模糊了室内外的界限。砖砌结构的开口为一系列隐蔽的天井提供了空气和阳光。

Salmtal Secondary School Canteen
萨姆塔尔中学食堂

Location/ 地点：Rhineland-Palatinate, Germany/ 德国，莱茵兰－普法尔茨
Architect/ 建筑师：SpreierTrenner Architekten (www.spreiertrenner.de)
Photos/ 摄影：Guido Erbring Architekturfotografie (www.guidoerbring.com)
Built area/ 建筑面积：552m²
Completion date/ 竣工时间：2012

Key materials: Façade – ceramic
Structure – concrete
主要材料：立面——瓷砖
结构——混凝土

Overview
The new school canteen of the Salmtal Secondary School in Germany was designed by SpreierTrenner Architekten (www.spreiertrenner.de) as a multifunctional building with the greatest possible flexibility. The space is used not only by children to eat every day, but also for special events such as music concerts, theatre plays or even Christmas fairs. This is why the main room was set out with a column-free square plan only subdivided by a mobile wall. The adjoining section at the rear contains all supporting facilities, such as a kitchen, storage space, toilets and staff facilities, etc. It has been set out with the option of extending it in the future.

The checked windows and the bright red Façade were inspired by the idea of creating a playful and interactive, yet efficient building. The starting point was the square plan, which allowed for maximum flexibility as it's multidirectional as well as easy to furnish for concerts uses, etc. Then all seats would be relatively close the stage.

To span a square plan most efficiently a two-directional grid was used. The grid got transferred to the façade as well, so the height of the room, the size of the windows and an efficient ratio for the wood trusses determined the grid proportion (exactly 1.575 m).

The big glazed entrance opens up the main canteen room to the outside and represents a welcoming gesture. The cantilevering canopy creates a transition zone between the interior and the playground.

Detail and Materials
All materials used were kept natural, robust, durable and simple. The main components were concrete, wood and ceramics.

The façade material draws its inspiration from the existing school building from the 70s, which has a red brick façade. So a red ceramics façade using 30mm thick panels was a contemporary close choice. SpreierTrenner Architekten

Concept Diagrams
概念示意图

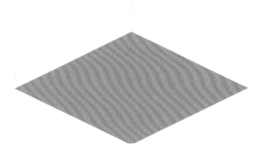

1. Square plan = flexibility
方形平面布局突显灵活性

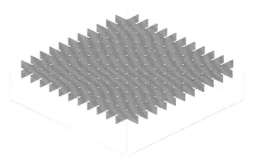

2. Grid structure creating column-free space
网格结构打造无支柱空间

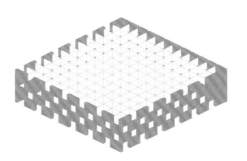

3. Grid transferred to exterior walls
网格结构延伸到外墙上

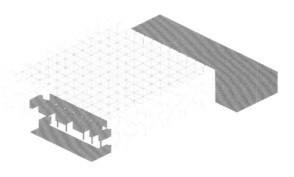

4. Kitchen and entrance added
增添的厨房和入口

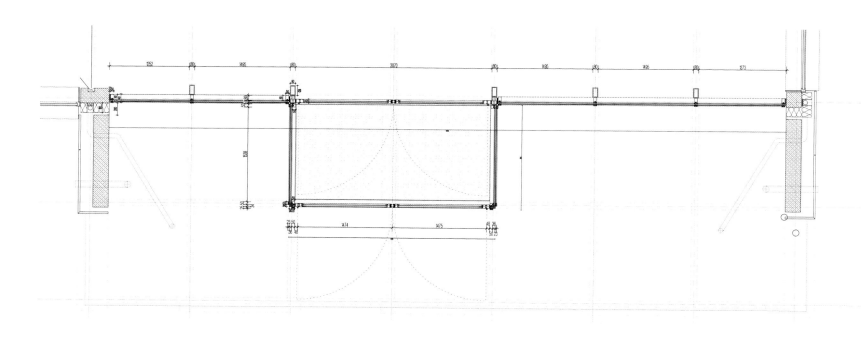

aimed to create something vivid and playful to engage the children, but also welcome any visitors. The building is the first thing you see when you arrive at the school complex, so it was important for the school to create a building that stands out and has a general welcoming and upbeat feel to it. That's why this vibrant shade of red was chosen. The glazed surface enhances the colour effect, makes the façade more robust to graffiti and easier to clean in general.

Although the façade is a bright red, no colours were used inside to allow the vibrant furniture to stand out. The concrete walls were constructed with a rough surface produced by OSB formwork that creates a warm texture. The floor shows the concrete screed surface, similar to the material used in car parks, covered with a transparent protective resin layer.

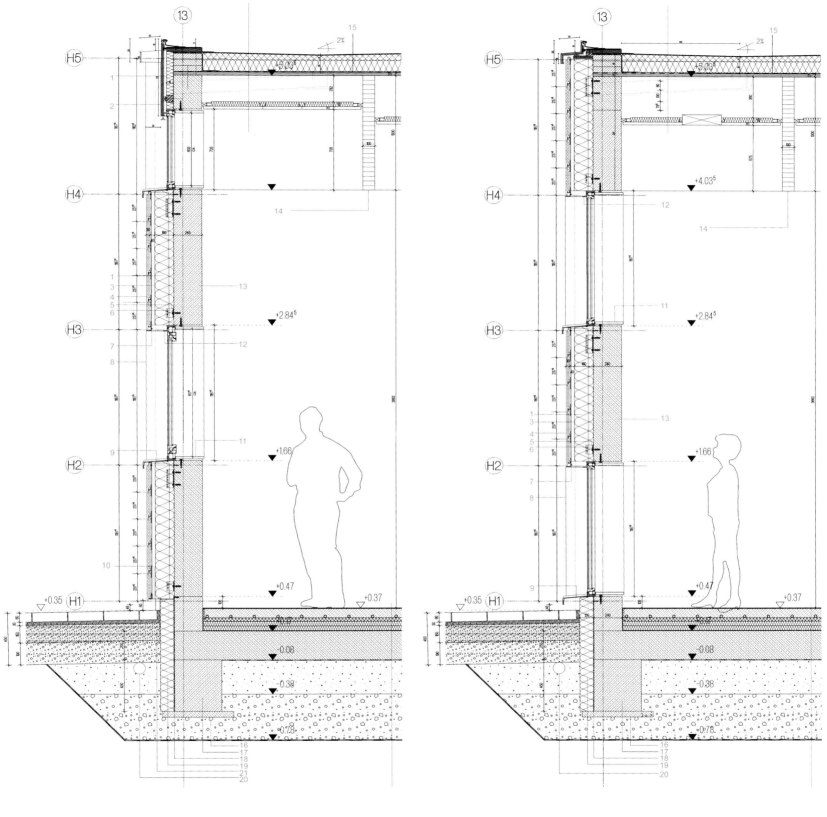

Façade Detail
1. Façade insulation 60mm
2. Aluminum – parapet
3. Aluminum substructure
4. Ceramic tiles
5. Wall fixing brackets
6. Insulation element
7. Aluminum soffit
8. Aluminum reveal
9. Aluminum window sill
10. Aluminum sheet
11. Timber frame
12. F-profile
13. 24cm in-site concrete
14. Bonded wood roof girders
15. Plywood-board on concrete wall / timber girders
16. Blinding layer 5cm
17. Ice wall concrete 40/45
18. Asphalt sheet
19. Insulation extruded Polystyrol 120mm
20. Drainage pipe 100mm
21. Base sheet aluminum

立面节点
1. 外墙隔热层 60mm
2. 铝栏杆
3. 铝制下层结构
4. 瓷砖
5. 墙面固定支架
6. 隔热元件
7. 铝底面
8. 铝窗侧
9. 铝窗台
10. 铝板
11. 木框
12. F型材
13. 24cm 现场浇筑混凝土
14. 黏合木屋顶梁
15. 混凝土墙上胶合板 / 木梁
16. 地基垫层 5cm
17. 冰墙混凝土 40/45
18. 沥青板
19. 挤塑聚苯乙烯隔热层 120mm
20. 排水管 100mm
21. 铝制底板

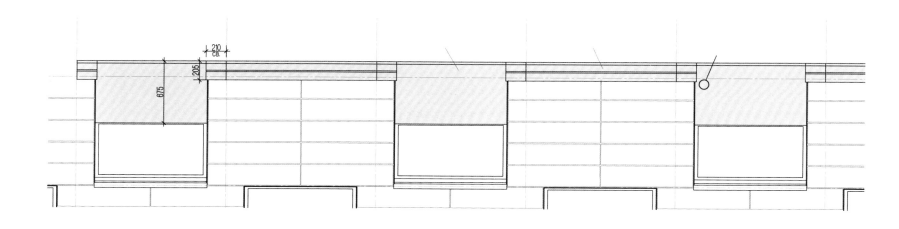

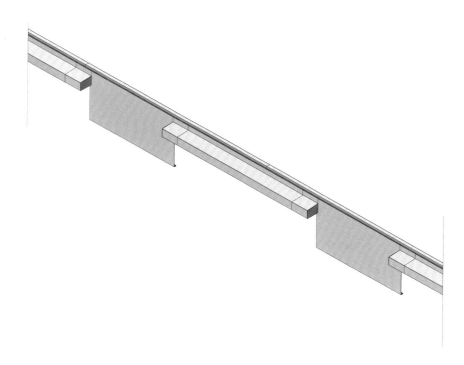

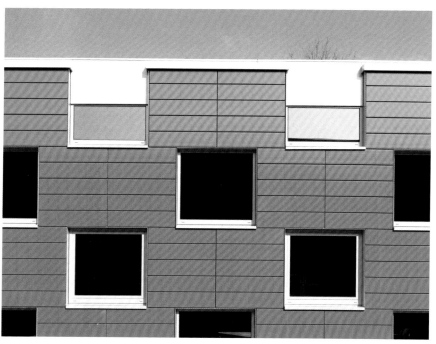

项目概况

德国萨姆塔尔中学新建的学校食堂由 SpreierTrenner 建筑事务所设计，是一座具有高度灵活性的多功能建筑。建筑空间不仅能满足每日的就餐需求，还能举办音乐会、戏剧表演、圣诞节派对等特殊活动。因此，主大厅采用无柱式方形平面布局，仅用移动墙隔开。食堂后方的附属结构内设置着所有的辅助设施，如厨房、储藏室、洗手间、员工设施等。食堂的设计已经为未来的扩建奠定了基础。

格纹式窗口和亮红色外墙的设计灵感来自于打造一个有趣的高效互动建筑的想法。设计的出发点是方形平面布局：多向等边空间保证了最大限度的灵活性，同时也便于进行音乐会等活动的布置，所有座位都可以更靠近舞台。

为了最高效地利用方形布局，设计采用了双向网格。同时，网格也被转移到了外墙上。因此，房间的高度、窗口的尺寸和木桁架的效率比共同决定了网格的比例和尺寸（精确地说，是 1.575 米）。

宽大的玻璃门入口将主食堂与外界连接起来，呈现出欢迎的姿态。悬臂式雨篷在室内和草场之间形成了过渡。

细部与材料

建筑所用的所有材料都保证了自然、坚固、耐用、简单的特性，其中最主要的材料是混凝土、木材和陶瓷。

立面材料的设计灵感来自于建于 20 世纪 70 年代的红砖面教学楼。因此，食堂建筑选择了更富现代感的 30 毫米厚瓷砖立面。SpreierTrenner 建筑事务所的目标是通过活泼生动的设计来吸引学生，同时也欢迎任何来访者。当人们走进校园，食堂是第一座映入眼帘的建筑，给人以热情乐观的感觉。因此建筑师选择了充满活力的红色系。抛光表面突出了色彩的效果，让立面不受涂鸦困扰且便于清洗。

尽管食堂外观呈亮红色，内部墙面却没有使用显眼的色彩，而是用亮色家具突出了氛围。定向刨花板模架为混凝土墙面营造了粗糙的表面，形成了温馨的纹理。地面露出了混凝土砂浆面，与停车场的铺装材料类似，并且在上方铺设了一层透明的树脂保护层。

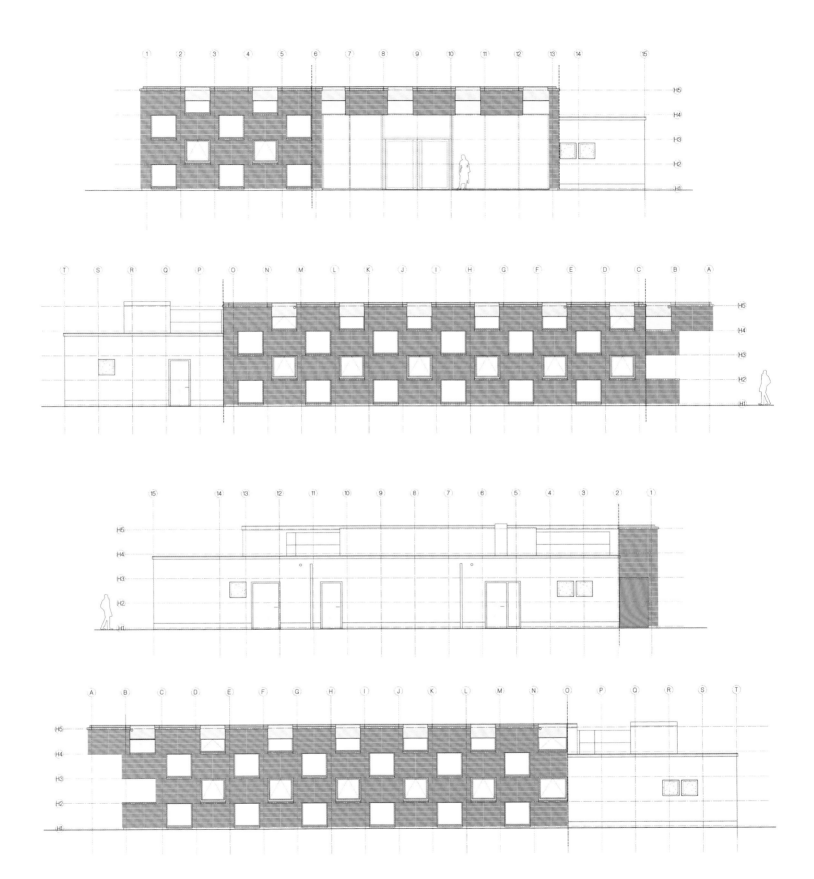

Kindergarten Parque Goya
戈雅公园幼儿园

Location/ 地点：Zaragoza, Spain/ 西班牙，萨拉戈萨
Architect/ 建筑师：Magén Arquitectos (Jaime Magén, Francisco Javier Magén)
Photos/ 摄影：Pedro Pegenaute, Jesús Granada
Built area/ 建筑面积：2,081.99m²
Budget/ 预算：2,435,489.00 €/2,435,489.00 欧元

Key materials: Façade – ceramic
主要材料：立面——陶瓷

Overview
The location, an urban edge at the north of Zaragoza, and the will to make a connection between interior and outdoor spaces led to approach the project as an open organisation of teaching linear elements colonising the area with an extensive layout, grouping around the outdoor spaces open to the landscape and shaping play areas. The strategy is based in the concepts of additive architecture and carpet-building, from the definition of the class as a unit or cell-types and their grouping into an open, flexible and extensive nature, setting up a rug or carpet on the ground, which defines the alternation between interior and exterior spaces.

The project approach also responds to the condition of very tight construction time (4 months) which involved the election of building solutions based on simplicity and quickness of execution. The programme defined by the Government of Aragon, which includes 12 classrooms, dining-hall and complementary spaces, breaks into an open layout that allows it to make a perceivable environment for children and ensure the lighting and natural ventilation in all spaces.

Detail and Materials
Outside, the building is defined by a continuous roof of inclined walls over a ventilated façade made by a large-format ceramic-extruded cladding. Ceramic fixed, anodised aluminium slats and the metallic finishing of the porches complete the exterior image of the building.

项目概况
项目位于萨拉戈萨北部的城市边缘，规划目标是在室内外空间内建立起积极的联系，打造开放型线性教学空间，将户外活动区集中起来面向景观开放并且塑造特色游戏区。项目的设计策略以"加法建筑"和"地毯式建筑"概念为基础：每间教室被设计成一个独立的单元，它们以灵活开放的形式聚集起来；在地面上配置一块"地毯"，明确划分出室内外空间的界限。

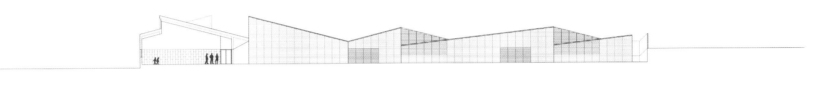

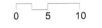

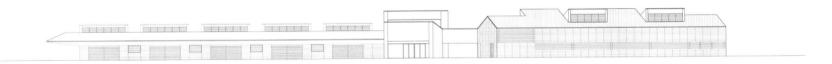

LOCATION

项目设计还充分考虑到了紧张的施工时限（4 个月），选择了简单快速的建造方式和施工方案。

项目配置由阿拉贡政府决定，包含 12 间教室、餐厅和辅助空间。开放式布局不仅让孩子们对环境一目了然，而且保证了各个空间的照明和自然通风。

细部与材料

建筑外观以连续的斜屋顶和斜边墙壁为特色，由大块瓷砖构成了通风立面。瓷砖、阳极氧化铝板和金属饰面的门廊共同组成了建筑的外部形象。

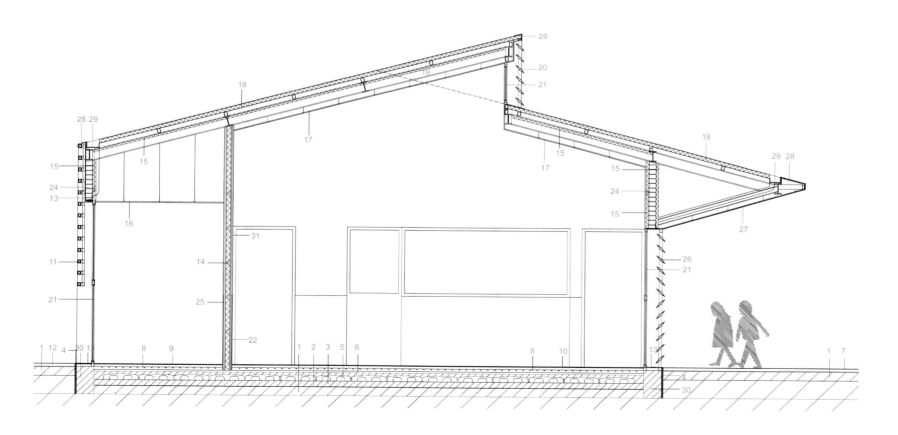

Detail Section – classroom 教室剖面节点

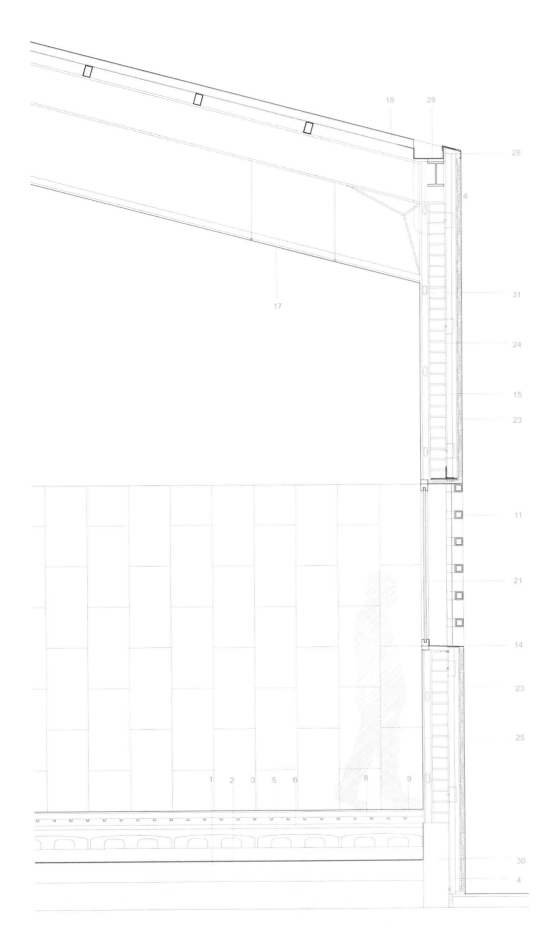

Detail
1. Gravel layer
2. Waterproofing membrane
3. Concrete cleaning base t=10cm
4. Extruded polystyrene joint t=3cm
5. Ventilated slab CUPOLEX t=15cm
6. Mortar cement leveling
7. Brown coloured concrete slab
8. Underfloor heating
9. Ceramic stone paving STONKER YUTE 90x45cm
10. Linoleum pavement
11. Lattice of squared ceramic slats FAVETON BRIOL
12. Rubber paving
13. Collapsible aluminium sheet
14. Plasterboard dividing wall
15. Insulation: extruded polystyrene panel t=40mm
16. Suspended ceiling of plasterboard
17. Suspended ceiling of plasterboard. Acoustical finish
18. Sandwich panel t=8cm
19. Wood suspended ceiling 2400x300x6mm
20. Adjustable aluminum slats UMBELCO UPO 250
21. Aluminium carpentry
22. Linoleum ZOCALO
23. Ceramic tile FAVETON CERAM 1200x200x28mm
24. Preformed brick pieces t=11cm
25. Ceramic stone ZOCALO STONKER YUTE 90x45cm
26. Adjustable aluminium slats UMBELCO UPO 250
27. Suspended ceiling of perforated plate
28. Coping of galvanised steel sheet
29. Steel perimetral gutter
30. Concrete wall
31. Plasterboard panel. Acoustical finish

节点
1. 碎石层
2. 防水膜
3. 混凝土底座 t=10cm
4. 挤塑聚苯乙烯接缝 t=3cm
5. 通风板 CUPOLEX t=15cm
6. 砂浆水泥找平
7. 棕色混凝土板
8. 地热供暖
9. 陶瓷地砖铺装 STONKER YUTE 90x45cm
10. 油毡铺面
11. 方形陶瓷条格架 FAVETON BRIOL
12. 橡胶铺装
13. 可折叠铝板
14. 石膏板隔断墙
15. 隔热层：挤塑聚苯乙烯板 t=40mm
16. 石膏板吊顶
17. 石膏板吊顶，隔音饰面
18. 夹心板 t=8cm
19. 木吊顶 2400x300x6mm
20. 可调节铝条 UMBELCO UPO 250
21. 铝窗框
22. 油毡 ZOCALO
23. 瓷砖 FAVETON CERAM 1200x200x28mm
24. 预成型砖块 t=11cm
25. 陶瓷地砖 ZOCALO STONKER YUTE 90x45cm
26. 可调节铝条 UMBELCO UPO 250
27. 穿孔板吊顶
28. 镀锌钢板顶盖
29. 钢排水槽
30. 混凝土墙
31. 石膏板，隔音饰面

Detail Section – dining hall 食堂剖面节点

Corte Verona Apartment Building
柯特维罗纳公寓楼

Location/ 地点: Wrocław, Poland/ 波兰，弗罗茨瓦夫
Architect/ 建筑师: Kazimierz Łatak, Piotr Lewicki/ Biuro Projektów Lewicki-Łatak
Photos/ 摄影: Piotr Łatak
Site area/ 占地面积: 16,247m²
Built area/ 建筑面积: 4,283m²

Key materials: Façade – brick slips
Structure – reinforced concrete
主要材料：立面——砖片
结构——钢筋混凝土

Overview
Corte Verona building continues a characteristic for Wrocław, multi-century use of brick in the process of construction. The building was given a well-known, well-functioning form of an urban block with a division of the space into: public (outside), private (flats) and semi-public (the courtyard).

Detail and Materials
The structure, of reinforced concrete, was designed as a set of transverse vertical walls, which, while separating particular flats, isolate them acoustically and horizontal slabs which are elements of the floors.

In a close vicinity there are examples of using brick both while constructing buildings of the same function (originated at the beginning of the twenties of the previous century, Grabiszyn housing estate, by Paul Heim and Albert Kempter) and in creating the so-called small architecture objects: walls, benches and fences. And it was the last of those objects, an openwork brick wall in the neighbourhood became a direct inspiration of shaping the structure of the new architecture by the authors. Bricks were (mentally) rescaled to a size of one flat and the gaps between one another became loggias.

The façade is made of brick slips (of two thicknesses and colours), placed interchangeably on each side of the quadrangle together with wooden windows and glass balustrades. On the ground floor level, there is silicate plastering (reinforced by a double polypropylene mesh) together with aluminium doors and windows.

项目概况
柯特维罗纳公寓楼延续了弗罗茨瓦夫的整体建筑风格，选择了砖作为主要建筑材料。建筑结构清晰、功能明确，将空间划分为三个层次：公共（外部）、私有（公寓）和半公共（庭院）。

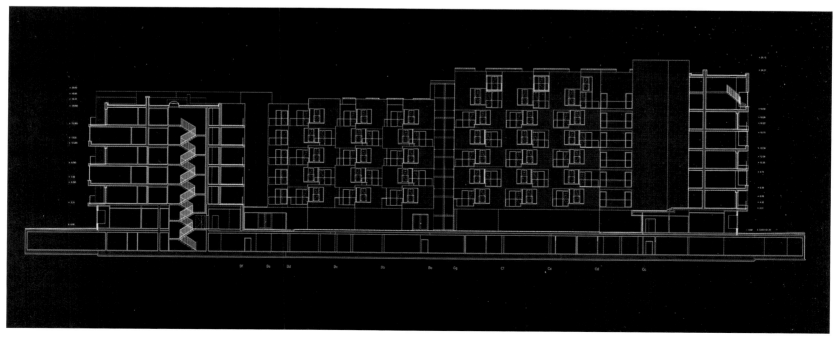

细部与材料

钢筋混凝土结构被设计成横断面的直立墙壁和水平楼板的组合。前者不仅隔断了不同的公寓,还能实现隔音效果;后者则是构成楼面的主要元素。

项目周边的同类建筑大多采用砖石作为建筑材料(大概从20世纪20年代开始,以保罗·海姆和艾伯特·坎普特的格拉比思辛住宅区为代表),其他的建筑也在墙壁、长椅、围栏中运用了砖元素。附近的一面镂空砖墙是项目设计的主要灵感。建筑师对砖块进行了重新组合,用一间公寓代表一块砖,而相邻公寓之间的镂空空间则形成了凉廊。

建筑立面由砖片(分为两种色彩和两种厚度)构成,它们在四边形的各面交替拼接,与木窗和玻璃围栏共同组成了丰富多变的外墙形式。一楼的立面由硅酸盐石膏墙面(双层聚丙烯网加固)以及铝门、铝窗构成。

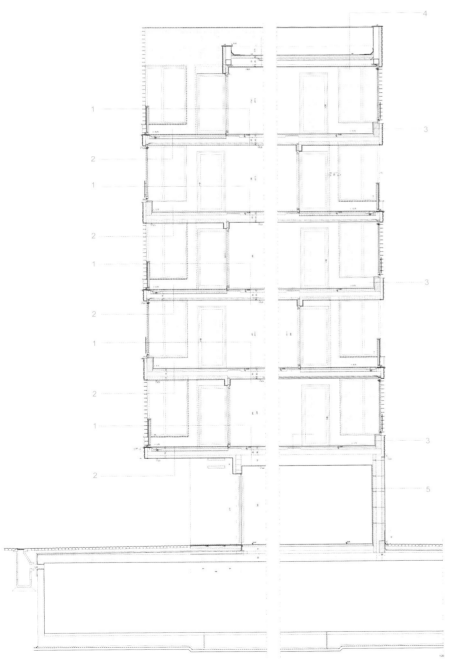

Detail

1. Parquet 2.0
   Screed concrete 4.0
   PE membrane -
   Sound insulation + styrofoam 7.0
   Reinforced concrete 20.0
   Gypsum board 1.25

2. Wooden deck 6.5
   Separation layer 6.5
   Waterproof layer -
   Styrofoam 10.0
   PE membrane -
   Reinforced concrete 20.0
   Styrofoam 10.0
   Plaster 0.5

3. Clinker tiles 2.0-4.0
   Polyurethane foam 12.0
   Porous concrete 12.0
   Gypsum board 1.25

4. Styrofoam 20.0
   PE membrane -
   Reinforced concrete 20.0
   Gypsum board 1.25

5. Plaster 1.5
   Ceramic brick 12.0
   Polystyrene 20.0
   Ceramic brick 12.0
   Gypsum board 1.25

节点

1. 护墙 2.0
   砂浆水泥 4.0
   聚丙烯膜 -
   隔音 + 聚苯乙烯泡沫 7.0
   钢筋混凝土 20.0
   石膏板 1.25

2. 木平台 6.5
   隔层 6.5
   防水层 -
   聚苯乙烯泡沫 10.0
   聚丙烯膜 -
   钢筋混凝土 20.0
   聚苯乙烯泡沫 10.0
   石膏 0.5

3. 熔渣釉面砖 2.0-4.0
   聚氨酯泡沫 12.0
   多孔混凝土 12.0
   石膏板 1.25

4. 聚苯乙烯泡沫 20.0
   聚乙烯膜 -
   钢筋混凝土 20.0
   石膏板 1.25

5. 石膏 1.5
   瓷砖 12.0
   聚苯乙烯 20.0
   瓷砖 12.0
   石膏板 1.25

Blood Centre
血液中心

Location/ 地点：Racibórz, Poland/ 波兰，拉齐布日
Architect/ 建筑师：FAAB Architektura Adam Białobrzeski | Adam Figurski
Photos/ 摄影：Bartłomiej Senkowski
Site area/ 占地面积：22,380m²
Built area/ 建筑面积：33,132m²

Key materials: Façade – NBK terracotta glazed ceramic panels, NBK terracotta Terrart-baguette, Appiani ceramic wall tiles, Schüco windows and doors
主要材料：立面——NBK 釉面陶瓷板、NBK 陶瓷条、Appiani 墙面瓷砖、Schüco 门窗

Façade material producer:
外墙立面材料生产商：
NBK（terracotta glazed ceramic panels & terracotta Terrart-baguette & Terrart-baguette）：(抛光陶瓷板 & 陶板)
Appiani (ceramic wall tiles)；(墙面瓷砖)
Schüco(windows and doors)（门窗）

Overview

The investment divided into three phases, including (1) construction of the new Regional Blood Centre building, (2) shelter for the mobile blood centre [bus] and (3) the first in Poland specialized Centre for Blood Cancer Diagnostics to be located within the existing building. At the moment, the first phase has been completed.

The first phase building houses modern cool rooms and storages, blood collecting unit, a complex of advanced medical laboratories and blood radiation laboratory, offices and conference centre. Within the building blood is being collected, tested, processed to divide into blood elements and purified with the technology based on radioactive materials.

Detail and Materials

The building elevation is finished with ceramics, including small ceramic tiles on the ground level and glazed ceramic panels on the upper floor levels. The upper levels consist of more than 2,000 ceramic elements. Among them 50x100cm flat panels, vertical blinds and (for the first time in Poland) curved panels of 145cm and 195cm radius. Three different shades of red were implemented to create irregularity on the surface of the all four elevations. These shades symbolize the variation of colour in blood. The colour of blood, oxygenated in the lungs and flowing in the arteries, is bright red, while blood flowing in the veins, containing carbon dioxide, is dark red.

The Ceramic pipes (TERRART®-BAGUETTE), placed at the uppermost level, protect the conference room from overheating as well as hide complicated and massive technological systems located at the roof level. These ceramic pipes add detail and make the enclosure lighter in its appearance.

The colour scheme, the irregularity of the elevation leitmotif and the panel's glossiness represent the richness of blood, often described as the gift of life.

The geometry is inspired by the function of the blood centre, the clash of biology with technology. Rounded elements represent biology, straight represent technology.

The intentional vivid colour scheme of the elevation, making the building visually surprising to the public, calls attention to the idea of the blood donation. Application of the glazed panels is inspired by the local Silesian building tradition, which is present in the historical building façades with the glazed bricks.

项目概况

项目投资分为三个阶段：（1）地区血液中心大楼的建造；（2）一栋采血中心（采血车）的停车亭；（3）波兰第一家专业血癌诊断中心的设立。目前，第一阶段的任务已经完成。

一期建筑内设有现代的冷藏室和仓库、血液采集室、高级医疗实验室、血液放射实验室、办公室和会议中心。在建筑内可以实现血液的采集、检测、血液成分分析、净化（通过放射性物质为基础的技术实现）等。

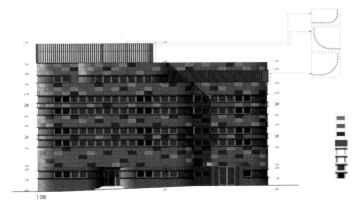
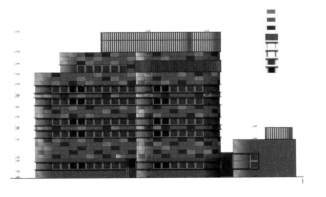
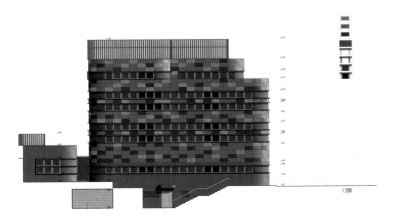
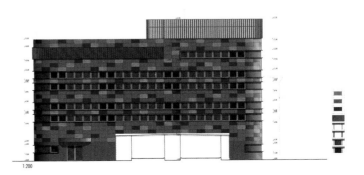

Façade Detail

1. Galvanised steel sheet: RAL 7035 (light grey)
2. Exterior vertical blinds, 50×150mm, ceramic, glazed, colour & material coordinated with wall panels
3. Glazed ceramic wall panels: NBK Terracotta
4. Aluminium window profiles: Schuco, RAL 7035
5. Glass façade: FW50 + HI Schueco
6. Windowsill: CORIAN SILVERITE
7. LED stripe (warm white light)
8. Rectangular steel tube: RAL 3003
9. Plaster & cement finishing painted acrylic-latex: Beckers Scotte 7, NCS 2000N
10. Ceramic mosaic tile: Appiani 2.5×5.0cm, RAL 3005
11. Building perimeter filled with washed gravel stones: (Φ1.5-2.5cm)

立面节点

1. 镀锌钢板：RAL 7035（浅灰色）
2. 外部垂直百叶条：50x50mm，釉面陶瓷，色彩与材料与墙面板相一致
3. 釉面陶瓷墙面板：NBK Terracotta
4. 铝窗框：Schuco, RAL 7035
5. 玻璃立面：FW50 + HI Schueco
6. 窗台：CORIAN SILVERITE
7. LED 灯带（暖白色光）
8. 矩形钢管：RAL 3003
9. 石膏水泥饰面，涂丙烯酸乳胶漆：Beckers Scotte 7, NCS 2000N
10. 陶瓷锦砖：Appiani 2.5 × 5.0cm, RAL 3005
11. 建筑外围填充着洗净砾石：直径 1.5~2.5cm

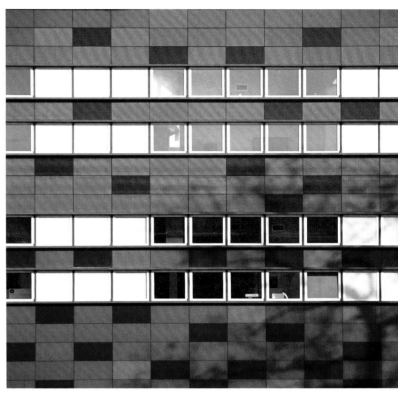

细部与材料

建筑立面采用瓷砖进行装饰,包含建筑底层的小型瓷砖块和上层的釉面陶瓷板两种形式。整个上层楼体外墙拥有超过 2,000 块陶瓷元件,其中包括 50x100cm 的平板、垂直百叶条以及半径为 145cm 和 195cm 的曲面板(在波兰尚属首创)。三种不同深度的红色在四面墙面上形成了不规则的感觉。这些色彩代表着血液色彩的变化。经肺部充氧后在动脉中流动的血流是鲜红色的,而静脉中的血流由于含有二氧化碳,呈现为深红色。

最顶层所使用的陶瓷管(TERRART®–BAGUETTE)能保护会议室不会过热,同时也能将屋顶复杂的大规模技术系统隐藏起来。这些陶瓷管能够增加建筑的精美感,让建筑外观看起来更轻盈。

外墙的色彩搭配、不规则的立面主题以及陶瓷板的光泽都体现了血液的丰富性,体现了血液作为"生命的馈赠"的价值。

建筑的结合造型从血液中心的功能中获得了灵感,即生物与技术的冲突。圆形元素代表着生物,直线元素则代表着技术。

建筑立面的鲜艳色彩使建筑让公众眼前一亮,从而呼吁人们关注献血。釉面板的应用参考了西里西亚地区的建筑传统,当地的历史建筑墙面通常均为釉面砖。

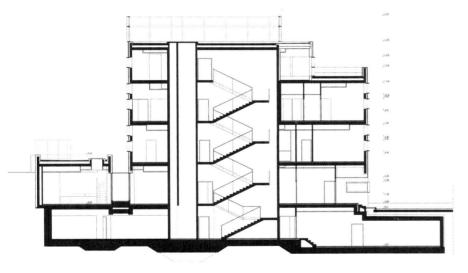
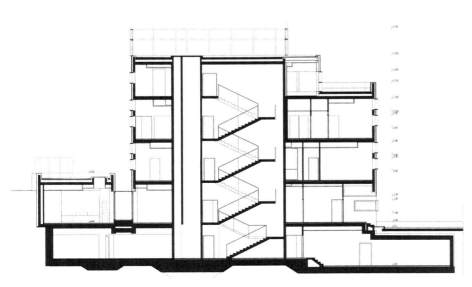

Teacher Training Centre, Archbishop of Granada

天主教格拉纳达总教区教师培训中心

Location/ 地点：Granada, Spain/ 西班牙，格拉纳达
Architect/ 建筑师：Ramón Fernández-Alonso and associated
Photos/ 摄影：Jesús Granada
Completion date/ 竣工时间：2012
Site area/ 占地面积：8,900m²
Built area/ 建筑面积：19,756.19m²

Key materials: Façade – ceramic
主要材料：立面——陶瓷

Façade material producer:
外墙立面材料生产商：
Cumella. Decorativa

Overview
The building is structured in four levels: two below the ground as parking and three above containing the program, which is divided into six areas: Common areas, located on the ground floor and composed mainly of following areas: vestibular areas, cafeteria, library, auditorium, gymnasium, stationery, photocopying; Teaching area, located on the first floor and part of the second floor, comprising classrooms and workshops for students; Departmental area. It consists on offices and meeting areas. It has a teacher consultation, on the second floor; Management and secretarial area, located on the ground floor; Parking area, located in the semi-basement and basement of the building, has 300 spaces for vehicles; Church with a separate entrance from the outside.

Detail and Materials
The initial idea is to provide a closer architecture, almost familiar, achieving it with the spaces composition as well as in the light's treatment and the texture's ceramic exterior skin. The materiality is linked from the beginning to the argument of the project process.

Ground floor is projected as a threshold space compressed by the building itself in its upper floors. The ceramic enclosure of the classrooms hangs over the city plan. This basic approach results from the structural solution adopted by this project based on the study of the section: a roof truss, containing the two upper floors of classrooms and departments, supported by two lines of brackets, saving a long distance. Under them, common areas are developed in continuity with the outside landscaped terraces and protected by powerful cantilever.

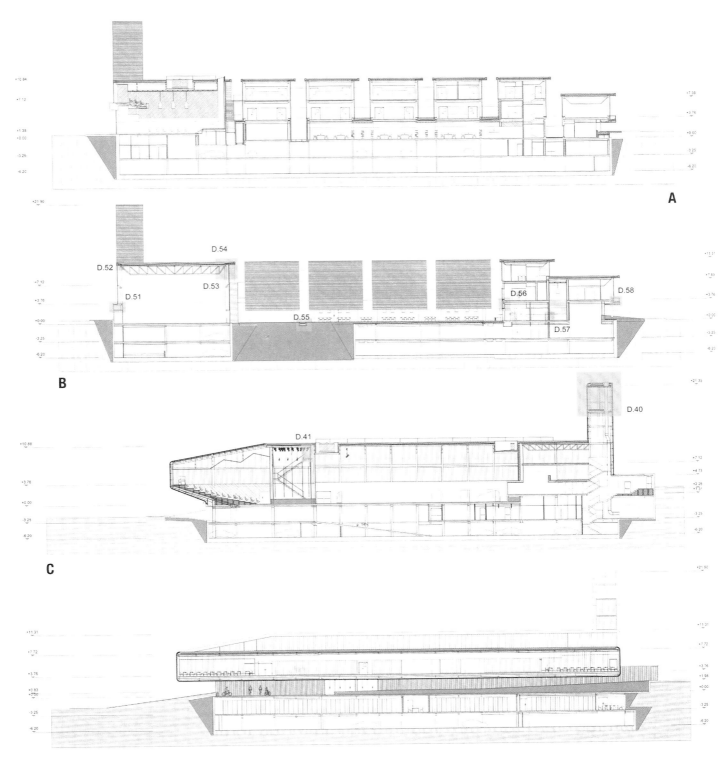

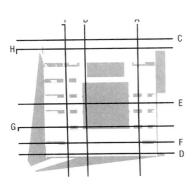

项目概况

建筑分为四层结构：地下停车场和三层地上空间。地上空间分为六个区域。公共区：位于一楼，由前庭区、自助餐厅、图书室、礼堂、健身房、文具店、影印室组成；教学区：位于二楼和三楼的部分区域，由教师和学生工作室组成；部门办公区：包含各种办公室和会议区，在三楼设有教师咨询室；管理及秘书区：位于一楼；停车区：位于地下室和半地下室，有300个车位；教堂：在外面设有独立入口。

细部与材料

项目的初始概念是通过光线处理和陶瓷表皮的质感设计出一座令人感到亲切的建筑。建筑师开始构思到项目过程中一直在研究材料的选择。

建筑一楼以入口空间的形式凸出，上方的楼层实现了一种压迫感。教室的陶瓷外墙仿佛笼罩在城市规划之上。这种基本设计来自于项目所选择的结构策略：屋架结构。该结构包含两层教室和部门办公楼层，由两条支架支撑，节约了很长的距离。下方的公共区域户外景观平台形成连续的空间，上方则有宏大的悬臂结构提供保护。

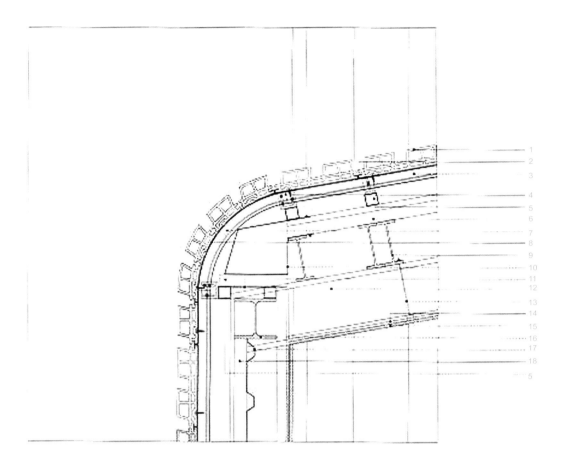

Façade Detail
1. Ceramic tile
2. PGR
3. Profile TV2010
4. Angel regulation
5. Horizontal auxiliary battens, Tubular 60x60mm
6. Sandwich panel
7. Auxiliary curved profile
8. Curved profiled sheet
9. Profile IPE 220
10. Galvanised steel gutter
11. Welded metal plate
12. Profile HEB 300
13. Threaded rod continually adjustable roof
14. Auxiliary profile securing continuous roof Pladur®
15. Continuous roof cardboard plate – plaster system Pladur®
16. Profile HEB 200
17. Galvanised corrugated sheet
18. Expanded polyurethane insulation foam

立面节点
1. 瓷砖
2. PGR
3. 型材 TV2010
4. 角度校准
5. 水平辅助管 60x60mm
6. 夹层板
7. 辅助曲式型材
8. 曲面板
9. 型材 IPE 220
10. 镀锌槽
11. 焊接金属板
12. 型材 HEB 300
13. 连续可调节屋顶的螺纹杆
14. 辅助型材，固定连续屋顶 Pladur®
15. 连续屋顶板，石膏系统 Pladur®
16. 型材 HEB 200
17. 镀锌波纹板
18. 泡沫聚氨酯隔热层

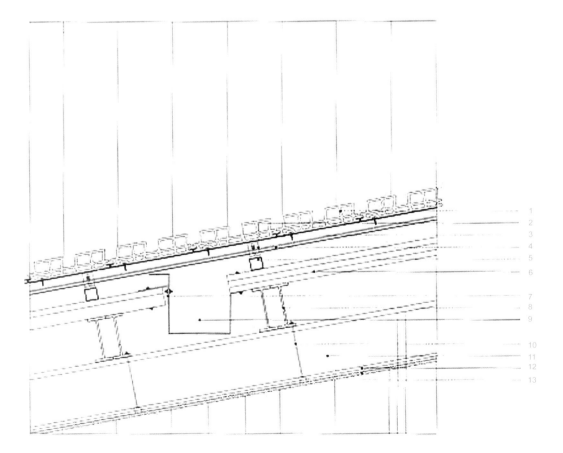

Façade Detail
1. Ceramic tile
2. Angel regulation
3. PGR
4. Profile TV2010
5. Horizontal auxiliary battens, Tubular 60x60 mm
6. Sandwich panel
7. Auxiliary anchor plate
8. Profile HEB 200
9. Rectangular galvanised steel gutter
10. Threaded rod continually adjustable roof
11. Profile HEB 300
12. Auxiliary profile securing continuous roof Pladur®
13. Continuous roof cardboard plate – plaster system Pladur®

立面节点
1. 瓷砖
2. 角度校准
3. PGR
4. 型材 TV2010
5. 水平辅助管 60x60 mm
6. 夹层板
7. 辅助锚固板
8. 型材 HEB 200
9. 矩形镀锌钢槽
10. 连续可调节屋顶螺纹杆
11. 型材 HEB 300
12. 辅助型材，固定连续屋顶 Pladur®
13. 连续屋顶板，石膏系统 Pladur®

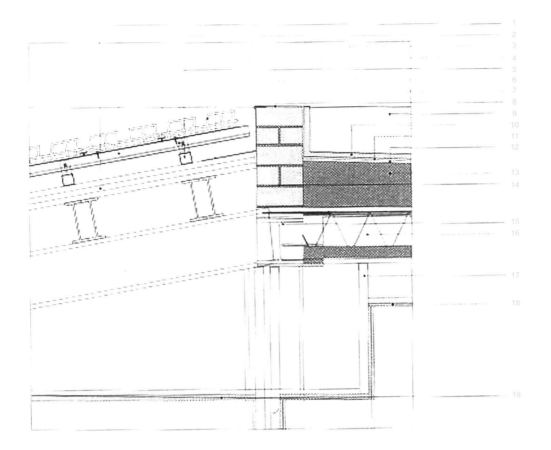

Façade Detail
1. Sandwich panel
2. Profile TV2010
3. PGR
4. Angle regulation
5. Horizontal auxiliary battens, Tubular 60×6 mm
6. Ceramic tile
7. Folded galvanised steel sheet
8. Top of galvanized steel sheet folded in 'U'
9. Ceramic protective layer of gravel recycled RCDs
10. Extruded polystyrene board Roofmate on film
11. Laminated polymeric bitumen waterproofing of $4.80 kg/m^2$
12. Regularisation mortar layer. e=20mm
13. Forming concrete slope celuar
14. Screed 1 foot perforated brick
15. Structural profile HEB 240
16. Cofradal wrought prefabricated plate 250
17. Structure Pladur® auxiliary system
18. Continuous roof system Pladur®
19. False ceiling

立面节点
1. 夹层板
2. 型材 TV2010
3. PGR
4. 角度校准
5. 水平辅助管 60x60mm
6. 瓷砖
7. 折叠镀锌钢板
8. 镀锌钢板顶盖，折成 U 形
9. 回收碎石的陶瓷保护层
10. 挤塑聚苯乙烯板 Roofmate
11. 夹层聚合沥青防水层 4.80 kg/m²
12. 校准砂浆层 e=20mm
13. 成形混凝土坡
14. 砂浆层，1 英尺预制砖
15. 结构型材 HEB 240
16. Cofradal 锻造预制板 250
17. 结构辅助系统 Pladur®
18. 连续屋顶系统 Pladur®
19. 假吊顶

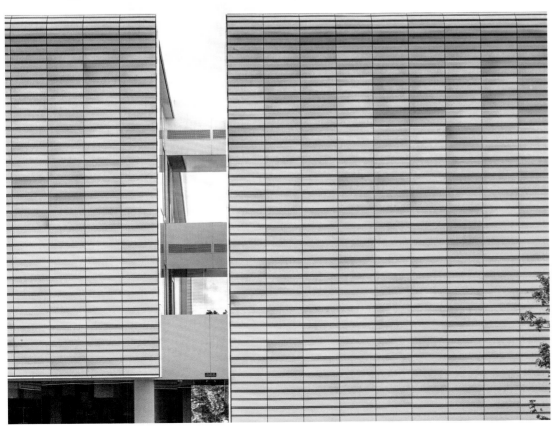

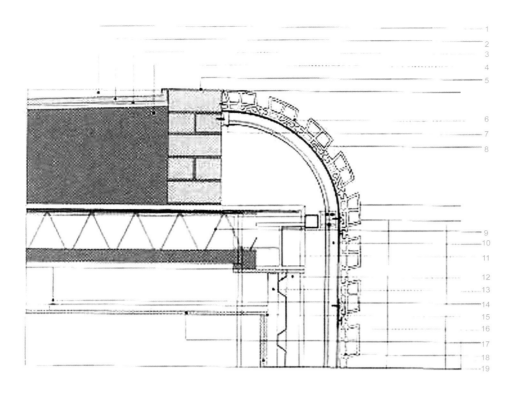

Façade Detail

1. Ceramic protective layer of gravel recycled RCDs
2. Extruded polystyrene board Roofmate on film
3. Laminated polymeric bitumen waterproofing of 4.80kg/m²
4. Forming concrete slope cellular
5. Top of galvanised steel sheet folded in 'U'
6. Convex curved ceramic tile
7. Profile 'L' for fastening aluminum profile parapet
8. Curved shaped plate
9. Cofradal wrought prefabricated plate 250
10. Profile TV2010
11. Galvanised corrugated sheet
12. Expanded polyurethane insulation foam
13. Upright in 'C' of 46 mm. Pladur® system
14. Threaded rod continually adjustable roof
15. Profile 'L' continuous roof fixing system Pladur®
16. PGR
17. Pladur® continuous ceiling hook system
18. Ceramic tile
19. Pladur® wall system consisting of two plates e=15

立面节点

1. 回收碎石的陶瓷保护层
2. 挤塑聚苯乙烯板 Roofmate
3. 夹层聚合沥青防水层 4.80 kg/m²
4. 成形混凝土坡
5. 镀锌钢板顶盖，折成 U 形
6. 凸面瓷砖
7. L 形型材，固定铝护栏
8. 弧形板
9. Cofradal 锻造预制板 250
10. 型材 TV2010
11. 镀锌波纹板
12. 泡沫聚氨酯隔热层
13. C 形支柱 46mm，Pladur® 系统
14. 连续可调节屋顶的螺纹杆
15. L 形连续屋顶固定系统 Pladur®
16. PGR
17. Pladur® 连续天花板挂钩系统
18. 瓷砖
19. Pladur® 墙面系统，由两层板材构成 e=15

Façade Detail

1. Protective layer of boulder gravel
2. Protective layer and mortar. 30mm. spacer layer on geotextile
3. Laminated polymeric bitumen waterproofing of 4.80kg/m²
4. Regularisation mortar layer. e=20mm
5. Folded galvanised steel sheet
6. Convex curved ceramic tile
7. Folded sheet of galvanised Steel with sealed joints
8. Wrought polystyrene vaults
9. Siphon drain
10. Corrugated galvanised steel sheet
11. Anodized aluminum 6060 T5 system Materia Granada
12. Horizontal anodized aluminum 6060 T5 system Materia Granada
13. Ceramic tile
14. Auxiliary profile securing continuous roof Pladur®
15. Continuous roof plate cardboard - plaster Pladur® system
16. Upright in 'C' of 46 mm. Pladur® system
17. Pladur® wall system consisting of two plates e=15

立面节点

1. 碎石保护层
2. 保护层和砂浆 30mm，土工布上方分隔层
3. 夹层聚合沥青防水层 4.80kg/m²
4. 校准砂浆层 e=20mm
5. 折叠镀锌钢板
6. 凸面瓷砖
7. 折叠镀锌钢板，密封接合
8. 锻造聚苯乙烯顶盖
9. 虹吸疏水
10. 波纹镀锌钢板
11. 阳极氧化铝 6060 T5 系统 Materia Granada
12. 水平阳极氧化铝 6060 T5 系统 Materia Granada
13. 瓷砖
14. 辅助型材，保护连续屋顶 Pladur®
15. 连续屋顶板，石膏板 Pladur® 系统
16. C 形支柱 46mm，Pladur® 系统
17. Pladur® 墙面系统，双层板 e=15

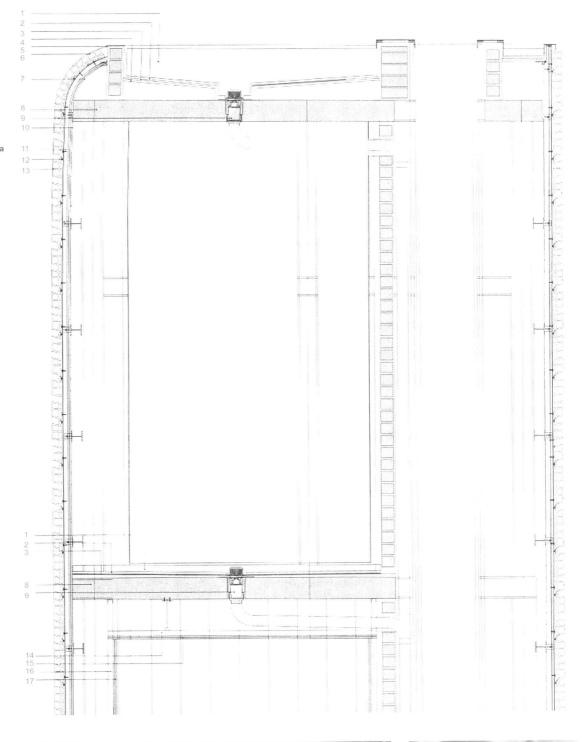

Fire Station, Vilnius
维尔纽斯消防站

Location/ 地点：Vilnius, Lithuania/ 立陶宛，维尔纽斯
Architect/ 建筑师：UAB "Laimos ir Ginto projektai"
Photos/ 摄影：Raimondas Urbakavičius
Area/ 面积：4,000m²
Completion date/ 竣工时间：2012

Key materials: Façade – tile "Argeton"
主要材料：立面——Argeton 砖

Overview

A new fire station was building built in Vilnius, capital of Lithuania, rising on the west bank of the river Neris. Building consists of 2 parts: emergency response centre (ERC) and a fire station. In the lower part, stretched along the river, are located premises of a fire station and a garage of the trucks. In the upper part, which is a dominant point in the architecture and is stretched along the street, are located premises of ERC. However, both these services do not intersect each other. Work in these structures is hard, tight and closed from the public. Therefore architects had a dual task between openness and closure.

Detail and Materials

Finish of a façade is coated with hanging tiles "Argeton". A dark colour blends the building with the surroundings. A white background of neighbouring volumes becomes a compositionally good frame for a fire station to live its independent, open and interesting life.

Moreover, the human factor was a basic when thinking about structure and architecture. Building is constructed opened to its surroundings and demonstrates close connection. Walls of glass with a view to the river help to relax and to calm. The dug lower part of the building with the grass roof creates and illusion of the hill. Thus other buildings do not lose the connection with the river. The grass roof is a space for a recreation, which is protected from the noise by the upper part.

项目概况

立陶宛首都维尔纽斯新建了一座消防站，就坐落在涅利斯河西岸。建筑由两部分构成：应急反应中心和消防站。低处沿着河流延伸的是消防站和消防车库。高处占据支配性地位且沿着街道延伸的是应急反应中心。然而，这两个机构所提供的服务并不交叉，机构内的工作困难、紧张且必须与公众隔离。因此，建筑师必须在开放与封闭之间把握好尺度。

细部与材料

建筑外墙面上覆盖着 Argeton 砖。深色让建筑融入了周边环境。旁边建筑的白色背景为消防站提供了良好的框架，充分烘托了消防站独立、开放而丰富鲜活的形象。

此外，人为因素是考虑设计结构和建筑的基础。建筑面向周边的环境呈现出开放的姿态，体现了紧密的联系。拥有河景的玻璃墙有助于消防员放松和平静心情。建筑的下半部分嵌入土坡，绿色屋顶让它看起来像一座小山。这样一来，其他建筑也能与河流建立起联系。绿色屋顶可以作为一个休闲空间，让人远离上方街道的喧嚣。

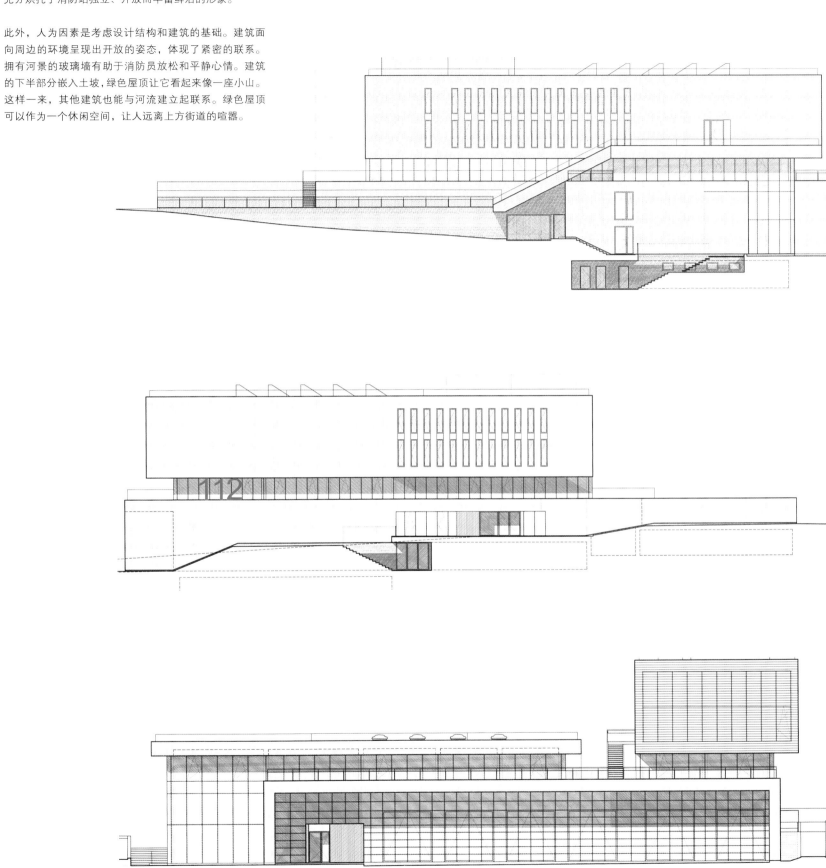

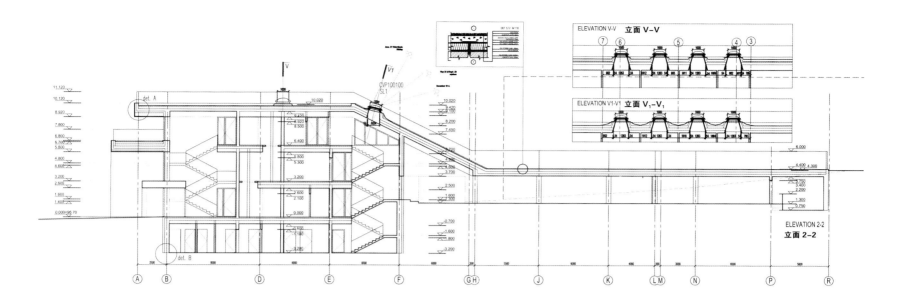

Parapet Detail
1. Heat insulation
2. Solid rock wool
3. Flashing 2 layers
4. Flashing
5. 60×200×6mm
6. Sheet metal cap
7. Concrete slab
8. Vapour barrier
9. Slope forming layer

护墙节点
1. 隔热层
2. 实心石棉
3. 双层防水板
4. 防水板
5. 60×200×6mm
6. 金属板顶盖
7. 混凝土板
8. 隔汽层
9. 造坡层

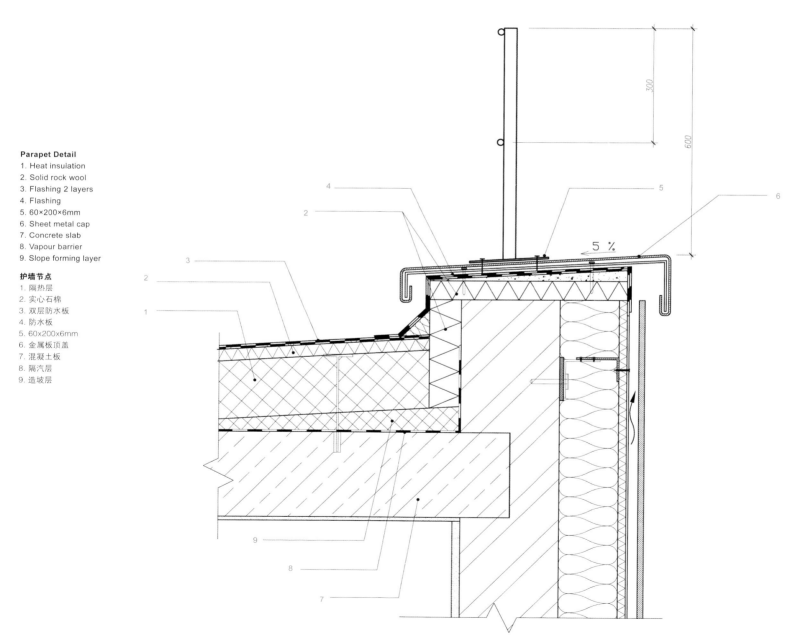

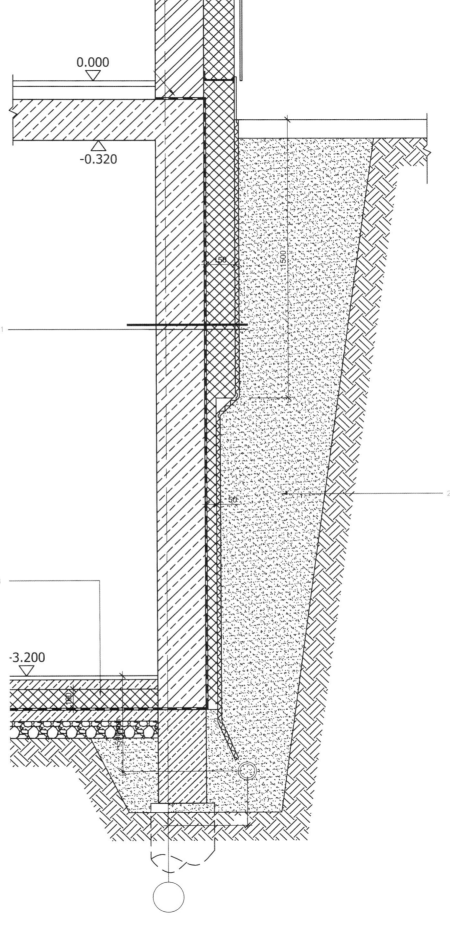

Detail
1. Monolithic concrete wall, 250 mm
   Flashing
   Polystyrene foam EPS100,150 IR 50mm
   Drainage sheets Iso-Drain P8 Geo, 8mm
2. Sand backfill
3. Polystyrene foam EPS150

节点
1. 整体浇灌混凝土墙，250mm
   防水板
   聚苯乙烯泡沫 EPS100，150 IR 50mm
   排水板 Iso-Drain P8 Geo，8mm
2. 砂土回填
3. 聚苯乙烯泡沫 EPS100

Apartments for Young People, Nursery and Park in San Sebastián

圣塞巴斯蒂安青年公寓、托儿所和公园

Location/ 地点：San Sebastián (Gipuzkoa), Spain/ 西班牙，圣塞巴斯蒂安（吉普斯夸省）
Architect/ 建筑师：Ignacio Quemada Arquitectos
Photos/ 摄影：Alejo Bagué
Built area/ 建筑面积：9,991m²
Cost/ 成本：6,363,601 €/6,363,601 欧元
Completion date/ 竣工时间：2011

Key materials: Façade – terracotta
主要材料：立面——陶瓦

Overview

Laid out along the line of maximum incline, the building provides a link between the lowest and highest areas of the estate, from Aizcorri Street at a height of +36, where it starts with only two floors, through to a height of +23 from which a walkway and lift connect to Benta Berri square at an elevation of +10. Another public lift built into the body of the building, and accessible from the walkway, bridges the entire height differential between the square and Aizkorri.

The nursery with 102 places occupies the ground floor and the semi-basement, while the apartments take up the upper four floors. The storage rooms are located in the areas built into the hill.

To soften its impact on the square, the building has a terraced profile. For this purpose, the floors are displaced backwards one relative to the next, with the apartments, and therefore the spaces in the façade, symmetrically staggered.

Detail and Materials

It was the geometry for the elevations which inspired the idea of conceiving each recess as a single prefabricated piece. The framework for the recesses having been placed in position, held by the slabs and the heavy brick leaf of the ventilated façade, the spaces between them are then covered with insulation and an outer leaf of terracotta pieces which are used throughout the building including exterior roofs of overhangs and passageways. The result is a construction system which accords with the sculptural composition of the elevations and body of the building.

To create the window and balcony elements, the precast frame system manufactured by

Composites Jareño was selected which is made from a single piece in a sandwich configuration, with two layers of reinforced thermo-stable resin and an injected polyurethane foam nucleus. The whole building is finished with four types of frame, the main one being the one which shapes the space for each apartment and which also incorporates the balcony overhang.

项目概况

建筑建在斜坡之上，在高低地势之间形成了连接：从 +36 高度的埃斯克里街（在这一高度层上仅有两层楼），穿过 +23 高度的通道和电梯，一路到达 +10 高度的本塔伯里广场。另一座位于建筑内部的电梯将广场与埃斯克里街之间的高度差连接起来。

托儿所可接收 102 名幼儿，占据着一楼和半地下室；公寓占据着上方的四层空间。储藏室位于建在山体内部的区域。

为了弱化建筑对广场的影响，建筑采取阶梯式造型。公寓所在的楼层从下至上依次后错，从而在外墙上形成了对称的错落感。

细部与材料

海拔高度差的几何形状激发了建筑师将各个后错楼板设计成单一的预制板的想法。先固定好后错空间的框架，然后添加墙板和通风立面的厚砖层，在二者之间填充隔热层，最后在建筑外层（包括突出的屋檐和通道）整体覆盖一层陶瓷砖。最终形成的结构系统与富有雕塑感的立面和建筑楼体都十分契合。

为了在楼体上制作窗户和阳台，建筑师选择了 Composites Jareño 制造的预制框架系统。该系统由单块夹心板结构制成，夹心板则由双层加固热稳定树脂和注入的聚氨酯泡沫内核构成。整座建筑共采用了四种框架形式，主框架为每间公寓塑造了空间，同时还合并了阳台的悬挑檐。

Window-balcony detail
1. Waterproof mortar
2. Self-adhesive SIKA MULTISEAL strip
3. KNAUFF ULTRAVENT BLACK insulation (50 mm)
4. Cavity wall
5. Terracotta panels with 1cm open and overlapped joints
6. Plasterboard closing
7. Plasterboard ceiling
8. Blind box rested in the interior jambas of the case
9. Window surround one-piece composite, including cantilever
10. Lacquered aluminum frame
11. Stainless steel balustrade
12. Porcelain tile

窗户 – 阳台节点
1. 防水灰浆
2. 自黏式密封条 SIKA MULTISEAL
3. KNAUFF ULTRAVENT BLACK 隔热层（50mm）
4. 空心墙
5. 陶瓷板，配 1cm 开口和搭接接缝
6. 石膏板封口
7. 石膏天花板
8. 百叶窗匣，位于窗侧内
9. 包围单块复合材料的窗户，包括悬臂梁
10. 涂漆铝框
11. 不锈钢栏杆
12. 瓷砖

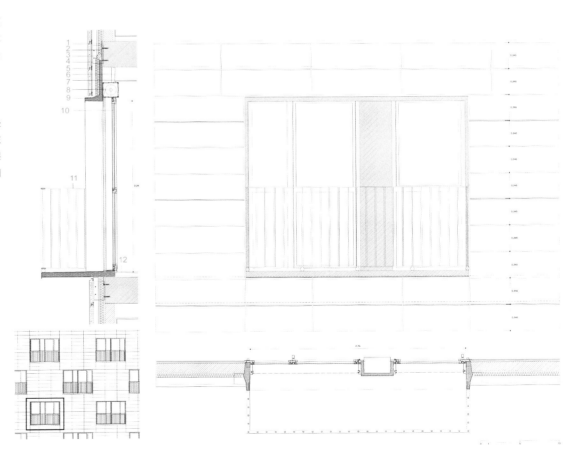

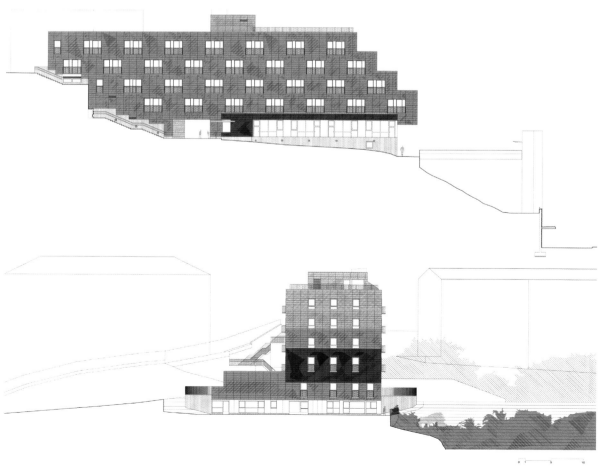

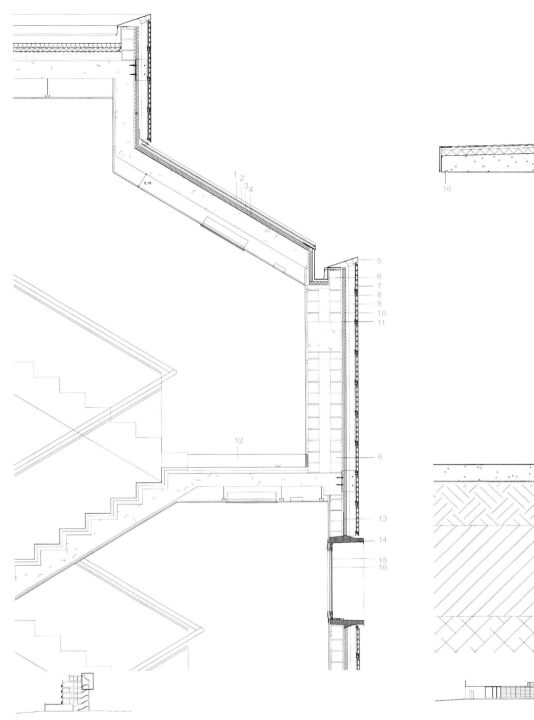
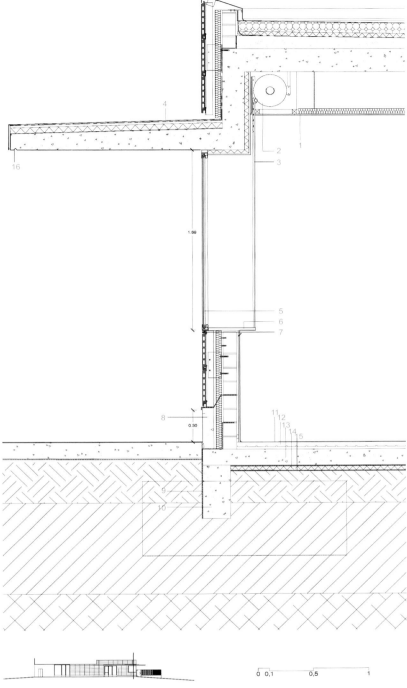

Façade Detail
1. Zinc (Quarz zinc) standing seam
2. PVC membrane VZM type
3. Waterproof chipboard (30 mm)
4. Fiberglass insulation
5. Composite coping
6. 1/2 foot of perforated brick
7. Terracotta panels with 1 cm open and overlapped joints
8. Ventilated cavity wall
9. Aluminum subframe
10. KNAFF ULTRAVENT BLACK insulation (50mm)
11. Waterproof mortar cement
12. Terrazzo skirting board in line with plasterboard
13. Self-adhesive SIKA MULTISEAL strip
14. Window surround one-piece composite, including cantilever
15. Lacquered aluminum frame
16. CLIMALIT glass 6/12/6

立面节点
1. 锌（石英锌）立缝
2. PVC 膜，VZM 型
3. 防水刨花板（30mm）
4. 纤维玻璃隔热
5. 复合顶盖
6. 1/2 英尺预制砖
7. 陶瓷板，配 1cm 开口和搭接缝
8. 通风空心墙
9. 铝制副框架
10. KNAFF ULTRAVENT BLACK 隔热层（50mm）
11. 防水灰浆水泥
12. 水磨石踢脚板，与石膏板对齐
13. 自黏式密封条 SIKA MULTISEAL
14. 包围单块复合材料的窗户，包括悬臂梁
15. 涂漆铝框
16. CLIMALIT 玻璃 6/12/6

Detail
1. False ceiling
2. Motorised blind
3. Blind rail
4. Microventilated zinc roof
5. Aluminum frame with thermal bridge breaking
6. Waterproof MDF sill
7. Lacquered aluminum sheet moulding
8. Reinforced concrete wall with vertical plank formwork
9. Waterproof sheet
10. Tie beam
11. Continuous pavement
12. Mortar
13. Concrete base with 20.20.6 welded wire mesh
14. Polystyrene insulation (4cm)
15. Polystyrene sheet
16. Runoff

节点
1. 假吊顶
2. 电动百叶窗
3. 百叶窗轨道
4. 微通风锌屋顶
5. 铝制框架，带热桥阻断
6. 防水中密度纤维板窗台
7. 涂漆铝板模架
8. 钢筋混凝土墙，带垂直木板模架
9. 防水板
10. 水平拉杆
11. 连续铺装
12. 灰浆
13. 混凝土底座，配 20.20.6 焊接钢丝网
14. 聚苯乙烯绝缘（4cm）
15. 聚苯乙烯板
16. 排水

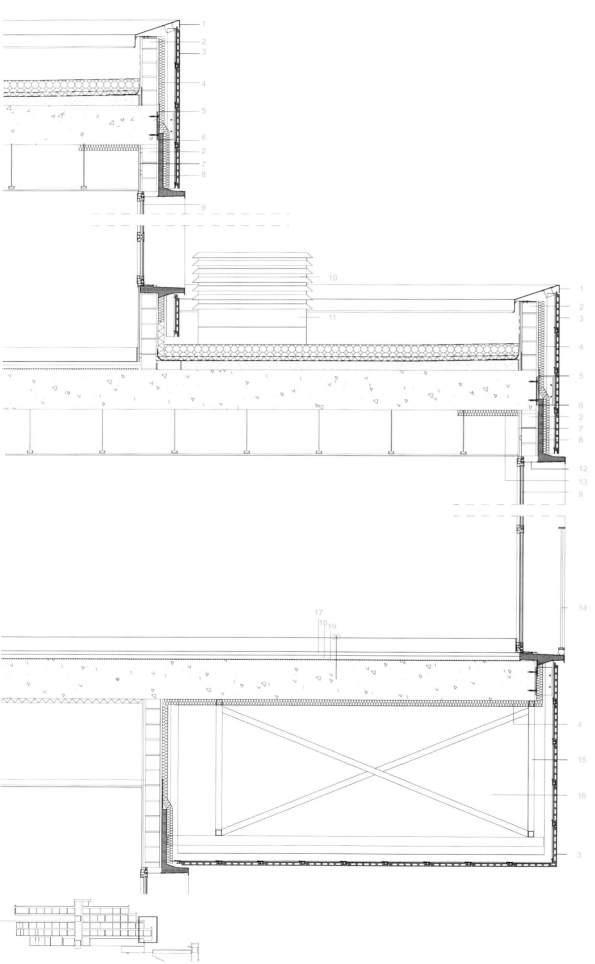

Façade Detail
1. Composite coping
2. 1/2 foot of perforated brick
3. Terracotta panels with 1 cm open and overlapped joints
4. KNAFF ULTRAVENT BLACK insulation (50 mm)
5. Self-adhesive SIKA MULTISEAL strip
6. Waterproof mortar cement
7. Plasterboard sheeting
8. Window surround one-piece composite, including cantilever
9. Lacquered aluminum frame
10. Lacquered steel sheet ventilation hat
11. Masonry + mortar plastering + self-protected bituminous waterproofing membrane
12. Aluminium lintel
13. 50cm wide extruded polystyrene strip, 50mm thick
14. Stainless steel balustrade
15. Painted steel subframe
16. Ventilated cavity wall
17. Ceramic tiles
18. Mortar
19. Anti-impact sheet
20. Flat beams, precast joists and concrete joist-filler block slab

立面节点
1. 复合顶盖
2. 1/2 英尺多孔砖
3. 陶瓷板，配 1cm 开口和搭接接缝
4. KNAFF ULTRAVENT BLACK 隔热层（50mm）
5. 自黏式密封条 SIKA MULTISEAL
6. 防水灰浆水泥
7. 石膏板
8. 包围单块复合材料的窗户，包括悬臂梁
9. 涂漆铝框
10. 涂漆钢板通风帽
11. 砌砖 + 灰泥石膏 + 自护式沥青防水膜
12. 铝过梁
13. 50cm 宽挤塑聚苯乙烯条，50mm 厚
14. 不锈钢栏杆
15. 涂漆钢制副框架
16. 通风空心墙
17. 瓷砖
18. 灰浆
19. 抗冲击板
20. 平梁，预制托梁和混凝土托梁填块板

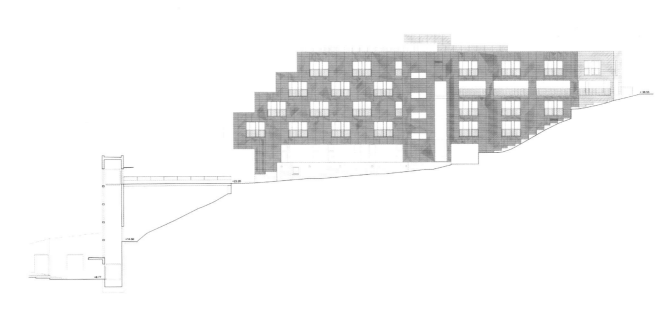

Agencia IDEA Headquarters
IDEA公司总部

Location/ 地点：Sevilla, Spain/ 西班牙，塞维利亚
Architect/ 建筑师：TRIANERA DE ARQUITECTURA S.LP
Photos/ 摄影：Víctor Sájara (www.victorsajara.com)
Site area/ 占地面积：3,364m²
Built area/ 建筑面积：10,769.90m²
Completion date/ 竣工时间：2011

Key materials: Façade – ceramic, glass, aluminum, inox steel
主要材料：立面——瓷砖、玻璃、铝、伊诺克斯不锈钢

Overview

The lot on which the building is located, was occupied by the Belgian pavilion at EXPO 1992, in Seville's Isla de la Cartuja. The square plan of the plot and its flat topography pose no isotropy or neutrality at its edges. There are important differences between the formal, spatial and visual characteristics of different fronts. To the south it opens onto the Newton Street, one of the most important avenues of the area. To the east, the Kepler St., with the rear façade of the French Pavilion. To north, the D'eluyar St., lower visual quality and finally to the west, the plot is right besides another new office building.

As for the distant views that the plot enjoys, interested diagonally ones, framed by ancient Portugal and Spain pavilions, allows the contemplation of a fragment of the Seville Historic Centre, Guadalquivir River and Barqueta Bridge. The first premise the architects started the proposal is that, for the development of office space, it is necessary to adopt the typological configuration that is best suited to achieve the desired flexibility and modularity, while a high level of environmental comfort (acoustic, lighting, heat and air quality), high energy efficiency and construction systems that responds to the conditioning factors of speed of execution and sustainability that are intended.

By aligning two blocks of its kind in the western and northern fronts, almost all the office space can be located in the top 5 floors getting much of their space to enjoy the interesting sights, and orientations that the climate allows the adoption of appropriate glazing systems, on the east and south façades that make the concave dihedral of the L form. The other façades, more closed, response to its accidental coincidence of worst orientations and views. In this case, treatment generates a contrast that makes expressive opening of the building diagonally to the south and east. Front the glass façades of first, these closed façades are formalised through a ceramic skin formed by great format pieces of FAVETON (400x1.350x30mm size), an extruded hollow-core tiles, in colour white. The dynamic diagonality that L plan generates is complemented by articulating a blind body on ground level on the south side, and a suspended canopy, with a large cantilever, that creates a large atrium access and covers the glass lobby of the main entrance. So, the ground floor is arranged as a confrontation between two parallel boxes to the south façade, including a garden area which is the centre of the building. These boxes are joined by the double height lobby that makes the transition between the atrium and courtyard.

Detail and Materials
The ceramic skin formed by great format pieces of FAVETON (400x1.350x30mm size), a extruded hollow-core tiles, in colour white, is a ventilated façade system and it does have an air gap between ceramic tiles and insulation exterior layer.

Section of tongue-and-groove piece can be placed horizontally and vertically. Horizontal position in this case. Height: 400 mm, between axes. Length: 1,350 mm. Thickness: 28 mm. Weight: 48 Kg/m². Standard junctions of 10 mm placed horizontally and vertically.

Assembly system. SAH anchorage system consists on an aluminium substructure made of vertical profile connections T type, fixed to the wall by L profiles called "Corbels". Two types available: Sustenance corbel (fixed point): They support the façade's weight and they are usually forged. Retention corbel (movable point): They hold up wind force.

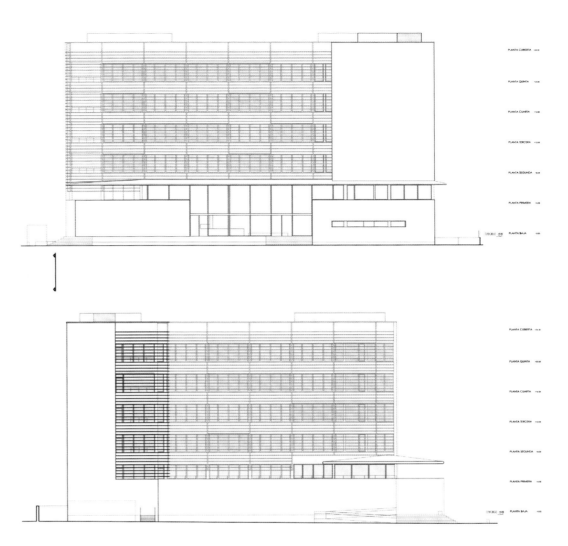

项目概况
建筑所在的场地曾经是1992塞维利亚世博会的比利时馆。场地呈正方形，地势平坦，各个边界都不尽相同。不同的朝向在形式、空间和视觉特征方面有着强烈的差异。场地南面是该地区最主要的街道之———牛顿街；东面是开普勒街，并且对着法国馆的后身；北面是德鲁耶街，视角效果较差；西面是一座全新的办公楼。

至于远景，项目的位置可谓得天独厚：古老的葡萄牙馆和西班牙馆首先映入眼帘，向后还可以瞥见塞维利亚历史中心、瓜达基维尔河以及巴克塔桥。项目的首要要求就是开发一个能够实现最大限度灵活性的模块化办公空间，该空间必须具备高等级的环境舒适度（音效、声学、供暖、空气质量等）、高能源效率，工程体系必须适应高速的执行需求以及可持续发展。

建筑的西、北两侧都是同类建筑，几乎所有办公空间都被设置在上面的五层楼，在L形凹面的东、南两面都享有上文提到的优美景色。根据当地气候，不同的朝向配有不同数量的玻璃装配。其他的立面更为封闭，因为它

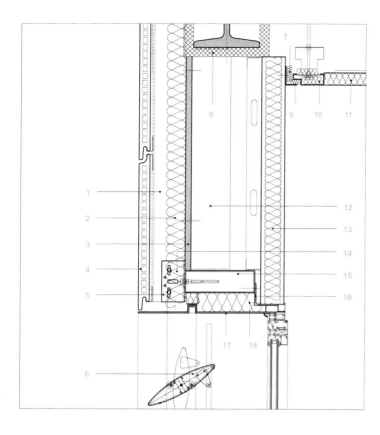

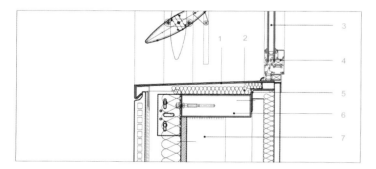

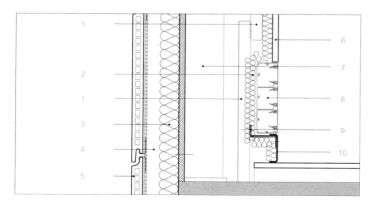

West façade – lintel of window
1. Ventilated air cavity, 40mm
2. Mineral wool thermal insulation, 60mm
3. Cement-wood panel, 15mm thick
4. Extruded ceramic tile, 1350x400x28mm, colour white
5. Aluminium supporting structure
6. Aluminium sun-shading motorised louvers, 200x40mm elliptical section, polyester paint finished
7. Perimeter ceiling joint with 80mm acoustic wool insulation infill
8. Plaster fire protection for steel structure
9. Bent to shape lacquered galvanised steel sheet, 2mm
10. 115. Perimeter joint suspended ceiling channel, with acoustic wool insulation infill
11. Metal suspended ceiling, 1350x1350mm plates
12. 100.100.3 mm galvanised steel vertical tube, with 120x120x10mm steel base plate
13. Acoustic plasterboard with mineral wool, 13+40mm
14. Aluminium supporting structure
15. Lintel: 180.60.3mm galvanised steel tube
16. L.60.3mm galvanised steel profile
17. Lintel formed by two bent to shape 4 mm lacquered galvanised steel sheets
18. Mineral wool thermal insulation, 60mm

西立面——窗梁节点
1. 通风气腔，40mm
2. 矿物棉隔热层，60mm
3. 水泥木板，15mm 厚
4. 挤制瓷砖，135x400x28mm，白色
5. 铝支撑结构
6. 铝制电动遮阳百叶窗，200x40mm 椭圆截面，喷涂聚氨酯漆
7. 外围天花板接缝，80mm 隔音棉填充
8. 钢结构石膏防火层
9. 弯曲成形涂漆镀锌钢板，2mm
10. 外围接缝吊顶槽，隔音棉填充
11. 金属吊顶，1350x1350mm 板
12. 100.100.3mm 镀锌钢垂直管，配 120x120x10mm 钢底板
13. 由两块 4mm 弯曲成形涂漆镀锌钢板构成过梁
14. 铝支撑结构
15. 过梁：180.60.3mm 镀锌钢管
16. L.60.3mm 镀锌钢型材
17. 由两块 4mm 弯曲成形涂漆镀锌钢板构成过梁
18. 矿物棉隔热层，60mm

West façade – window opening detail
1. Bent to shape lacquered galvanised steel sheet, 4mm
2. Mineral wool thermal insulation, 60mm
3. Double glazing: 6mm toughened glass + 12mm cavity + 4+4mm lam. safety glass
4. Aluminium frame
5. L.60.3mm galvanised steel profile
6. Lintel: 180.60.3mm galvanised steel tube
7. 100.100.3 mm galvanised steel vertical tube, with 120x120x10mm steel base plate

西立面——窗口节点
1. 弯曲成形涂漆镀锌钢板，4mm
2. 矿物棉隔热层，60mm
3. 双层玻璃：6mm 钢化玻璃 +12mm 空气层 +4+4mm 夹层安全玻璃
4. 铝框
5. L.60.3mm 镀锌钢型材
6. 过梁：180.60.3mm 镀锌钢管
7. 100.100.3mm 镀锌钢垂直管，配 120x120x10mm 钢底板

West façade – vertical elevation corner
1. Socket support structure: two 30.30.3mm steel tubes
2. Mineral wool acoustic insulation, 50mm
3. Mineral wool thermal insulation, 60mm
4. Ventilated air cavity, 40mm
5. Extruded ceramic tile, 1350x400x28mm, colour white
6. Plasterboard, 15mm
7. 100.100.3 mm galvanised steel vertical tube, with 120x120x10mm steel base plate
8. Facilities aluminum channel
9. Z.70.30mm galvanised steel profile
10. Socket: bent to shape lacquered galvanised steel sheet, 2mm. 73-221mm high

西立面——垂直立面墙角节点
1. 插槽支撑结构：两根 30.30.3mm 钢管
2. 矿物棉隔音层，50mm
3. 矿物棉隔热层，60mm
4. 通风气腔，40mm
5. 挤制瓷砖，135x400x28mm，白色
6. 石膏板，15mm
7. 100.100.3mm 镀锌钢垂直管，配 120x120x10mm 钢底板
8. 槽型铝材
9. Z.70.30mm 镀锌钢型材
10. 插槽：弯曲成形涂漆镀锌钢板，2mm，73~221mm 高

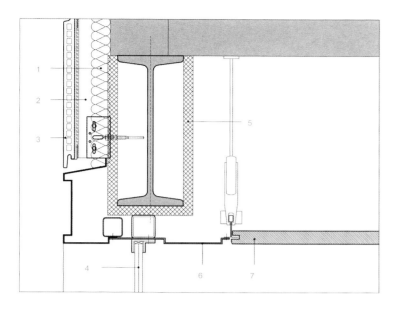

West façade – porch lintel on the ground
1. Mineral wool thermal insulation, 60mm
2. Ventilated air cavity, 40mm
3. Extruded ceramic tile, 1350x400x28mm, colour white
4. Laminated safety glass 10+10mm, with neoprene profiles
5. Plaster fire protection for steel structure
6. Lintel formed by two bent to shape 4 mm lacquered galvanised steel sheets
7. Solid wood wool panels suspended ceiling, 600x600x35mm

西立面——一层门廊横梁节点
1. 矿物棉隔热层，60mm
2. 通风气腔，40mm
3. 挤制瓷砖，135x400x28mm，白色
4. 夹层安全玻璃 10+10mm，配氯丁橡胶剖面
5. 钢结构石膏防火层
6. 由两块 4mm 弯曲成形涂漆镀锌钢板构成过梁
7. 实心刨花板吊顶，600x600x35mm

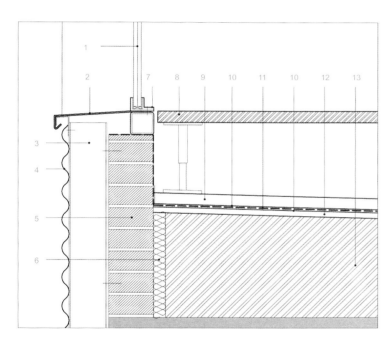

West façade – porch corner on the ground floor
1. Laminated safety glass 10+10mm, with neoprene profiles
2. Bent to shape lacquered galvanised steel sheet, 4mm
3. Z.50.30.3mm galvanised steel profile, for metal sheet support
4. Undulating metal sheet, 0.5mm PL18/76 type, polyester lacquered finish
5. Brick wall, 115mm
6. Expanded polystyrene, 30mm plate
7. Inox steel frame, with L.60.35.5 and L.35.5 plates
8. Granite tiles flooring, 600x300x30mm grey colour, flamed finish, above adjustable PVC supports
9. Reinforced mortar layer, 40mm, with fibre glass grid
10. Geotextile layer, 150 gr/m²
11. Bituminous sealing layer
12. Regulation mortar layer, 15 mm, 2% minimum slope
13. Cellular concrete layer, 100-200mm, 2% minimum slope

西立面——一层门廊角部节点
1. 夹层安全玻璃 10+10mm，配氯丁橡胶剖面
2. 弯曲成形涂漆镀锌钢板，4mm
3. Z.50.30.3mm 镀锌钢型材，支撑金属板
4. 波形金属板，0.5mm PL18/76 型，聚氨酯漆饰面
5. 砖墙，115mm
6. 挤制发泡聚苯乙烯板，30mm
7. 伊诺克斯不锈钢架，配 L.60.35.5 和 L.35.5 板
8. 花岗岩砖地面，600x300x30mm，灰色，火烧面，铺在可调节 PVC 支撑上
9. 加固砂浆层，40mm，配纤维玻璃网格
10. 土工布层，150 gr/m²
11. 沥青密封层
12. 找平砂浆层，15mm，2% 最小坡度
13. 多孔混凝土层，100~200mm，2% 最小坡度

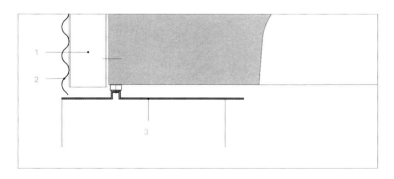

West façade – lintel of entrance to basement
1. Z.50.30.3mm galvanised steel profile, for metal sheet support
2. Extruded galvanised steel sheet, 2mm
3. Lintel formed by two bent to shape 4 mm lacquered galvanised steel sheets

西立面——通往地下室入口横梁节点
1. Z.50.30.3mm 镀锌钢型材，支撑金属板
2. 挤制镀锌钢板，2mm
3. 由两块 4mm 弯曲成形涂漆镀锌钢板构成过梁

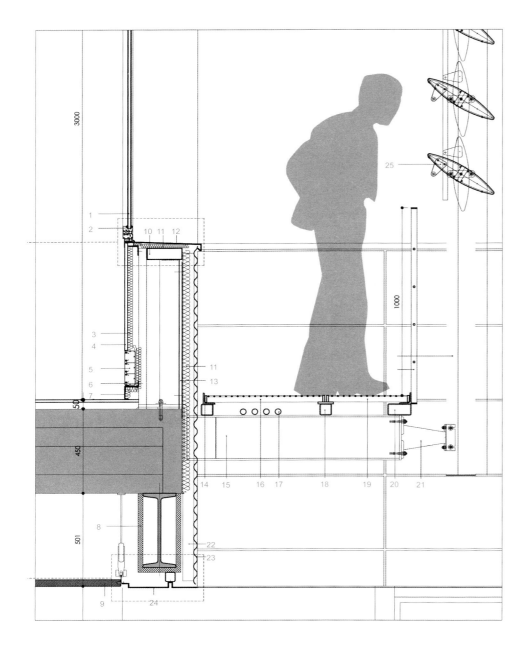

East façade – roof corner of porch
1. Double glazing: 6mm toughened glass + 12mm cavity + 4+4mm lam. safety glass
2. Aluminium frame
3. Mineral wool acoustic insulation, 50mm
4. Plasterboard, 15mm
5. Facilities aluminum channel
6. Z.70.30mm galvanised steel profile
7. Socket: bent to shape lacquered galvanised steel sheet, 2mm. 73-221mm high
8. Plaster fire protection for steel structure
9. Solid wood wool panels suspended ceiling, 600x600x35mm
10. Lintel: 180.60.3mm galvanised steel tube
11. Mineral wool thermal insulation, 60mm
12. Bent to shape lacquered galvanised steel sheet, 4mm
13. Cement-wood panel, 15mm thick
14. L.60.6mm galvanised steel profile, for metal grid support
15. Façade steel bracket, each 2700 mm, with L.250.300.15 mm steel base plate
16. Galvanised steel grid, 30x30/25mm and Ø6 mm, with 35x3mm plates steel frame
17. Drainage pipe, Ø30mm
18. 60.60.4mm galvanised steel tube
19. Galvanised steel grid, 30x30/25mm and Ø6 mm, with 35x3mm plates steel frame
20. 120.60.4mm galvanised steel tube, welded to the steel bracket for grid support
21. Galvanised steel bracket for sun-shading support
22. Z.50.30.3mm galvanised steel profile, for metal sheet support
23. Undulating metal sheet, 0.5mm PL18/76 type, polyester lacquered finish
24. Lintel formed by two bent to shape 4 mm lacquered galvanised steel sheets
25. Aluminium sun-shading motorised louvers, 400x80mm elliptical section, polyester paint finished

东立面——门廊屋顶一角节点
1. 双层玻璃：6mm 钢化玻璃 +12mm 空气层 +4+4mm 夹层安全玻璃
2. 铝框
3. 矿物棉隔音层，50mm
4. 石膏板，15mm
5. 槽型铝材
6. Z.70.30mm 镀锌钢型材
7. 插槽：弯曲成形涂漆镀锌钢板，2mm，73-221mm 高
8. 钢结构石膏防火层
9. 实心刨花板吊顶，600x600x35mm
10. 过梁：180.60.3mm 镀锌钢管
11. 矿物棉隔热层，60mm
12. 弯曲成形涂漆镀锌钢板，4mm
13. 水泥木板，15mm 厚
14. L.60.6mm 镀锌钢型材，支撑金属格栅
15. 立面钢支架，每个 2700mm，配有 L.250.300.15 mm 钢地板
16. 镀锌钢格栅，30x30/25 mm，Ø 6 mm，配有 35x3mm 钢板架
17. 排水管，Ø30mm
18. 60.60.4mm 镀锌钢管
19. 镀锌钢格栅，30x30/25 mm，Ø 6 mm，配有 35x3mm 钢板架
20. 120.60.4mm 镀锌钢管，焊接在钢支架上，支撑金属格栅
21. 镀锌钢支架，支撑遮阳装置
22. Z.50.30.3mm 镀锌钢型材，支撑金属板
23. 波形金属板，0.5mm PL18/76 型，聚氨酯漆饰面
24. 由两块 4mm 弯曲成形涂漆镀锌钢板构成过梁
25. 铝制电动遮阳百叶窗，400x80mm 椭圆截面，喷涂聚氨酯漆

们的朝向或正对的景色并不合适。在本项目中，东、南两侧的对角空间享有大面积的玻璃窗口。封闭的立面采取陶瓷表皮的形式，采用 FAVETON 型白色挤制空心瓷砖。

建筑 L 形内凹的空间在南侧建造了一个底座结构和悬臂式天篷，形成了一个大型中庭通道，将主入口的玻璃大厅覆盖起来。这样一来，一楼空间就在建筑南立面形成了两个平行盒状结构的对置，而在建筑中央则设有一个花园区域。两个盒状结构由双高大厅连接起来，实现了中庭和庭院之间的过渡。

细部与材料
采用 FAVETON 型白色挤制空心瓷砖构成的陶瓷表皮是一个通风立面系统，它在瓷砖和隔热板外层之间有一个空气腔。

舌榫件可以采用水平和垂直两种安装方式，本项目采用了水平安装。高度：400 毫米（轴间）。长度：1,350 毫米。厚度：28 毫米。重量：48 千克/平方米。标准水平和垂直接缝均为 10 毫米。

装配系统。ASH 锚定系统由铝制下层结构（T 型垂直型材连接件）通过 Corbels L 型材固定在墙面上。有两种类型可供选择：支持托梁（固定点）——它们支撑立面的重量，通常采用锻造的形式。保持托梁（移动点）：它们拦截风力。

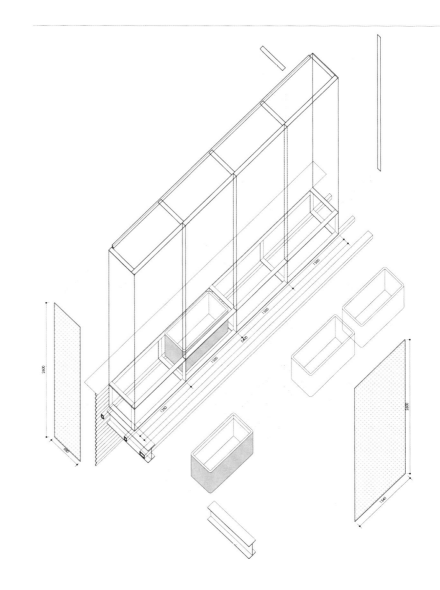

Na Vackove Residential Houses in Prague
纳瓦克夫住宅区

Location/地点: Prague, Czech Republic/ 捷克，布拉格
Architect/建筑师: UNIT architekti, Jiran Kohout architekti/Michal Kohout, David Tichý, Zdeněk Jiran
Photos/摄影: Filip Šlapal
Site area/占地面积: 9,306m²
Built area/建筑面积: 3,843.05m²
Completion date/竣工时间: 2012

Key materials: Façade – ceramic façade cladding – dark brick tiles, thin-layer façade plaster (facing public space)
主要材料：立面——深色瓷砖覆面、薄层立面石膏（朝向公共空间）

Overview
The housing complex Vackov represents project which emerged from long term research of residential qualities of the built environment. Utilising low-rise/high density principles, Vackov is an example of compact urban development with attractive common spaces and as such represents a viable alternative to suburban housing. The goal was to design contemporary urban housing which reflects human scale, offers appropriate hierarchy of privacy and foster social interaction. The main emphasis was put on creating a legible environment with an intuitive layout.

The design integrates a variety of demands placed on urban housing today: services and comfort which correspond to the traditional urban density, but also a level of privacy and the pleasure of living in a private house with garden. It borrows its scale, legibility and a hierarchy of public spaces from the traditional city and combines it with neighbour relations of local community of garden cities.

At the core of the design is an understanding of social interactions between the residents followed by designing of corresponding outdoor spaces whose purpose and usage is determined by varying levels of intimacy on a scale between "public" and "private".

Detail and Materials
The material and spatial design of the façades follows their orientation either to the public space or to the semi-private courtyards.

Façades facing the semi-private courtyards are covered with vertical wooden boards. These light-coloured façades convey a sense of home and direct us to the world inside.

Ventilated façade is made of vertical planks of softwood placed on the horizontal grid of wood slats. The grid is anchored into the reinforced concrete wall with galvanised anchors. The wooden pine planks "Thermowood" are 12 cm width without surface finishing. They are folded on the façade running side by side using the tongue and groove principle. Every board of the façade is anchored to the grid with the steel screws.

On the contrary, more formal dark façades representing the "front" of the buildings outward are covered with ceramic façade cladding of dark brick tiles and they are structured with balconies.

项目概况

纳瓦克夫住宅区的建成是对居住品质长期研究的成果。项目采用低层高密度原则，是紧凑型城市开发项目的便饭，配有出众的公共空间，完全可以与城郊住宅的质量相媲美。项目的目标是设计一套具有人性尺度的现代城市住宅，提供合适的隐私等级，促进社交互动。项目的重点在于打造一个清晰直观的居住环境。

设计融合了当今城市住宅的多种要求：既保证与传统城市高密度住宅的服务和舒适度，又提供了私人花园住宅的私密感和宜居性。它从传统城市中借鉴了规模、直观性和公共空间的层次感，并且将其与花园城市社区联系起来。

设计的核心是对居民之间的社交活动的理解。建筑师设计了一个具有多层次私密感的户外空间，在"公共"与"私密"之间建立起相应的联系。

细部与材料

建筑立面的材料选择和空间设计分别与立面的朝向相对应，分为公共空间朝向和半私密庭院朝向两种。

朝向半私密庭院的立面上覆盖着垂直木板，这些浅色立面传递出一种居家感，引导着人们走进去。

通风立面由垂直软木板放置在水平木条格栅上构成。格栅通过镀锌锚件固定在钢筋混凝土墙上。Thermowood松木板条12厘米宽，表面未经任何处理。它们利用凹凸缝拼法叠加铺装在墙面上。每块外墙板都通过钢螺丝固定在格栅上。

另一方面，建筑正面采用更为正式的深色墙面，由深色瓷砖覆盖，它们与阳台构造在一起。

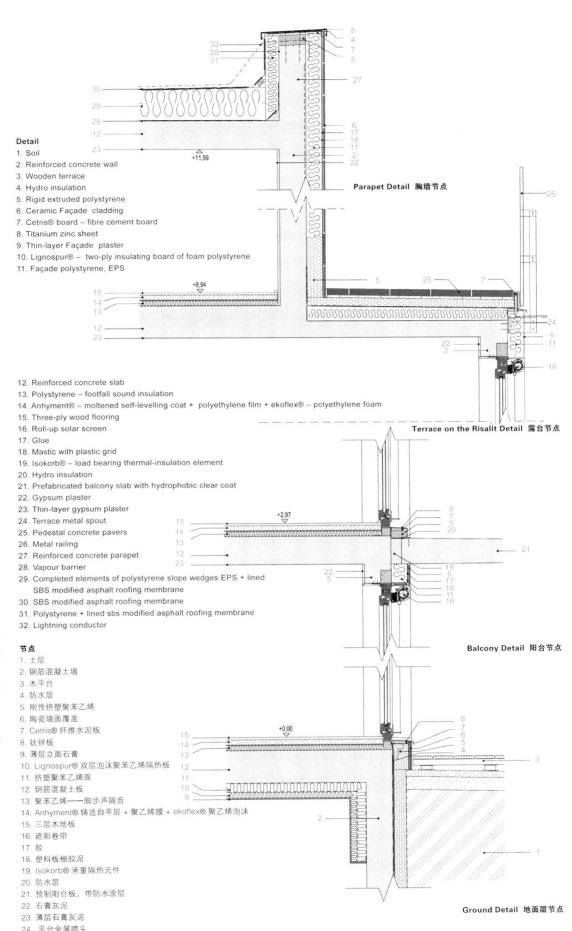

Detail
1. Soil
2. Reinforced concrete wall
3. Wooden terrace
4. Hydro insulation
5. Rigid extruded polystyrene
6. Ceramic Façade cladding
7. Cetris® board – fibre cement board
8. Titanium zinc sheet
9. Thin-layer Façade plaster
10. Lignospur® – two-ply insulating board of foam polystyrene
11. Façade polystyrene, EPS
12. Reinforced concrete slab
13. Polystyrene – footfall sound insulation
14. Anhyment® – moltened self-levelling coat + polyethylene film + ekoflex® – polyethylene foam
15. Three-ply wood flooring
16. Roll-up solar screen
17. Glue
18. Mastic with plastic grid
19. Isokorb® – load bearing thermal-insulation element
20. Hydro insulation
21. Prefabricated balcony slab with hydrophobic clear coat
22. Gypsum plaster
23. Thin-layer gypsum plaster
24. Terrace metal spout
25. Pedestal concrete pavers
26. Metal railing
27. Reinforced concrete parapet
28. Vapour barrier
29. Completed elements of polystyrene slope wedges EPS + lined SBS modified asphalt roofing membrane
30. SBS modified asphalt roofing membrane
31. Polystyrene + lined sbs modified asphalt roofing membrane
32. Lightning conductor

Parapet Detail 胸墙节点
Terrace on the Risalit Detail 露台节点
Balcony Detail 阳台节点
Ground Detail 地面层节点

节点
1. 土层
2. 钢筋混凝土墙
3. 木平台
4. 防水层
5. 刚性挤塑聚苯乙烯
6. 陶瓷墙面覆盖
7. Cetris® 纤维水泥板
8. 钛锌板
9. 薄层立面石膏
10. Lignospur® 双层泡沫聚苯乙烯隔热板
11. 挤塑聚苯乙烯面
12. 钢筋混凝土板
13. 聚苯乙烯——脚步声隔音
14. Anhyment® 铸造自平层 + 聚乙烯膜 + ekoflex® 聚乙烯泡沫
15. 三层木地板
16. 遮阳卷帘
17. 胶
18. 塑料板栅胶泥
19. Isokorb® 承重隔热元件
20. 防水层
21. 预制阳台板，带防水涂层
22. 石膏灰泥
23. 薄层石膏灰泥
24. 平台金属喷头
25. 托架混凝土铺装
26. 金属栏杆
27. 钢筋混凝土护墙
28. 隔汽层
29. 聚苯乙烯楔形元件 + 苯乙烯改性沥青屋面膜内衬
30. 苯乙烯改性沥青屋面膜
31. 聚苯乙烯 + 苯乙烯改性沥青屋面膜内衬
32. 避雷针

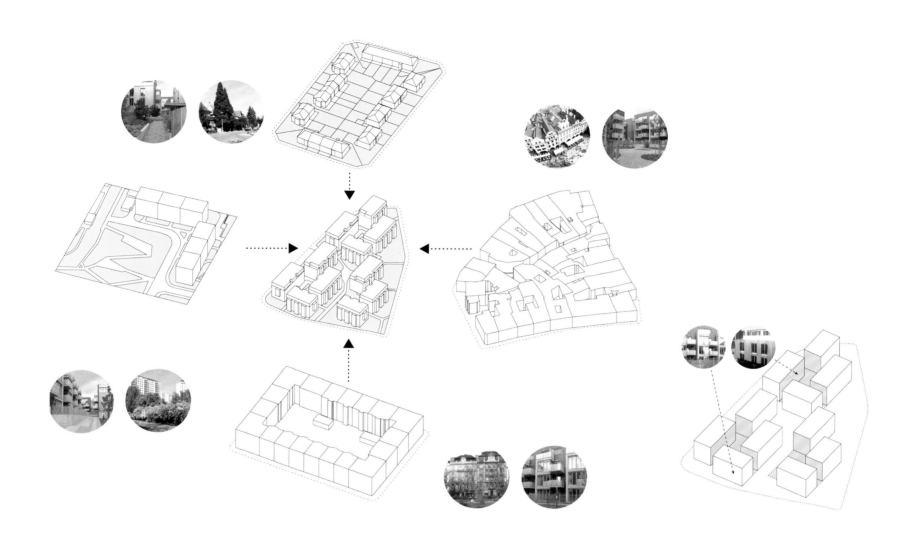

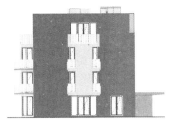

North 北立面

East 东立面

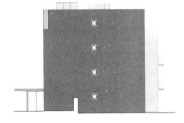

South 南立面

West 西立面

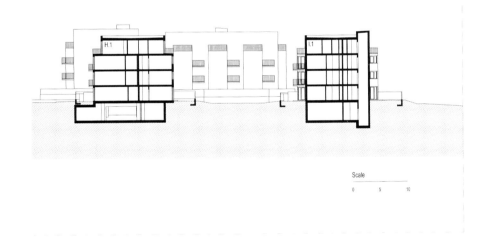

Scale
0 5 10

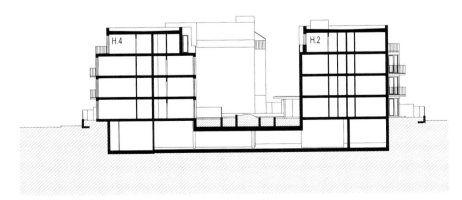

Scale
0 5 10

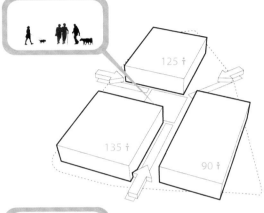

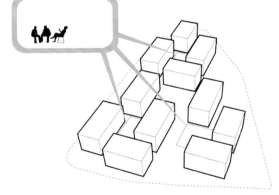

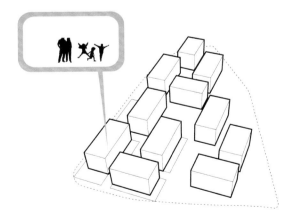

Building 18 of Getafe Campus
赫塔费校园18号楼

Location/ 地点：Madrid, Spain/ 西班牙，马德里
Architect/ 建筑师：juan beldarrain/ estudio Beldarrain
Photos/ 摄影：Francisco Berreteaga
Site area/ 占地面积：8,610m²
Built area/ 建筑面积：21,153m²
Completion date/ 竣工时间：2013

Key materials: Façade – terracotta
Structure – concrete
主要材料：立面——瓷砖
结构——混凝土

Overview
The building design is sustained on the engagement with innovation and sustainability, as well as on the conviction that these two go hand in hand.

On the one hand, as opposed to an educational architecture that is traditionally rigid, systematic and repetitive, the architects take sides for a flexible, diverse and humanised architecture. An architecture tending less to be a reflection of the institution's authority and more attentive to provide services to teachers and students, creating spaces for diverse encounters where a great number of new models of learning and research can be accommodated.

On the other hand, the whole building responds to the growing commitment of all the parties towards the environmental sustainability. During the project and construction phases, countless sustainability measures have been incorporated, with the result that a degree of excellence has been achieved that had not been reached in Spain until now by an educational centre. It will be the first one of its kind to obtain a LEED Certificate and it will be in the highest grade: PLATINE LEED.

The process of conceiving the building has constituted a deliberate pendular movement between the confidence in the system and the repetition, on the one hand, and the search for exception and diversity on the other hand. System and repetition, on their own, provide a reassuring coherence and an economy of means, at the same time as they offer an intense flexibility. On the other hand, exception and diversity humanise architecture and make the user free to choose among multiple possibilities of use, which awards him a leading role.

Detail and Materials
The structure is systematically translated to the façade so as to provide flexibility to the bays. Concrete pillars are placed at a distance of

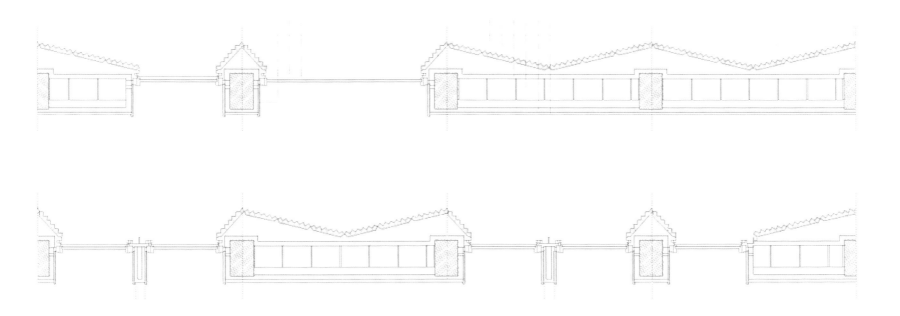

2.20m from each other and the 11.40 span is solved by means of posttensioned slabs. Nevertheless, the pillars disappear behind the folds of the ventilated ceramic façade, to the point that in some places, such as in the principal access, the plates seem to float weightless.

The façade, extraordinarily systematic, repeats the same plates, ceramic pieces and windows throughout all the building. However, the windows are scattered in an apparently disorderly manner solving with efficacy the illumination of the different uses and enhancing the desire to show the diversity proposed by the design.

The extensive use of few materials in all the building is explored up to their last possibilities, requiring qualities that will allow them to respond in an attractive and effective manner to very different situations.

In a campus of buildings made of white concrete and exposed bricks, a façade has been chosen that reproduces these two materials reinterpreting them in sustainability terms. Between elongated bands of decontaminating precast concrete that project shade on the façades in summer, a ceramic skin has been extended having the same colour as that of the exposed bricks of the adjacent buildings. It is composed of extruded terracotta pieces that form a

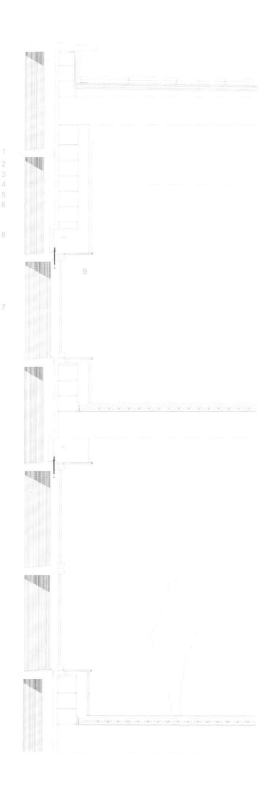

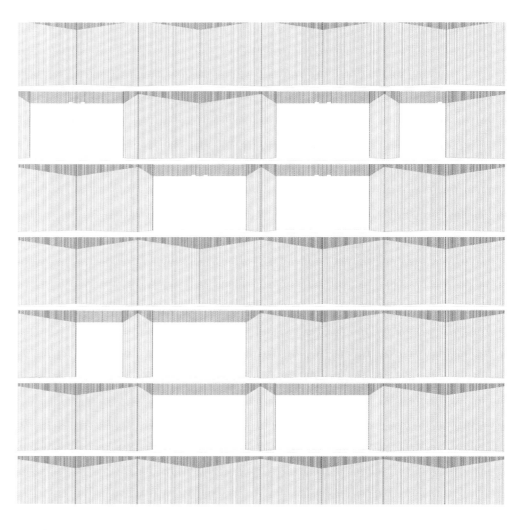

ventilated façade having an extraordinary thermal efficacy that, at the same time, conceals the structure between the geometry of its folds. The repetition of a same piece extended to 6,800m² of façade made it possible to design the geometry of the piece and its fluting, creating a façade of vibrant texture. On the upper part of the building, the façade becomes a lattice that conceals the roof installations allowing these spaces to be ventilated. To this end, some modules are substituted by lattices made up of baguettes of the same material, continuing the irregular texture given to the façades by the window scattering.

Ventilated Ceramic Façade
1. Horizontal precast concrete slats with decontaminating photocatalytic effect
2. Extruded ceramic pieces anchored to the concrete slats
3. Internal insulation using mortar and rock wool (5cm)
4. Thermo-clay block (24cm)
5. Semi-rigid mineral fiber thermal insulation with an inner layer of polyethylene
6. Plasterboard
7. Aluminium frames in lacquered aluminium and double glazing with thermal break
8. Roller aluminium shutter
9. Window frame made of maple wood

通风陶瓷立面
1. 水平预制混凝土板条，具有净化光触媒效应
2. 挤制瓷砖，锚固与混凝土板条上
3. 内置隔热层，灰浆 + 石棉（5cm）
4. 隔热黏土砌块（24cm）
5. 半刚性矿物纤维隔热层，内层为聚乙烯
6. 石膏板
7. 涂漆铝框架和双层断热玻璃
8. 卷帘铝百叶
9. 枫木窗框

项目概况

本项目的建筑设计建立在创新和可持续发展的基础上,二者相互结合,共同实现了项目的成功。

一方面,与传统教学建筑刻板、系统化、重复的印象相反,建筑师选择了灵活、多样、人性化的建筑形式。建筑力求避免教学机构的形象,将更多注意放在为师生提供服务、为不同的使用者创造空间上,打造了大量全新模式的教学研究空间。

另一方面,整座建筑都充分注意了环境可持续发展的方方面面。在规划和施工阶段,项目实施了多种可持续策略,从而实现了西班牙本土教学机构从未有过的优异成绩,是西班牙同类项目中第一个获得 LEED 绿色建筑认证最高顶级白金认证的项目。

在项目的构思过程中,建筑师不断在系统而重复的设计和特殊而多样的设计之间摇摆不定。系统和重复本身能保证项目的协调性和经济实用,同时也能增加项目的灵活性。另一方面,特殊而多样能让建筑更人性化,让使用者可以自由选择多种功能空间,从而让项目成为同类中的领导者。

细部与材料

建筑结构被系统化地转移到建筑立面上,从而保证了内部空间的灵活性。混凝土柱的排列间隔为 2.20 米,而后拉力楼板则解决了 11.40 米跨度的问题。柱子隐藏在通风陶瓷立面的褶皱后面,在主要通道等处,墙板似乎轻快地飘浮在空中。

建筑立面非常系统化,利用相同的板材、瓷砖和窗户塑造了整个建筑表面。窗户呈不规则的方式分散开,不仅满足了不同功能空间的不同照明需求,而且优化了建筑的多样化设计。

建筑大量使用了单一的材料。有限的材料以无限的应用方式呈现出来,实现了富有吸引力且高效的建筑设计。

在遍布着白色混凝土建筑和砖砌建筑的校园中,建筑师决定用一种更富可持续性的方式来实现一种呼应。在细长的净化预制混凝土条带(条带凸出墙面之外,在夏日能起到一定的遮阳作用)之间的陶瓷墙面呈现为与附近砖砌建筑同样的色彩。墙面由挤制瓷砖构成,具有卓越的热能功效,同时还能将建筑结构隐藏在墙面的褶皱之间。6,800 平方米的单一材料立面实现了瓷砖形状及其槽缝设计的多样化,从而形成了富有动感的纹理。建筑的上半部分立面成格栅状,将屋顶装置隐藏起来,实现了空间的通风。一些模块被长条瓷砖所构成的格栅所取代,延续了窗口赋予立面的不规则纹理。

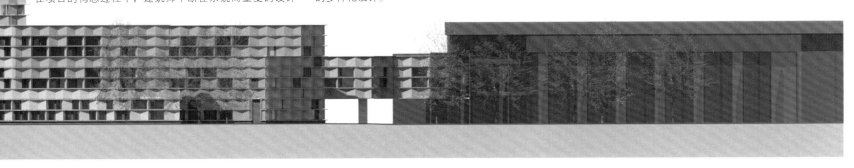

Centre of Air Navigation Services
空中导航服务中心

Location/ 地点: Budapest, Hungary/ 匈牙利, 布达佩斯
Architect/ 建筑师: László Kalmár, Zsolt Zsuffa/ ZSK Architects (Zsuffa és Kalmár Építész Műterem)
Photos/ 摄影: Tamás Bujnovszky
Site area/ 占地面积: 5,072m²
Floor area/ 楼面面积: 10,480m²

Key Materials: Façade – NBK Keramik (exterior Sun Control System), Schüco windows
主要材料：立面——NBK 陶瓷材料（外部遮阳系统）、Schüco 窗

Overview
The design of ZSK Architects continued the scenario of its neighbour building in a way that made them able to fit this significantly different spatial programme into the system determined by the previous building to be extended.

The surroundings do not require any new architectural character, preferably the existing one needs to be reinforced. Instead of brick façade cladding, the architects applied a ceramic lamella system installed on a separated load bearing structure as a secondary skin acting as an energetically sufficient sunscreen. This solution – according to their plans – took over the main characteristic of the neighbour building, namely the architectural use of burnt clay and at the same time radically reinterpreted it: the familiar connotations of brick, its massiveness, sensuality and the handcraft was replaced by the lightness (associated with flying) and industrial style of ceramic lamellas.

Associating this with the Faraday cage effect might not lead too far, it essentially means to enclose the space to be protected with metal mesh in order to shield it against exterior electromagnetic influence.

Detail and Materials
The most important — though at the same time most abstract – architectural development of this building lays in the architectural effect and meaning of ceramic lamellas, which solution is not without antecedents.

On one hand the use of lamellas provides a sculpturesque/object-like interpretation since seeing from different view points, the recognised massiveness of surfaces is changing too and leads to the differentiation of surfaces. However, the lamella system is completely homogeneous and undifferentiated at the same time, this way presenting a surface-like impression too. A good example of the former case can be the Central Signal Box building of Herzog and de Meuron

from 1999, where – covering the simple mass – the statuesqueness of metal cladding and its transition between solid and openwork surfaces strengthens primarily the form itself and its mass-like character, and results in an unusual, dematerialising impact. At the same time, Renzo Piano's buildings (Daimler building in Potsdamer Platz, Berlin at the turn of 2000 or the New York Times headquarter building from 2007) and their ceramic façades are approaching from the surface, since they use lamellas in their solid and open nature too in this way they project the façade behind them to the surface.

Transparent skin or mass with structured surface: these two features are present on the building at the same time. The dilemma of the design is the matter of openings: in case of spaces of typically circulation function (where shaded façade may be allowed) the lamellas run in front of the windows, but at offices with permanent working environment the ceramic shades, otherwise consistently placed everywhere around the building, are stopped. These large openings are framed with concrete rim in terracotta colour, further strengthening the surface-like character of the façade skin, and in this surface windows are articulated either as openings from a "curtain" drawn aside or cut out by using elements of a certain depth.

Besides the dematerialisation of mass and surface, in this case ceramic lamellas mean a radical brick façade transformation too. The façade elements refer to the building of ANS II only in their material and colour; all the other connotations offer drastically different interpretations. The ceramic elements are manufactured by the German company NBK, and in this very situation the components were produced with the length of 180 cm resulting from a development done in co-operation with the architects. The span quite large compared to the cross section dimension made it possible to install the supports rarely; and by this the conceptional intention, namely aerial appearance was emphasised. The ceramic beams have an end only at places where they meet an opening, and in this way they whip round the building like a never-ending bandage. It is felt this effect the strongest at spots where – looking through the lamellas of the corners or the ones running in front of the ground floor or upstairs court – due to the absence of any building façade behind the shades we can see a floating, nearly dematerialised, independent building skin. At these places the contours of the form are dissolved, the building is surrounded by a peculiar vibration.

项目概况

ZSK建筑事务所的设计继承了项目周边建筑的特色，使这个截然不同的空间项目融入了已有的建筑系统之中。

项目所在的环境并不需要任何新增的建筑特色，而是希望能够对已有的建筑环境进行强化。建筑师放弃了砖砌墙面，而选择将瓷砖薄层系统安装在独立的承重结构上，使其作为具有遮阳功能的第二层建筑表皮。这种设计既参考了周边建筑的烧陶材料，同时又对其进行了大胆的改造：砖块的传统内涵、它的厚重感、质感以及手工艺已经被陶瓷薄片的轻盈感和工业风格所取代。

或许可以将这种设计与法拉第笼效应联系起来，即：将空间包围在一个金属网内，使其抵御外界的电磁干扰。

细部与材料

项目将最重要也是最抽象的建筑设计置于建筑效果和陶瓷薄片的意义中，这在同类建筑中尚属首创。

一方面，陶瓷片的运用为建筑带来了一种雕塑感。随着视角的变换，建筑表面也会呈现出不同的效果。然而，

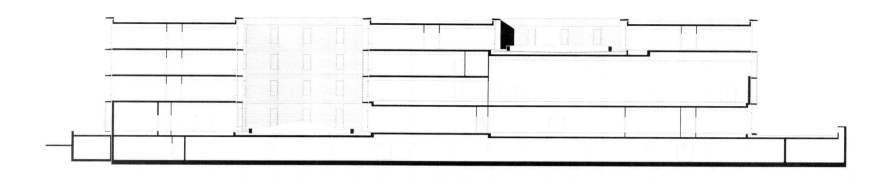

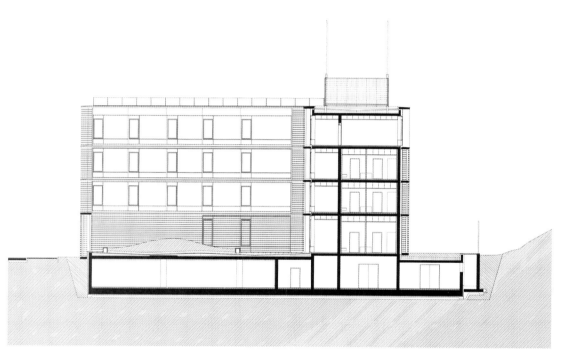

陶瓷片系统是完全均匀而无差别的，呈现出一种建筑表面的感觉。由赫尔佐格和德梅隆在1999年设计的中央信号楼就是此种设计的典范：简单的建筑主体被雕塑般的金属网包裹起来，虚与实表面之间的过渡突出了建筑的造型和体块，形成了独特的消失效果。此外，伦佐·皮亚诺的建筑（例如2000年的波茨坦广场戴姆勒大楼、2007年的纽约时报总部大楼）和它们的陶瓷立面也与该项目类似，它们所采用的陶瓷片也虚实结合，从而将后方的墙壁凸显出来。

半透明表皮与带有结构化表面的体块：这两种特征同时体现在一座建筑里。设计的困难在于窗口设计：在典型的交通空间中（可以使用遮阳立面），陶瓷片从窗前穿过；但是在永久性的办公环境中，前方的陶瓷片则应当被移除。这些大面积的窗口以陶土色混凝土为边框，进一步突出了立面表皮的平面感。窗户呈现为两种形式，一种可以向侧面拉开，另一种则向内切入一定的深度。

除了体块和表面的非物质化设计之外，在本项目中，陶瓷片还意味着对砖砌立面的彻底改造。立面元素仅在材料和色彩上参考了旁边的ANS II大楼，其他方面则是彻底的颠覆。陶瓷元件有德国NBK公司制作，长达180厘米。与横截面相比，元件的跨度极大，可以尽量减少支撑结构，这样一来，就能形成更具空气感的外观。陶瓷梁仅在窗口处设有端点，像一条不间断的绷带把建筑包裹起来。从转角、一楼正前方或楼上庭院看去，这种效果更加强烈，因为这些地方没有建筑外墙，陶瓷遮阳条就像飘浮在空中一样。建筑造型的轮廓消失了，建筑被一种特殊的震动环绕起来。

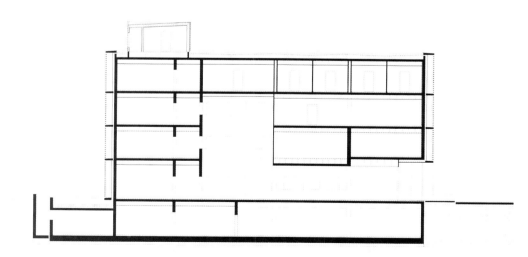

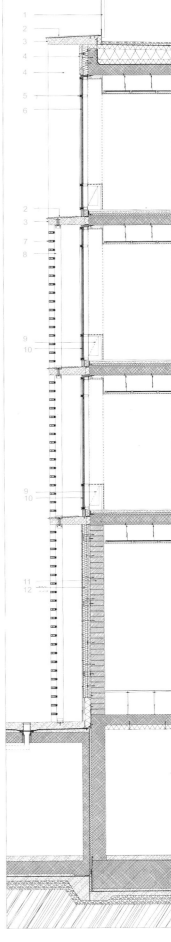

Section Façade
1. Galvanised steel safety barrier
2. VM Antrazinc tining
3. Coloured concrete moulding
4. Coloured concrete façade-panel
5. Schuco curtain wall
6. Schuco curtain wall (transparent glass)
7. Ceramic lamella (NBK Keramic sunscreen)
8. Lamellas' Alu support system
9. Fan-coil
10. Schuco curtain wall (translucent white glass)
11. Brick masonry
12. Thermal insulation

立面剖面
1. 镀锌钢安全栅
2. VM Antrazinc 锌顶盖
3. 彩色混凝土模架
4. 彩色混凝土立面板
5. Schuco 幕墙
6. Schuco 幕墙（透明玻璃）
7. 陶瓷薄片（NBK 陶瓷遮阳）
8. 陶瓷片的铝制支架系统
9. 风机盘管
10. Schuco 幕墙（半透明白玻璃）
11. 砖砌结构
12. 隔热层

Małopolska Garden of Arts
小波兰艺术花园

Location/ 地点：Krakow, Poland/ 波兰，克拉科夫
Architect/ 建筑师：Ingarden & Ewý Architects (IEA) www.iea.com.pl
Built area/ 建筑面积：1,579.3m²
Usable floor area/ 可用楼面面积：4,330.76m²
Photos/ 摄影：Krzysztof Ingarden
Completion date/ 竣工时间：2012

Key materials: Façade – ceramic tiles
主要材料：立面——瓷砖

Overview

The Małopolska Garden of Arts is a functional cross between two cultural institutions in Krakow: the JuliuszSłowacki Theatre and the MałopolskaVoivodeship Library. The wing on Szujskiego Street holds a modern art and media library, with multimedia books and music, while the section standing on RajskaSreet, has been developed by the theatre. The new hall – operating, as a studio theatre, conference space, concert hall, and venue for banquets and exhibitions – holds retractable stages for 300 people. Altogether, the space houses a theatre together with a cosy cinema with 98 seats, a café, and premises for the organisation of educational, art-related activities.

The main idea of the architects was to develop a method of contextual design to achieve a free and innovative form. It was to be autonomous so that it could be considered an individual form with a modern expression of its own but at the same time not obscuring the character of the place and following the legal regulations in Poland, laid down in the Planning and Spatial Development Act. According to these rules, a new development is expected: "to continue the function, parameters, features and indicators of development and land use, including the dimensions and the architectural form of structures, the building alignment and the intensity of land use" (which means that legislative provisions are aimed at ensuring contextual continuation).

The design method adopted in the Małopolska Garden of Arts project is based on two contrary principles of interpretation of its neighbourhood. On the one hand it respects the morphological continuity by using signs based on iconic representations of the general type – in other words a geometrical code, as well as a comparable scale of neighbouring structures. On the other hand, one abstracts from the context and interprets it. The historical forms and materials, in this case brick, have been processed. The façades

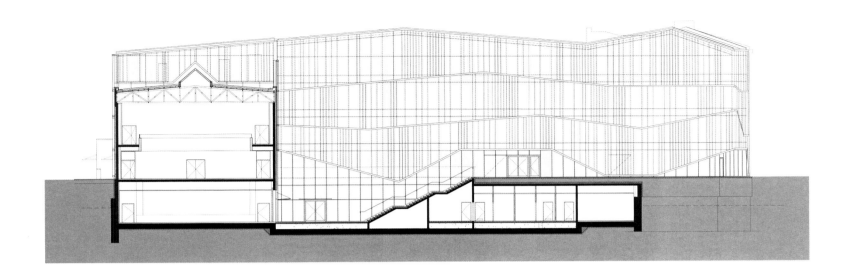

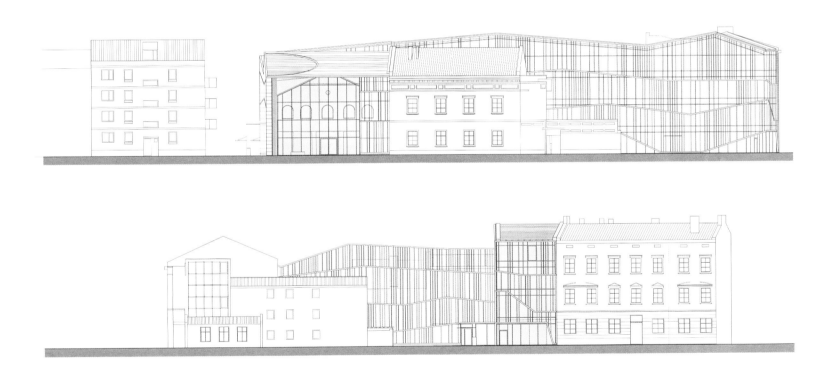

have a ceramic facing following the author's original design. It is an attempt to create contemporary architecture resulting from the analysis of the site and from the intention to understand the relation between the traditional and the contemporary language of architecture.

Detail and Materials

The method used in the Małopolska Garden of Art can be defined as a method of a type and stylistic interpretation of the neighbouring development – reinterpreting the local codes in terms of their geometrical features and material. As far as the approach to design is concerned it can be characterised as an attempt at a creative interpretation of the reality (mimesis), while at the same time radically straying from the context in terms of abstract and free composition of the body and the façades, and giving traditional materials (e.g. brick) new forms.

The characteristics of the old building material, particularly of the former hall – ceramic brick – are continued as well. Having reinforced the construction, the front wall of the hall was also recreated using brick from the dismantled structure itself. It was impossible to save the original walls and foundations considering their poor condition and inadequate load-bearing capacity.

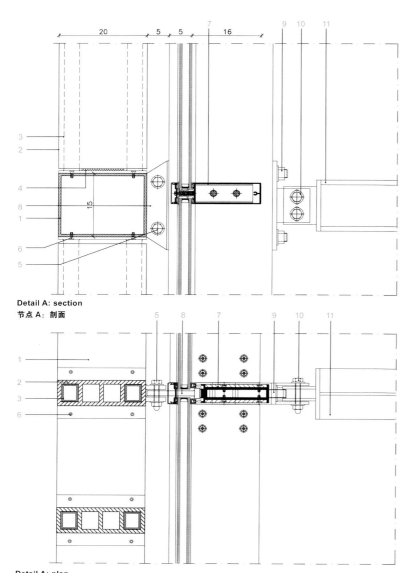

Detail A: section
节点 A：剖面

Detail A: plan
节点 A：平面

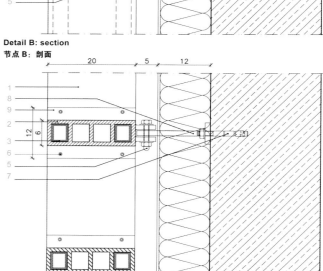

Detail B: section
节点 B：剖面

Detail B: plan
节点 B：平面

Detail A	节点 A
1. Main structure: steel RHS 200x150x6	1. 主结构：矩形钢管 RHS 200x150x6
2. Ceramic panel in 3 different colours	2. 三色陶瓷板
3. Aluminum RHS 40x35x4	3. 矩形铝管 RHS 40x35x4
4. Cushioning layer	4. 缓冲层
5. Screw, class 8.8	5. 螺钉，8.8 级
6. Self-tapping screw	6. 自攻丝螺钉
7. Façade system beam	7. 立面系统梁
8. Zinc coated steel flat bar	8. 镀锌扁钢条
9. Steel threaded rod Φ20mm	9. Φ20mm 钢螺纹杆
10. Screw, class 8.8	10. 螺钉，8.8 级
11. Façade bracing spot-bolted to interior slab	11. 立面支架，点栓于室内板上

Detail B	节点 B
1. Main supporting structure: steel RHS 200x150x6	1. 主支撑结构：矩形钢管 RHS 200x150x6
2. Ceramic panel in 3 different colours	2. 三色陶瓷板
3. Aluminum RHS 40x35x4	3. 铝 RHS 40x35x4
4. Cushioning layer	4. 缓冲垫
5. Screw, class 8.8	5. 螺钉，8.8 级
6. Self-tapping screw	6. 自攻丝螺钉
7. Steel bolt	7. 钢螺栓
8. Zinc plated steel flat 10mm	8. 镀锌钢板 10mm
9. Steel flat bar bolted to support main structure with welded vertical RHS	9. 扁钢条，螺栓固定，利用焊接矩形钢管支承主结构

项目概况

小波兰艺术花园位于克拉科夫的两座文化设施——朱丽斯斯洛瓦齐剧院和小波兰图书馆之间。苏杰斯克果街一侧的建筑内设置着现代艺术与媒体图书馆，收藏着多媒体图书和音乐；拉吉斯卡街一侧则由剧院负责开发。新建的大厅集实验剧场、会议空间、音乐厅、宴会厅、展览厅等功能于一身，它的可伸缩舞台可容纳 300 人。此外，项目还设有 98 个坐席的小型电影院、咖啡厅以及举办艺术教学相关活动的场所。

建筑师的主要理念是开发一套能够实现自由创新形式的环境化设计方案。设计是独立的，可以被看作是现代表达的一种独立形式，同时又不会影响其所在场所的特色、遵循波兰的《规划与空间开发法》的规定。根据以上原则，新项目预期的要求为："延续开发及土地使用的功能、参数、特色和指数，其中包括建筑结构的尺寸和形式、建筑队列以及土地使用的强度。"（即法律条款规定必须保证周边环境的延续性）

小波兰艺术花园所采用的设计策略基础是对周边环境的两种截然不同的诠释方法。一方面，它延续了建筑形态，采用了普遍的几何造型和与周边建筑结构相适应的体量规模。另一方面，它又抽离于环境之外。建筑选择了传统的形式和材料——砖，并对其进行了加工处理。建筑的陶瓷立面遵循了设计师的原始设计。项目是一次尝试，试图在环境分析及了解传统与现代建筑语言的关系的前提下打造一座现代建筑。

细部与材料

小波兰艺术花园所采用的设计方法可以看作是对周边开

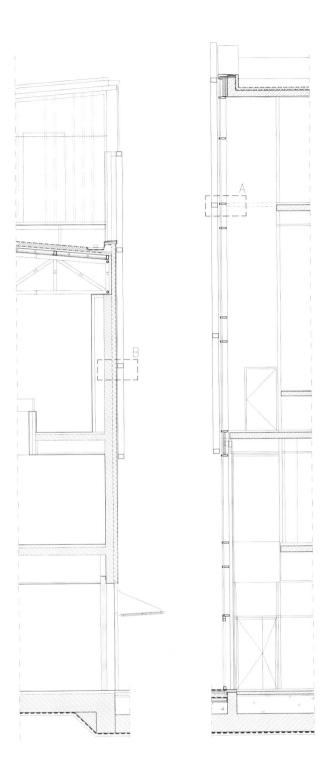

发环境的类型化和风格化诠释，它在几何特征和材料方面都对当地的设计传统进行了重新诠释。从设计层面上来讲，它是对现实创新解读的一次尝试。然而，从体块和立面的抽象而自由的构成上来讲，它彻底偏离的环境，赋予了传统材料（砖）全新的形式。

设计延续使用了旧建筑材料——瓷砖。经过结构加固之后，大厅的前墙用拆卸下来的砖块进行了重新塑造。因为考虑到材料的不良状况和不足的承重能力，基本无法保留原有的墙壁。

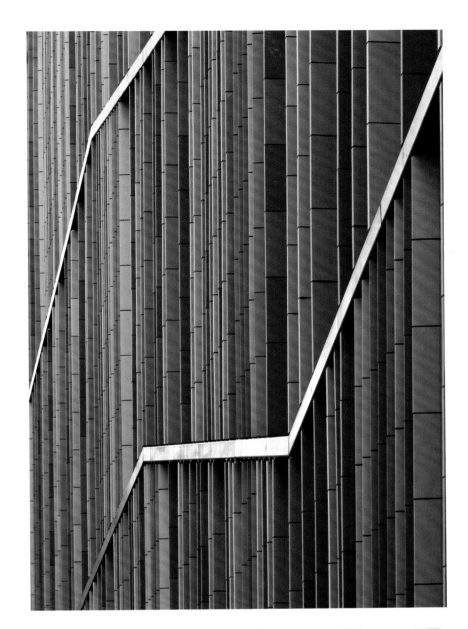

Martinet Primary School
马丁内特小学

Location/ 地点: Barcelona, Spain/ 西班牙，巴塞罗那
Architect/ 建筑师: Humbert Costas Tordera, Manuel Gómez Triviño, Jaime Blanco Granado, Carlos Durán Bellas, Josep M. Estapé i de Roselló/Mestura architects
Photos/ 摄影: Pedro Pegenaute
Gross area/ 总面积: 3,448.71m²

Key materials: Façade – ceramic
主要材料：立面——陶瓷

Overview

The building takes the form of a "U" shape on the small plot of plan, centred around the daycare, which frees the rest of the un-built space for the elementary school's patio. The school is located in the neighbourhood of Almeda in Cornellà de Llobregat, Barcelona.

The support services, including a dining room, kitchen, gym, changing rooms and teachers rooms, are located in one volume of the ground floor. The classrooms are in a three-level volume with the day¬care on the ground floor. The daycare faces south, and has direct access to its own private patio via a porch. The elementary school classrooms are located on the first and second floors and are accessed by a corridor from the south, but receive light from the north. A small part of the programme is located as an annex between the two other volumes, holding the lobby, concierge and services, and the library with separate outdoor access.

Detail and Materials

There is a latticework made from ceramic paste facing south and offering passive solar protection to the corridors. It is made of ceramic pieces enameled in various appropriate colours: the side facing west is done in three shades of green representing spring and the east-facing section in shades of brown representing autumn. The latticework fulfils various purposes. The main façade of the school is visible from Ronda de Dalt, a highway, and the ceramic latticework is an eye-catching element that nonetheless inte¬grates it into the environment of neighbouring industrial containers, while the three-dimensional ge-ometry of the pieces stands out on closer inspection. From the inside, the latticework forms a double façade, which creates a play of light and shadow that alters with the changing seasons. The use of colour provides a changing landscape aimed at achieving an appropriate and cheerful environment for children face a fairly inhospitable urban place.

Why the designers choose the façade material?

All schools built in Catalonia by the local government meet the same standards as the number and size of all the spaces that make up, however the constraints of the place and environment produce always different projects. In this school, a lot of smaller surface than usual and unfriendly environment, conditioned building is very compact and introverted, in which the treatment of light and colour come to prominence thanks to the use of the ceramic pieces.

The use of ceramics has always enjoyed a leading role in Catalan architecture and its use to form celosia allows to create interesting effects with the power of the Mediterranean light.

Features of the constructional materials product?

- Glazed ceramic tile screen made with stoneware pasta, created to afford passive protection from the sun on the south side of the corridors that lead to the primary classrooms.
- Assembling the vertical and horizontal joints with stainless steel rods 6 mm in diameter.
- Galvanised steel forged anchor for resisting the horizontal thrust
- Anti-rust primer and two coats of synthetic enamel paint for all metal elements.

Technology of construction materials?
The criteria applied to the construction solution for the screen have assured the structural

stability and durability of the materials that have been used. The tiles have a groove all around their sides that means the vertical and horizontal joints were able to be strengthened by means of 6mm diameter stainless steel rings. A total of 7 vertical expansion joints were created, with horizontal bracing in the middle framework, 3 vertical bracings through the full height of the screen and side bracings at the two ends. The joints are 2cm thick and have been made in lime mortar with a plasticized aeration additive applied semi-dry. The additive provides the enough plasticity to accommodate possible shrinkages.

项目概况

建筑呈U形，坐落在一块小型场地里，中央环绕着日托班，其他的空白空间则可以作为小学的天井。学校位于巴塞罗那的阿尔梅达社区。

餐厅、厨房、体育馆、更衣室、教师办公室等辅助服务设施全部聚集在一楼的一个空间内。教室占据了三层楼的空间，其中日托班设在一楼。日托班方向朝南，通过门廊直接与专属天井相连。小学教室分设在二三楼，通过朝南的走廊进出，从北面获取阳光。一小部分功能区以附属楼的形式建在两个主楼之间，内部设置着大厅、门房、服务区和图书馆，其中图书馆设有单独的外部通道。

细部与材料

建筑朝南的外墙有一面由陶瓷坯泥制成的格架，为走廊提供了被动遮阳。构成格架的瓷砖涂上了不同色彩的瓷釉：朝西的一侧由三种绿色调构成，代表着春天；朝东的一侧由灰色调构成，代表着秋天。格架具有多重功能。学校的正面正对一条环形高速路，陶瓷格架十分显眼，同时又能使学校融入周边的工业集装箱环境。走近学校，瓷砖的三维立体造型会变得十分突出。在内部，格架构成双层立面，营造出随着季节变迁而变化万千的光影效果。色彩的运用赋予了学校变化的外观，在相对荒凉的城市空间内为孩子们打造了一个愉悦的环境。

设计师为什么选择这种立面材料？

所有建在加泰罗尼亚地区（西班牙东北部）、由当地政府筹建的学校都必须符合同样的空间规划和规模标准，但是由于场地和环境因素的限制，不同学校的设计各不相同。在本案中，项目场地面积较小，周边环境也不甚友好，因此建筑必须十分紧凑且内向。瓷砖的应用让建筑对光线和色彩的处理变得十分突出。

陶瓷的运用在加泰罗尼亚地区的建筑中十分常见，而格架的形式在地中海阳光的沐浴下形成了有趣的光影效果。项目所使用的各种建筑材料有什么特点？

- 釉面瓷砖幕墙由陶瓷坯泥制成，能为建筑南侧与主要教室区域相连的走廊提供被动遮阳。
- 用直径6毫米的不锈钢条装配瓷砖的水平和垂直接缝。
- 用镀锌钢锻造锚件稳定水平推力
- 用防锈底漆和双层合成瓷漆喷涂所有金属件。

建筑材料的技术应用？

幕墙的建造标准保证了材料的结构稳定性和耐久性。瓷砖的各面都有凹槽，因此可以用直径6毫米的不锈钢环对水平和垂直接缝进行加固。总共有7个垂直伸缩接缝，中间框架上带有水平支撑。3个垂直支架贯穿了幕墙的全高，两侧也分别设有侧面支架。接缝为2厘米厚，用石灰砂浆和增塑充气添加剂在半干时填充。添加剂提升了砂浆的塑性变形能力，能够应对未来可能的收缩。

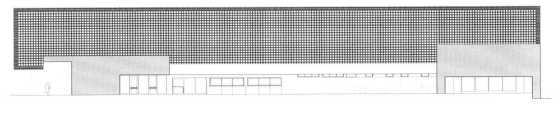
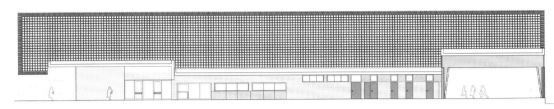
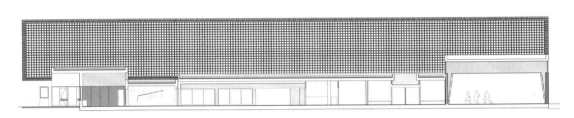
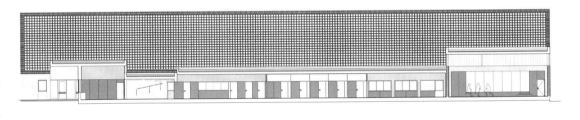

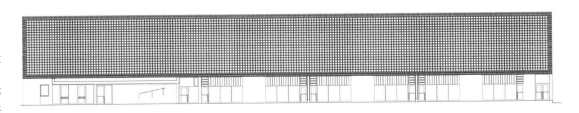

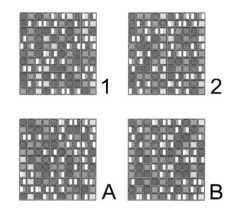

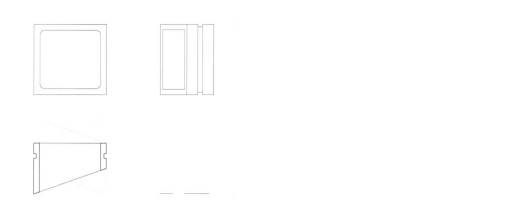

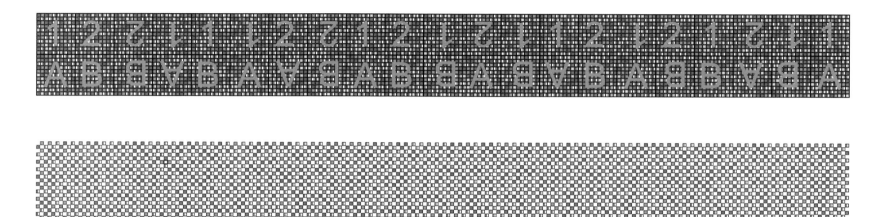

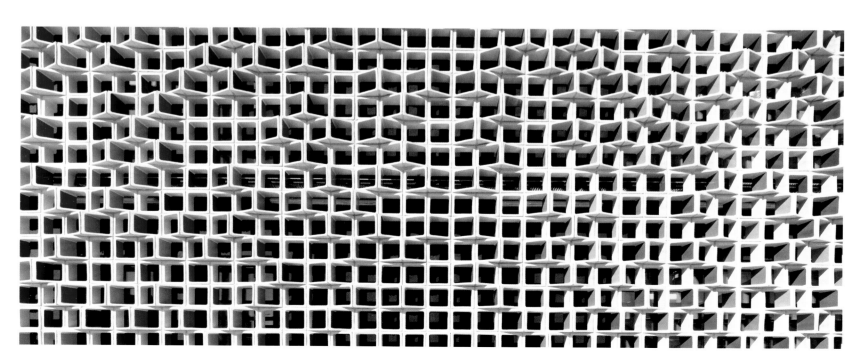

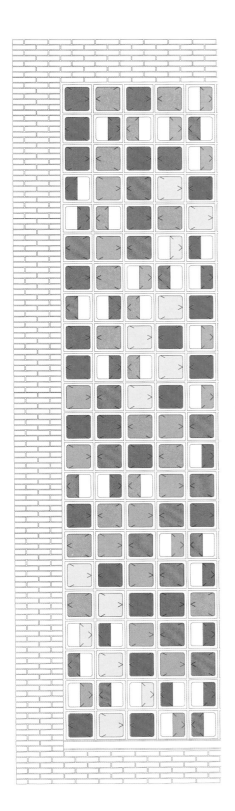
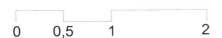
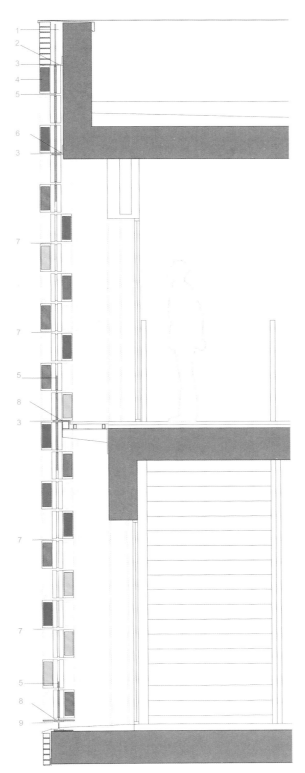

Façade Section Detail

1. Mortar fill and fixing by strip
2. Metal profile LD 48
3. Neoprene coro
4. Ceramic tile
5. Stainless steel rod
6. Metal profile L 100.100.10
7. Longitudinal reinforcement
8. Galvanised steel plate
9. Metal profile HEB 100

立面剖面节点

1. 砂浆条缝填充和固定
2. 金属型材 LD 48
3. 氯丁橡胶
4. 瓷砖
5. 不锈钢条
6. 金属型材 L 100.100.10
7. 纵向钢筋
8. 镀锌钢板
9. 金属型材 HEB 10

Mediatheque in Carballo
卡瓦略图书馆

Locaiton/ 地点: Carballo, Spain/ 西班牙，卡瓦略
Architect/ 建筑师: Mr. Óscar Pedrós, PhD. Architect
Photos/ 摄影: Óscar Pedrós, Héctor Santos – Díez (Santos-Díez | BISimages)
Area/ 面积: 157,000m²
Budget/ 预算: 1,199,588 €/1,199,588 欧元
Completion date/ 竣工时间: 2013

Key materials: Façade – ceramic
Structure – concrete
主要材料：立面——瓷砖
结构——混凝土

Overview
Thinking on a 21th Century library, as a friend of the architect would say, is like wondering about a book's cemetery. At that point, all efforts must conciliate architectonic non-variable values as sequence, promenade and staying with immateriality brought by new technologies. There is no other solution than sharing jealously material space with vitality to avoid Carballo's newest building ending as book warehouse.

The building is placed in Carballo, a medium-sized Spanish village, 35 km far away from A Coruña, Galicia, Northwest Spain. Urban atmosphere is strongly rude, full of party walls and unfinished housing blocks. The building becomes quite hermetic and opens itself to an inner patio acting as an open space for reading.

A U-shaped ground floor surrounds a central patio with controlled access from outside. This patio behaves as an extension of future public space northwards.

Floor plans are designed following the complicated plot's geometry to get a better ratio of profit in relationships between inner and outer space, melting the threshold between them. In terms of construction, the building shows two important features; the first one is a structural effort in designing all singular corners of the building: due to the rotation of first floor looking for urban views (from inside) and the proper building's presence (from outside), some cantilevered elements appear as part of its language. Special presence gains the hanging auditorium and the covered entrance.

Detail and Materials
There's yet other significant feature is in terms of materialisation: one of the building's aims is to get as much natural lighting as possible, both to save energy and for a comfortable reading. Then an off-white façade allows to get lots of light by reflection without shining. 3x1 m ceramic tiles were chosen to fit the scale of an institutional building, all along the rain-screen façade.

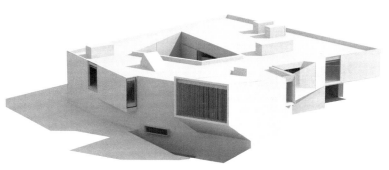

Their thickness (only 6 mm) makes possible a bigger space exploitation as inner perimeter gets bigger. Shelves for books will hide the visible structure thrusting out of the inner walls. Working with a really thin façade (25 cm in total), obliges also to set all installations out of the perimeter (running through the floor or inside walls) to protect holes in the stand-alone brick exterior party-wall. The most specific and significant thing in the design of it is the fact that the tiles are laid in a staggered shape giving continuity to vertical joints. It's an unusual configuration for rain-screen façades and it is rather unseen. The final budget was 764.00 €/sqm, a definitely low-cost architecture.

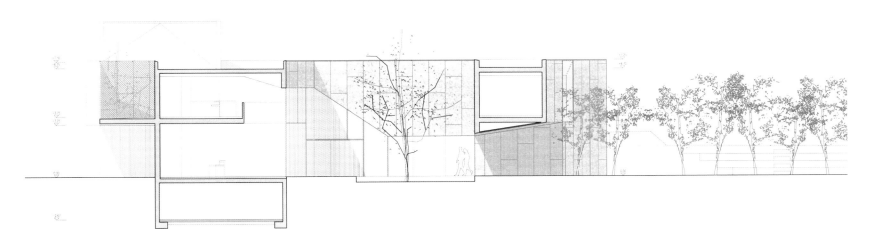

项目概况

建筑师的一位友人曾说道，"21世纪的图书馆就好像是图书的墓地。"毫无变化的建筑标准和新技术让图书馆变得死气沉沉。只有通过让丰富的物质空间充满活力才能避免让卡瓦略这座最新的建筑沦为图书仓库。

建筑坐落在卡瓦略——一座位于西班牙西北部，距离拉科鲁尼亚35千米远的中型城镇。这里的城市氛围十分粗犷，到处是界墙和未完成的住宅楼。建筑比较封闭，内部有一个可用作露天阅读空间的天井。

一楼的U形平面布局环绕着中央天井展开，与外界的联系有限。这个天井是未来公共空间向北扩张的一部分。

为了获得更好的室内外空间比例并使二者融为一体，建筑的楼面布局遵循了复杂的地块形状。在结构上，建筑有两个重要特征：一是建筑所有的棱角设计；因为二楼空间渴望寻求城市景象，并且为了呈现合适的建筑外观，建筑采用了一些悬挑结构。特殊的外观设计实现了悬挑的礼堂和带顶的入口。

细部与材料

建筑的另一个显著特征是它对材料的运用：建筑设计的目标之一是尽量使用自然采光，这样既能节约能源，又能让阅读更舒适。因此，灰白的外墙能够通过反射获得大量光线。3x1米的瓷砖十分适合建筑的学院派风格，它们排列构成了雨幕外墙。薄薄的瓷砖（仅有6毫米）让内部空间更大。书架将内墙上突出的结构元素隐藏起来。由于外墙极薄（总厚度仅有25厘米），室内装置必须离开建筑外围以保护对砖砌外界墙上的孔洞。设计最特别的一点在于瓷砖是错列排列的，保持垂直接缝的连续性。对雨幕外墙来说，这种配置十分罕见。项目最终的预算是764欧元/平方米，绝对是一座低成本建筑。

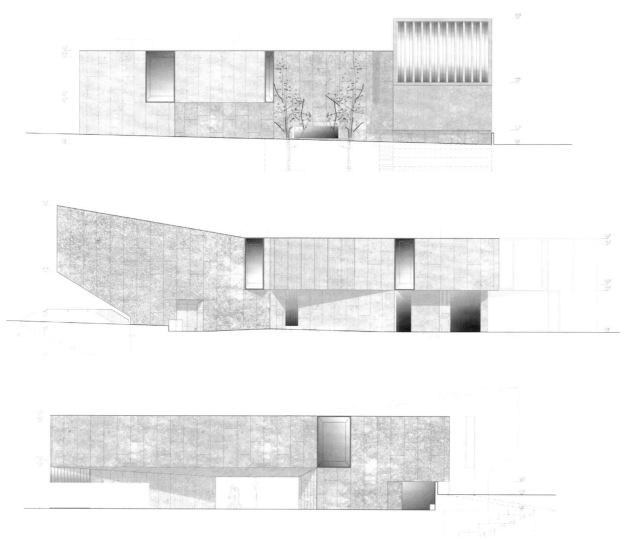

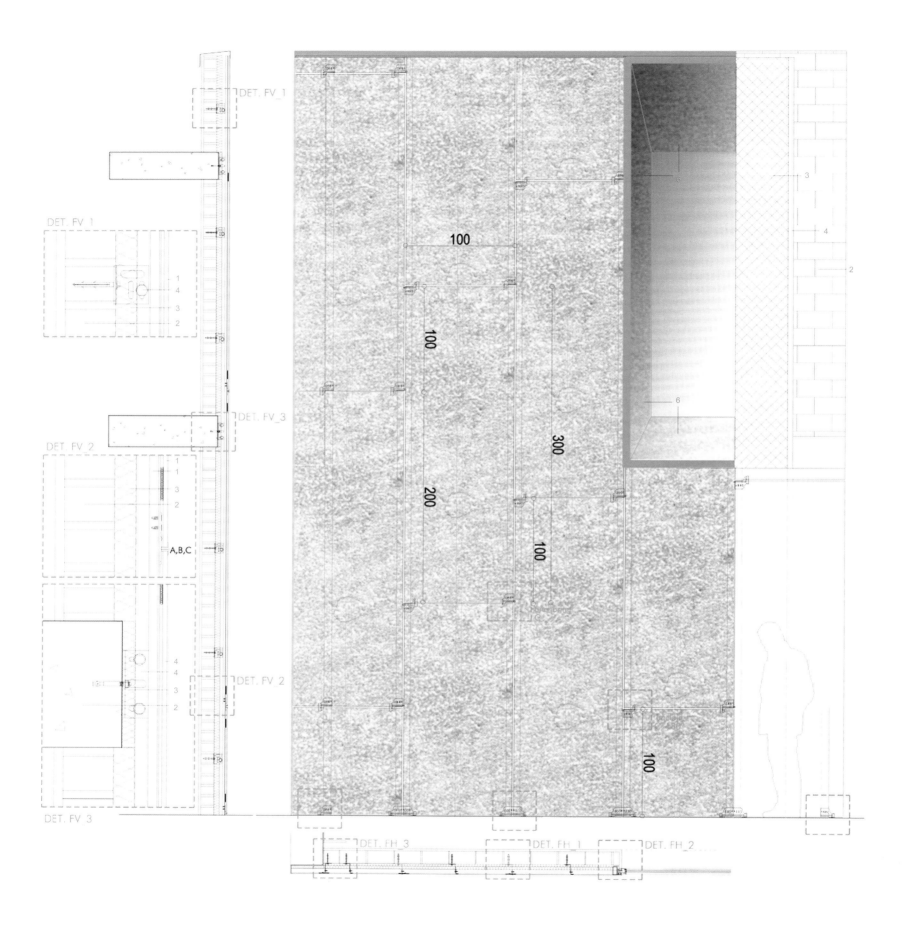

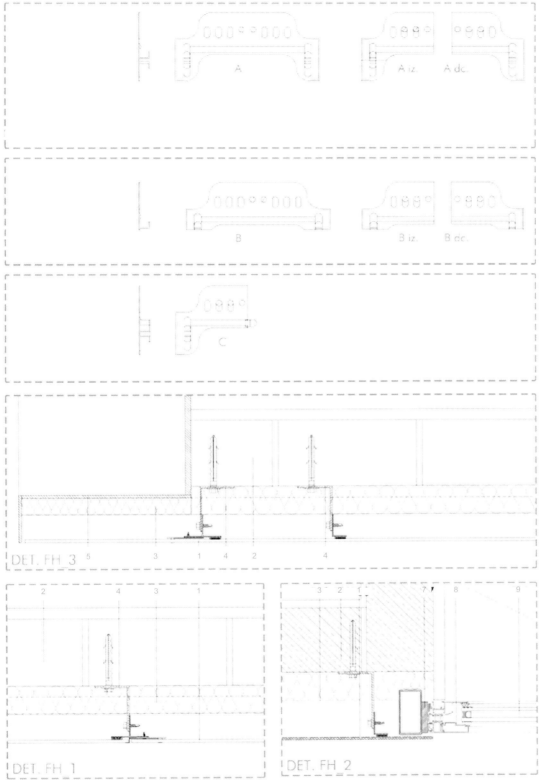

Rain Screen Façade System
1. Rain screen façade of 3.00x1.00m, ceramic tiles, 3+3mm thick reinforced inside with mat, off-white coloured, staggered-pattern, holders in RAL 9001-lacquered aluminium
2. Perforated brick in partition wall, 25x12x8cm lying flat (half-foot)
3. 60mm thick mineral wool (MW) layer with protective layer outwards
4. Supporting substructure in black lacquered aluminium
5. Concrete filling under flashing
6. Inside ceramic flashing 3+3mm thick
7. Neoprene strip to avoid galvanic coupling
8. Standing carpentry of extruded aluminium, lacquered colour X1 8808 VT(S0342) exterior
9. Double laminated glazing 4+4/12/6+6 with acoustic polyvinyl butiral (low-emissivity glass in patio's north glazing)

雨幕外墙系统
1. 雨幕外墙由 3.00x1.00m 瓷砖构成，3+3mm 厚，内部以衬板加固，灰白色，错列图案，由 RAL 9001 色涂漆铝架支撑
2. 隔断墙的多孔砖，25x12x8cm，平铺
3. 60mm 厚矿棉层，外覆保护层
4. 下层支承结构，黑色涂漆铝
5. 防水板下混凝土填充
6. 内部瓷砖防水板 3+3mm 厚
7. 氯丁橡胶条，避免电阻耦合
8. 挤制铝材立式窗框，X1 8808 VT(S0342) 色外部涂漆
9. 双层隔热玻璃 4+4/12/6+6，配聚乙烯隔音条
（天井的北窗采用低辐射玻璃）

Chapter 2
Stone

第二章 石材

As a high-class construction material, stone is extensively used in exterior and interior decoration, curtain wall decoration and public facility construction. (See Figure 2.1 and Figure 2.2)

石材（Stone）作为一种高档建筑装饰材料广泛应用于室内外装饰设计、幕墙装饰和公共设施建设。（见图2.1、图2.2）

2.1 Categories & Features

Now, stone products on the market mainly belong two categories: natural stone and artificial stone. Natural stone used in building façade generally includes granite, slate, sandstone and marble. (See Figure 2.3 to Figure 2.5)

Granite is a common type of igneous rock. With high density and mar-proof and anticorrosion qualities, it is a perfect decorative material for public places and exterior walls. Granite is easily embrittled so only small parts will fall off when it is damaged, which won't affect the overall flatness.

Slate is type of metamorphic rock. With medium rigidity and low water absorption rate, its life span is approximately 100 years. The colour of slate varies in grey, yellow, green, blue, black and red, most in solid colours. It expresses a natural and elegant feeling without polishing, ideal for decorations of cultural places.

Sandstone is a type of sedimentary rock. Sandstone is sound-proof, radiation-free, damage resistant and weather resistant, and doesn't fade easily. It absorbs moisture and is insoluble in water. As a type of matte stone material, sandstone looks natural and expresses a simple yet elegant style, decent and warm at the same time. It is commonly used in exterior and interior walls, sculptures and garden

constructions.

Marble is a non-foliated metamorphic rock composed of recrystallised carbonate minerals, most commonly calcite or dolomite. The grains of marble are generally irregular and the rigidity is low. Marble is radiation-free, colourful and abrasive resistant. Its life span is generally 50 years. Marble is commonly used for interior walls and floors.

Artificial stone used in architectural field includes artificial marble, vitrified stone and stone composite panel. (See Figure 2.6)

Artificial marble has a high density and its colour is adjustable, suitable for decorative finish. Its disadvantages lie in low rigidity and uneven smoothness.

Vitrified stone is a stable material with high anti-corrosion and anti-pollution ability and high bending strength. Its colour is bright and soft, without obvious colour distortions.

Stone composite panel is composed of stone and structural reinforced panel (The surface usually uses 2-10mm natural stone, while the reinforced panel uses aluminium honeycomb panel, aluminium-plastic panel, plywood or safety glass (THK≤20mm). The two are bonded together by epoxy resin or polyester resin.) On one hand, stone composite panel remains natural stone's grain and colour; on the other hand, it is light-weighted, anti-bending and easy to install.

2.1 分类与特性

目前市场上常见的石材主要分为天然石和人造石。建筑常用饰面天然石材大体分为花岗岩、板岩、砂岩、大理石。（见图2.3~图2.5）

花岗岩是一种非常坚硬的火成岩岩石。密度很高，耐划痕和耐腐蚀，具有良好的装饰性能，可适用公共场所及室外的装饰。花岗岩具有脆性，受损后只是局部脱落，不影响整体的平直性。

板岩是一种变质岩。硬度适中，吸水率小，其寿命一般在100年左右。颜色多以单色为主，如灰、黄、绿、青、黑、红等，在装饰上无需磨光，给人自然大方之感，适于富有文化内涵的场所的装饰。

砂岩是沉积岩的一种。具有隔音、吸潮、抗破损、耐风化、耐褪色、水中不溶化、无放射等特点。砂岩属于亚光石材，自然形态十足，在装饰上呈现素雅、温馨风格，但同时不失高贵大气，是一种暖色调石材，常用于室内外墙面装饰、雕刻艺术品、园林建造等。

大理石是由沉积岩和沉积岩的变质岩形成，表面条纹分布一般较不规则，硬度较低。其不含辐射，色彩丰富、耐磨性好，使用寿命一般在50年，常用于室内墙及地面装饰。

建筑用人造石材通常包含人造大理石、玻化石、石材复合板。（见图2.6）

人造大理石结构致密，可调节色彩，适于饰面装饰。但其硬度较小，光度不一致。

玻化石性能稳定、耐腐蚀、抗污性强、抗折强度高、色彩艳丽柔和、无明显色差。

石材复合板是由石材与结构增强板黏结而成（表面常采用2~10mm天然石材，结构增强板使用少于20mm厚的蜂窝铝板、铝塑板、胶合板或安全玻璃，黏结剂多为环氧树脂和聚酯树脂）。保留天然石材自然纹理、色彩、较轻、抗折强度高、便于安装。

2.6

2.2 在建筑中的应用

石材是人类历史上应用最早的建筑材料之一，并一直为人们所青睐，世界著名的古埃及金字塔、古希腊雅典卫城等都是石头建筑杰出而不朽的代表作。如今，石材很少作为结构材料使用，多用于建筑内外装饰材料，呈现自然气息，带来浓重的历史感和文化感。

石材用于墙面装饰最为常见的形式即为板材和块材。其中板材作为饰材的构造形式通常包括湿挂法（钢筋网挂贴法）、干挂法（现代石板幕墙）和室内装饰的黏结固定法等。而块材通常采用石材垒砌方式。（见图2.7~图2.9）

2.2 Application in Architectural Field

As one of the oldest building materials in human history, stone is favoured by people all the time. The world-known ancient Egypt pyramids and Acropolis of Athens are both excellent and lasting master works of stone structure. Today, stone is more used as decorative material rather than structural material. Besides natural feelings, stone also bring strong senses of history and culture.

Two most common forms of stone wall decoration are panel and block. The configuration method of stone panel includes wet hanging (reinforcing mesh hanging and sticking), dry hanging (modern stone curtain wall) and sticking fixation of interior decoration. Stone blocks are often built as masonry. (See Figure 2.7 to Figure 2.9)

· Wet Hanging

First, fix the reinforced mesh to the structural wall and punch holes at the back of the stone panel. Then tie the stone panel and reinforced mesh together by steel wire, and pour cement mortar into the joints between the panel and the wall layer by layer. This method is slow and the quality of construction is hard to ensure. Therefore, it is rarely used except for the construction of small project or wall base.

· Dry Hanging

Modern stone curtain wall usually uses dry hanging. Supporting frame and stone panel compose a non-load-bearing architectural enclosure system. The gap between wall and stone panel is approximate over 80mm which is easy for construction, adjustment of surface flatness and installation of thermal insulation layer.

The joint between panels needs careful treatment and is often sealed with sealant. This sealing method can reduce heat loss but its detailed visual effect is unfavourable. With the development of technology and update of ideas, open seam structure with more delicate visual effect is adopted by more and more architects. However, you should pay attention to the rainwater entering through the joints and make precautions. Some architects combine metal line or metalwork similar to that used in glass wall with stone panel to create more details.

· Bonding Fixation

Bond the stone panel on the wall base with epoxy resin or other adhesives. This method features fast construction and low cost, perfect for interior wall with secure wall base and limited height.

· Stone Laying

Lay the stone blocks together as self-supported structure. This method is apt for thick and low wall. If the stone wall is not used as interior structural wall, it is often fixed with the structural wall with metal hook or steel mesh. The gap is filled with cement mortar. The size, dimension, shape and laying method matter a lot to the final effect of the wall. (See Figure 2.10)

Besides these basic construction methods mentioned above, architects also create some unique methods by themselves, such as infill stone block wall. An architect should determine appropriate stone type, surface texture, stone thickness, curtain wall thickness, colour and overlapping mode with other materials according to the building's overall style.

2.10

• 湿挂法

将钢筋网固定于结构墙体，石板背部穿孔，然后通过钢丝将石板和钢筋网绑在一起，石板及墙体间分层浇筑水泥砂浆。这种方式施工速度较慢，质量不易保证，因此除一些小型工程或墙体基座部分外以较少采用。

• 干挂法

现代石板幕墙通常采用干挂法，由支承构架与石板组成不承受主体结构载重的建筑围护结构体系。墙体与石板间距一般在80mm以上，施工方便、易于调节墙面平整度并有利于墙体的保温。

石板间缝隙的处理非常重要，最为常见的是采用密封胶密封。其具有自身的优势，如减少热能流失，但同时也存在一定的缺点，尤其是细节的视觉效果。随着技术的进步和观念的更新，开缝式构造越来越多地被建筑师采用，视觉效果更为精致美观，但同时要注意进入板缝的雨水对幕墙的影响，采取一定的措施进行预防。一些建筑师在设计中将金属线条或类似于点支玻璃幕墙的金属件和石板结合在一起，营造更多的细节。

• 黏结固定法

用环氧树脂或其他黏胶将石板黏结在墙面的基层，施工速度快、造价低，适合室内空间基层牢固、高度有限的墙面处理。

• 垒砌法

将石块砌筑在一起，通过石材自身承重，适用于相对较厚且高度有限的墙体。如石块墙体不作为结构内墙，通常用金属搭钩、钢网片等将石块与结构墙固定，中间浇筑水泥砂浆。石块的规格、比例、形状以及垒砌方式对于墙面的最终效果具有重要的影响。（见图2.10）

除这些基本构造方式之外，建筑师根据自己的爱好创作出特殊的做法，如填充石块墙体。建筑师应根据建筑的整体风格选择石材的品种、表面质感、石材厚度、幕墙厚度、色彩以及与其他材质交叠部位的处理方式。

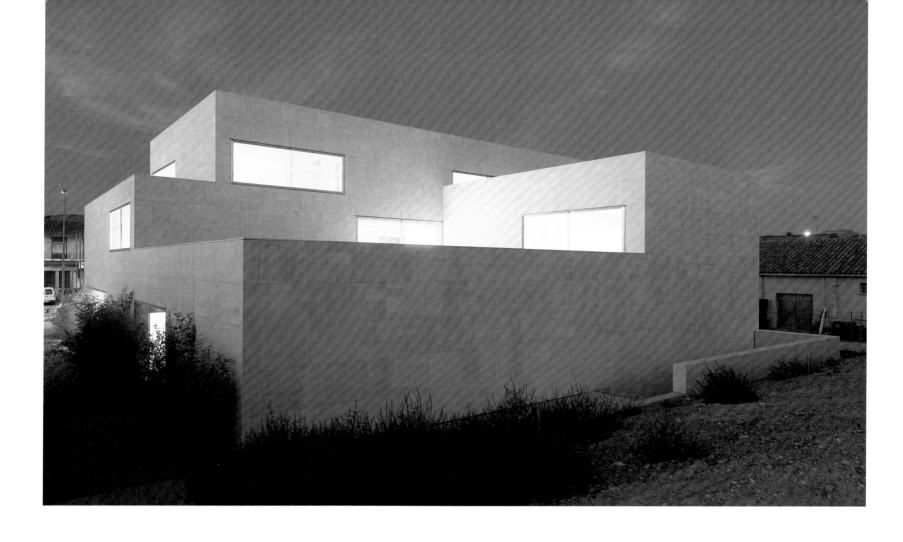

Bajo Martin County Seat
巴乔马丁郡政府

Location/ 地点：Teruel, Spain/ 西班牙，特鲁埃尔
Architect/ 建筑师：Magén Arquitectos (Jaime Magén, Francisco Javier Magén)
Photos/ 摄影：Pedro Pegenaute
Site area/ 占地面积：1,286.10m²
Completion date/ 竣工时间：2011

Key materials: Façade – stone, alabaster
主要材料：立面——石材、雪花石膏

Overview
The site is located on the outskirts of Hijar, capital of the county, along the national highway N-232 and the old abandoned silo. It was a dysfunctional urban environment, including existing industrial buildings, and the front of residential townhouses, just across the road.

The absence of urban qualities in the environment legitimises a certain autonomous condition of the building, rising from the land to form a unified solution, clear and compact. Therefore, the necessary link of building and place, reinforced by its institutional character, not articulated from urban relationships with the immediate environment, but from references to geographical landscape, history and culture, present in their external configuration. The group of carved volumes on local materials – stone and alabaster, alludes, in an abstract and geometric way, to stone groups that occur in quarries in the area.

Detail and Materials
The stone surfaces, opaque or translucent, exhibit materials and expressive features of alabaster in relation to the day or night lighting.

The ordered group volumes in the outside, compact, heavy and massive, is poured inside. The space pierces and perforates the solid volume, producing a dynamic system of voids, connected visually and spatially, diagonally, linking the three floors and articulating the circulation spaces, access and meeting. The continuity with the outside material and the presence of natural light into the interior through various gaps, strengthen the condition of the interior space as empty excavated, drawn from the section as a fundamental tool of the project.

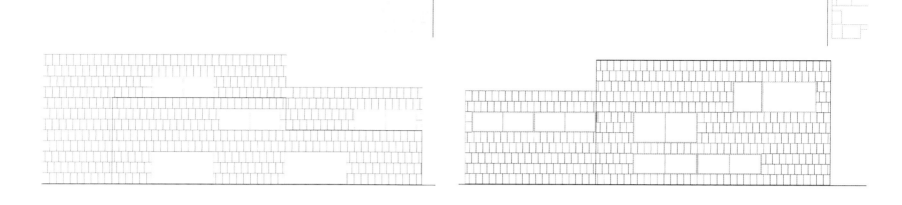

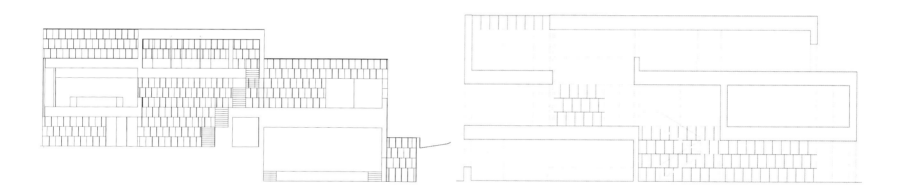

项目概况

项目场地位于巴乔马丁郡首府海加尔的市郊，紧邻 N-232 国家高速公路和一座废弃的筒仓。这是一个功能失调的城市环境，容纳着各种工业建筑，路对面还有许多联排别墅。

建筑环境中城市风格的缺失赋予了建筑一种独立感。建筑从地面上破土而出，整个外观显得清晰而紧凑。此外，由于建筑本质上明显的机构特征，十分有必要将建筑与环境联系起来。这种联系不是与直接环境的城市关系，而是在外部配置上体现出该地区的地理景观、历史和文化。由石材、雪花石膏等本土材料所塑造的雕刻体块以抽象图形的方式暗示着该区域内采石场中的石堆。

细部与材料

随着日夜的光照，透明或不透明的石材表面充分展示着材料的特质，并且呈现出雪花石膏富有表现力的纹理。整齐堆放的砌块从外面看起来紧凑、厚重、沉稳，是由内部浇灌而成。空间穿透实心体块，形成了富有动感的孔洞系统，在视觉和空间上将三个楼层连接起来，同时也连接了内部交通流线、通道和会面空间。外部材料的连续性以及从各种开口进入室内的自然光突出了室内空间的挖掘效果。

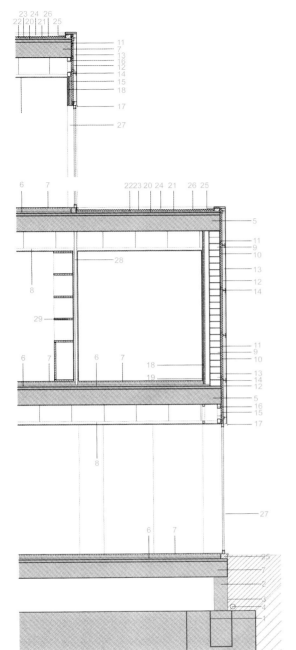
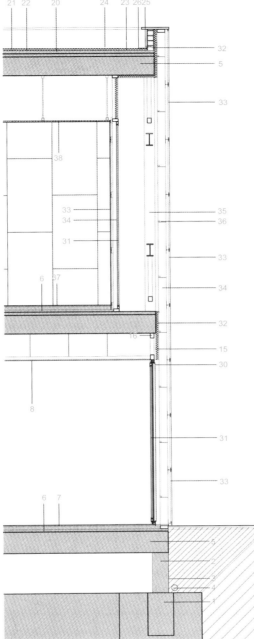
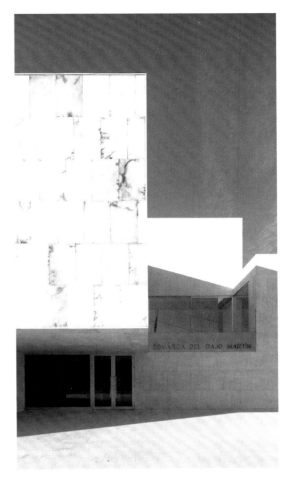

Detail
1. Concrete beam
2. Preformed brick pieces wall
3. Waterproofing membrane
4. Linear drainage
5. Pre-stressed self-supporting termacol slab 27+3+5
6. Mortar cement leveling
7. Stone paving limestone moca
8. Suspended ceiling of plasterboard
9. Performed brick pieces wall
10. Cement mortar wall
11. Insulation of extruded polystyrene panel
12. Air chamber
13. Natural limestone moca cladding
14. Adjustable anchorage type halfen
15. Pine wooden board
16. Tube section 80.6.3 of galvanised steel
17. Flashing of galvanised steel sheet
18. Plasterboard dividing wall
19. Plinth made of DM board colour white
20. Slope forming concrete
21. Geotextile sheet
22. Waterproofing membrane
23. Insulation of extruded polystyrene panel
24. Gravel layer
25. Performed brick pieces wall
26. Mortar cement layer
27. Aluminum carpentry reynaers
28. Plasterboard dividing wall
29. Wooden board finished in iroko wood
30. Aluminum carpentry reynears sliding door
31. Climalit glazing extrawhite 4+4.10.4
32. Plasterboard panel with insulation N 10+30
33. Alabaster glass: alabaster stone with extrawhite glazing attached
34. Horizontal substructure to anchor alabaster
35. Structure: steel lattice beam
36. Lighting: fluorescent tube
37. Paving: board parklex bamboo
38. Suspended ceiling: board parklex bamboo

节点
1. 混凝土梁
2. 预成型砖砌墙
3. 防水膜
4. 线形排水
5. 预应力自承式隔热板 27+3+5
6. 砂浆灰泥找平层
7. 石灰石铺砌
8. 石膏板吊顶
9. 预成型砖砌墙
10. 水泥砂浆墙
11. 挤塑聚苯乙烯隔热层
12. 空气腔
13. 天然石灰石覆面
14. 可调节式锚件,halfen 型
15. 松木板
16. 镀锌钢管型材 80.6.3
17. 镀锌防水板
18. 石膏板隔断墙
19. 白色 DM 板底座
20. 成坡混凝土
21. 土工布
22. 防水膜
23. 挤塑聚苯乙烯隔热层
24. 碎石层
25. 预成型砖砌墙
26. 砂浆水泥层
27. reynaers 铝窗框
28. 石膏隔断墙
29. 木板,绿柄桑木饰面
30. reynears 铝框拉门
31. Climalit 超白玻璃 4+4.10.4
32. 石膏板,带隔热层 N 10+30
33. 雪花玻璃:雪花石膏配超白玻璃
34. 水平下层结构,用于固定雪花石膏
35. 照明:荧光灯管
36.
37. 地面铺装: parklex 竹板
38. 吊顶: parklex 竹板

NUK College of Humanities and Social Sciences
高雄大学人文社科学院

Location/ 地点: Kaohsiung, Taiwan, China/ 中国台湾，高雄
Architect/ 建筑师: MAYU architects+
Photos/ 摄影: Guei-Shiang Ke
Gross floor area/ 总楼面面积: 12,020m²
Completion date/ 竣工时间: 2012

Key materials: Façade – wash stone
Structure – reinforced concrete, steel frame for arcade
主要材料: 立面——水洗石
结构——钢筋混凝土、钢架（拱廊）

Overview

The architectural plan is formed by two perpendicular, interlocking courtyards. Symbolically, courtyard-type college has many precedents in architectural history and can be referred to the liberal and humanity value that forms the core of university education.

In this arrangement, an ecological pond becomes a focal point of the project, and it is visible from most corners of the college through two arcades in different height. The courtyards extend horizontally and create typological relationship to the earth. Semi-outdoor arcades and semi-enclosed courtyards are usable in this harsh and extreme tropical weather, because of the shadow created and flexible spaces defined.

In order to ensure fluid movement between courtyards, public programmes such as entrance lobby, exhibition gallery, and vertical circulation are deployed around where courtyards meet. Accordingly, playful articulation of architectural form creates multiple paths and viewpoints. Here is also the social and interdisciplinary node of the college.

Under the extreme tropical climate, several measures are taken to ensure the sustainability of the project. On the outer boundary of the architecture, punctured windows articulated with sun shading devices respond to the need to prevent incident direct sunlight. The foundation water reservoir is installed in order to retain excess water and decrease runoffs caused by typhoons. The stored water is filtered and reused

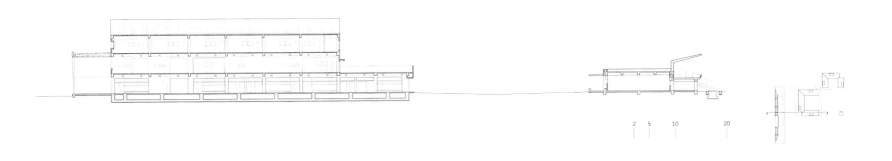

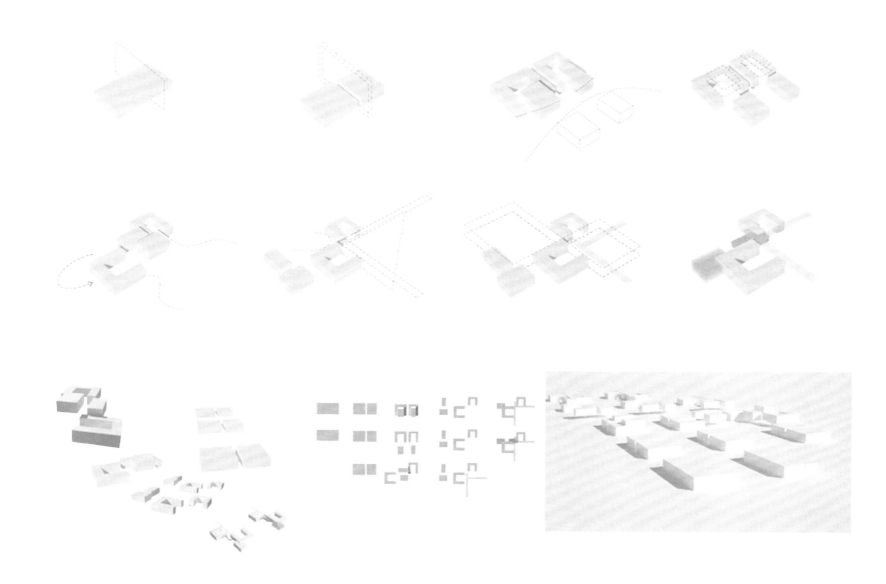

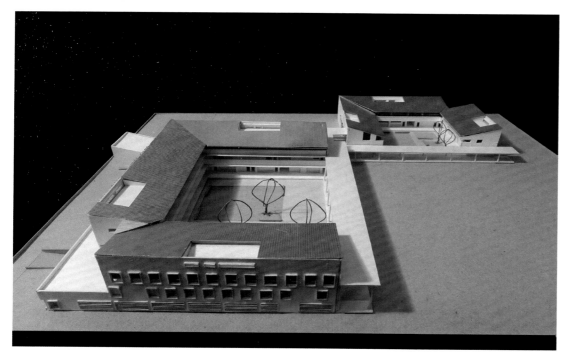

for irrigation. The open foyer, courtyards, and arcade ensure unimpeded circulation of fresh air through major public spaces in the building. The natural ventilation reduces the reliance of air conditioning in milder seasons of spring and autumn. In order to preserve the flourishing indigenous biological system, a central ecological pond is planned to provide diverse habitat for species. Endemic plants of Taiwan with different bloom season are chosen for new vegetation to further diversify the biological environment.

Detail and Materials

The low budget of this project necessitated architects to apply the locally most economical structural system and material: reinforced concrete column and beam system. Thus they faced the dilemma: on the one hand, the courtyard typology required monolithic outer façade to foster the outside-inside distinction; on the other hand, the

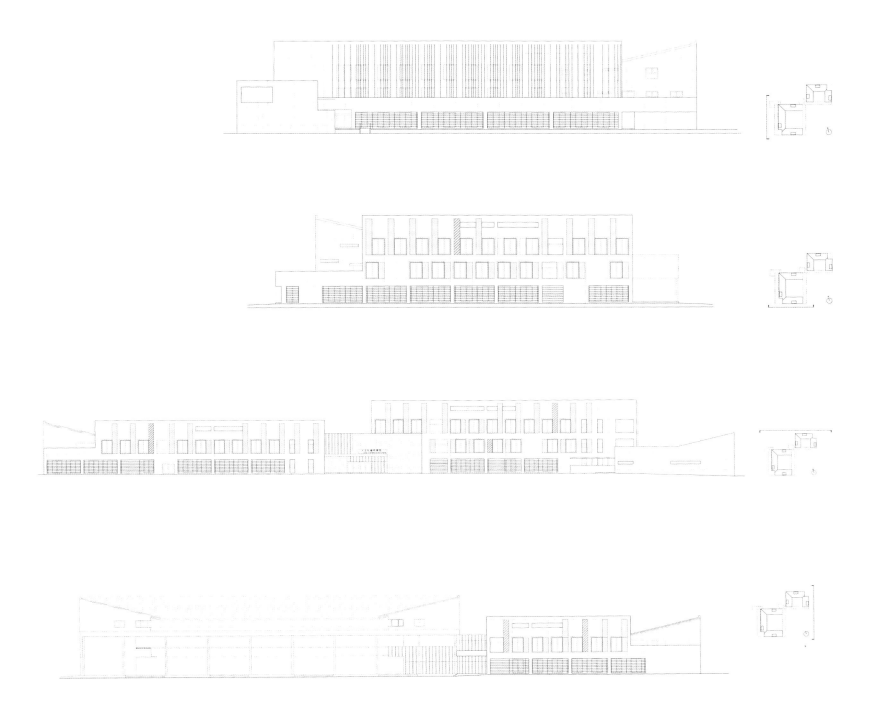

repetitive arrangement of classrooms and faculty studies would put many punctured windows on the façade and created unfavorable grid and hierarchy.

First step to solve the dilemma is focusing on the opening. Rather than puncturing the façade with "windows", they transformed them into "transparent walls". The windows are proportioned vertically and aluminum sun shading panels are installed according to the opening recesses. Secondly, they deployed the openings in a shifting manner on the façade, avoiding any possibility of grid perception. The pronounced metal panels gave the architecture depth and shadow that is welcoming in the tropics.

Colour palettes played important role in this project. Low cost wash stone rendering- usually in grey- is common in Taiwan. A special beige wash stone was chosen to give the architecture a livelier background. Aluminum panels were painted in crimson red and grey to from an even more abstract façade composition. One exception to this colour palette was at the main entrance, where wood stripe panels were installed on the façade, and their similarity to the beige wash stone gave subtle harmony and variation.

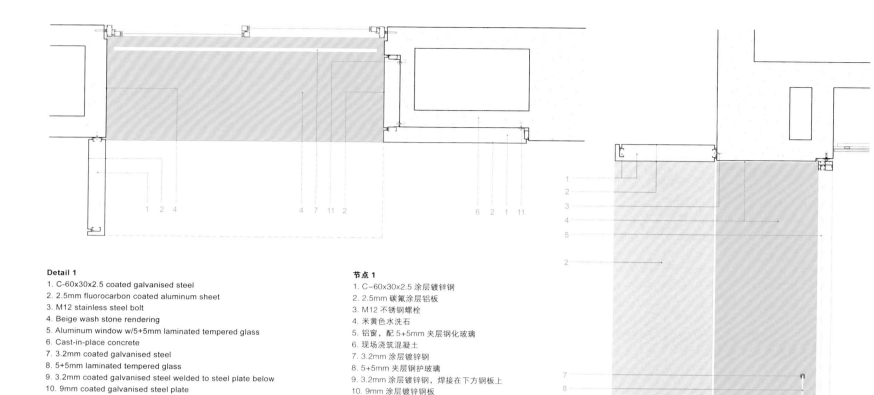

Detail 1
1. C-60x30x2.5 coated galvanised steel
2. 2.5mm fluorocarbon coated aluminum sheet
3. M12 stainless steel bolt
4. Beige wash stone rendering
5. Aluminum window w/5+5mm laminated tempered glass
6. Cast-in-place concrete
7. 3.2mm coated galvanised steel
8. 5+5mm laminated tempered glass
9. 3.2mm coated galvanised steel welded to steel plate below
10. 9mm coated galvanised steel plate
11. L-40x40x3 coated galvanised steel angle, bolted by M12 stainless steel bolt

节点 1
1. C-60x30x2.5 涂层镀锌钢
2. 2.5mm 碳氟涂层铝板
3. M12 不锈钢螺栓
4. 米黄色水洗石
5. 铝窗，配 5+5mm 夹层钢化玻璃
6. 现场浇筑混凝土
7. 3.2mm 涂层镀锌钢
8. 5+5mm 夹层钢护玻璃
9. 3.2mm 涂层镀锌钢，焊接在下方钢板上
10. 9mm 涂层镀锌钢板
11. L-40x40x3 涂层镀锌角钢，M12 不锈钢螺栓固定

项目概况

建筑的平面布局由两个相互垂直的庭院构成。庭院式学院在建筑史中有许多先例，象征着自由和人性价值，即大学教育的核心。

在这种布局中，生态池塘成为了整个项目的焦点，透过两条拱廊，从学院的大多数角落都能看到池塘的景色。庭院在水平方向延伸，与地面形成了紧密的联系。在严酷而极端的热带环境中，半开放拱廊和半封闭庭院提供了怡人的阴凉和灵活的空间。

为了保证两个庭院之间流畅的交通，入口大厅、展览厅、垂直交通等公共设施都环绕着两个庭院的交叉点而建。同样的，建筑造型的有趣连接也形成了多样化的通道和视角。这里同样是学院的社交互动和跨学科互动中心。

在极端的热带气候条件下，建筑师采用了集中方式来保证项目的可持续性。建筑外围的窗口配有遮阳装置，可以避免阳光直射。蓄水池能留住多余的雨水并减少台风造成的雨水径流。存储起来的经过过滤被用于植物灌溉。开放的门厅、庭院和拱廊保证了楼内公共空间的自然通风，提供了新鲜的空间。同时，自然通风也减少了气候较为温和的春秋两季的空调使用率。为了保护本地生态系统，中央生态池塘为多种生物提供了栖息地。景观绿化选择了在不同季节开花的台湾本土植物，丰富了学院内部的微生态环境。

细部与材料

由于项目预算较低，建筑师必须选择最经济的本土结构系统和材料：钢筋混凝土梁柱系统。这样一来，他们就陷入了两难的困境：一方面，庭院式设计要求使用庞大的外立面来突出内外的区别；另一方面，大量重复的教室和教员研究室将会在立面上形成大量窗口，从而形成不恰当的网格状窗口布局。

建筑师将解决问题的第一步放在开窗设计上。他们并没有在立面上打开过多的窗口，而是将窗口转化成"透明

Detail 2
1. Plain roof tiling
   Roof batten
   Moisture-diffusing membrane
   t=15cm concrete slab
2. Beige wash stone rendering
3. 3.2mm coated galvanised steel
4. 5+5mm laminated tempered glass
5. 3.2mm coated galvanised steel welded to steel plate below
6. 9mm coated galvanised steel plate
7. Aluminum vertical louver, fluorocarbon coated
8. Cast-in-place concrete
9. L- 75x75x6 coated galvanised steel angle fixer
10. 20mm terrazzo tile
    60mm reinforced concrete pressure-distribution slab
    25mm rigid thermal insulation
    3mm heat sealed bituminous sheet membrane
    Bituminous roof sealing layer
    150mm reinforced concrete slab
    Suspended ceiling:
    9mm plasterboard
    12mm acoustic mineral fiber board
11. 200mm reinforced concrete parapet
    Bituminous roof sealing layer
    3mm heat sealed bituminous sheet membrane
    1/2b brick w/cement mortar coating
    Polyurethane sealing layer
12. Stainless steel fixer, fluorocarbon coated
13. Φ=15mm stainless steel tension cable, fluorocarbon coated
14. Aluminum horizontal louver, formed by 2mm fluorocarbon coated aluminum sheet

节点 2
1. 平屋顶瓦
   屋顶木条
   散湿膜
   t=15cm 混凝土板
2. 米黄色水洗石
3. 3.2mm 涂层镀锌钢
4. 5+5mm 夹层钢化玻璃
5. 3.2mm 涂层镀锌钢，焊接在下方钢板上
6. 9mm 涂层镀锌钢板
7. 铝制垂直百叶，碳氟涂层
8. 现场浇筑混凝土
9. L– 75x75x6 涂层镀锌角钢固定件
10. 20mm 水磨石砖
    60mm 钢筋混凝土分压板
    25mm 刚性隔热层
    3mm 热封沥青膜
    沥青屋面密封层
    150mm 钢筋混凝土板
    吊顶：
    9mm 石膏板
    12mm 隔音矿物纤维板
11. 200mm 钢筋混凝土护墙
    沥青屋面密封层
    3mm 热封沥青膜
    1/2b 砖，配水泥砂浆涂层
    聚氨酯密封层
12. 不锈钢固定件，碳氟涂层
13. Φ=15mm 不锈钢拉索，碳氟涂层
14. 铝制水平百叶，由 2mm 碳氟涂层铝板构成

的墙壁"。直立的窗口外部安装了铝制遮阳板。然后，他们用不同的形式在立面上设计窗口，避免了重复的网格感。显眼的金属板赋予了建筑深度和阴凉，后者在热带气候环境中是深受欢迎的。

色彩搭配在项目中起到了重要作用。灰色的低成本水洗石在台湾地区十分常见。而特殊的米黄色水洗石为建筑提供了更为活泼的背景。铝板被漆成深红色和灰色，形成了对比更强烈的墙面组合。唯一跳脱于这一色彩搭配之外的是建筑的主入口。建筑师在主入口处安装了木板条，与米黄色类似的原木色赋予了建筑微妙的和谐感和变化感。

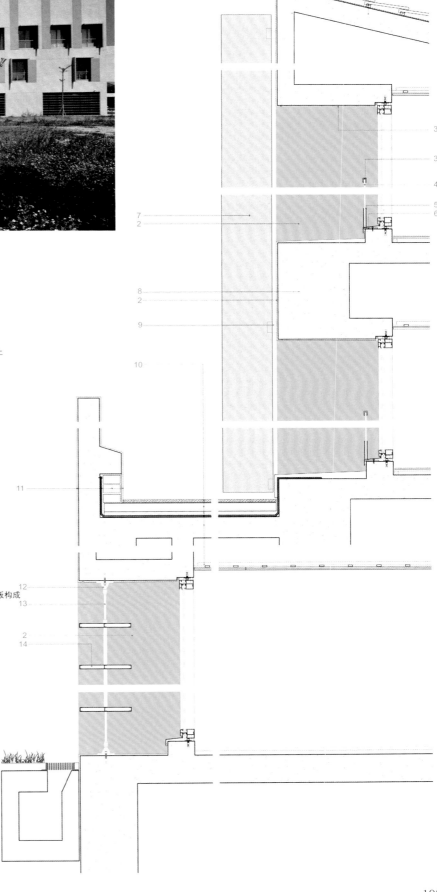

Rheinfels III, Chur
莱茵岩大厦

Location/ 地点 : Chur, Switzerland / 瑞士，库尔
Architect/ 建筑师 : NocasaBaumanagement
Area/ 面积 : 34,500m²
Completion date/ 竣工时间 : 2013

Key materials: Façade – travertine
主要材料：立面——石灰华

Overview
Innovation and Perspective
The new and generous building development in the most popular urban development area of Chur combines innovation and modernism. Good accessibility and the proximity to various local recreational areas, a lot of sun and unimpaired visibility as well as fascinating architecture make the building an attractive residential and commercial location.

Smooth Transitions and a Park
The eight-storey hybrid building is designed for versatile functionality. The ground floor on street level consists of generous retail areas. The heart of the building development is the vivid park located on the first floor. It is designed as a patio and "encounter zone" for the adjoining residential buildings and is connected with the ground floor from the outside by large flight of stairs.
There is a smooth transition to the exclusive living area with its 5 storeys. The apartments embrace the park with an angular design and can be reached comfortably by the patio as well as by a generously built internal development.

Detail and Materials
Waves and Travertine
The façade made of travertine reminds of historical industry buildings and creates a new urban esthetics. It generates a warm background for the inbuilt park's lush planting and stands as an impulse for the further urban development in the area Kleinbruggen/Rheinfels.

The apartments fascinate with their wave-like balconies, establishing a gentle link between apartment and patio and giving the building its unmistakable expression.

The complex isolates itself markedly from the outside. The balconies are highlighted like observation decks and ensure an impressive view.

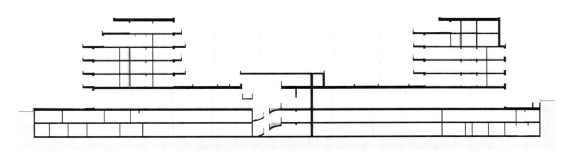

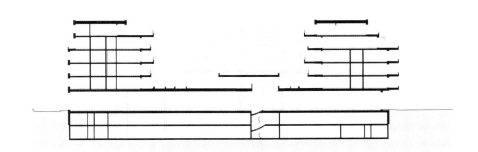

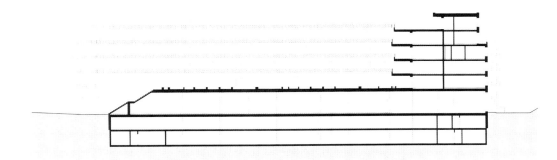

项目概况
创新与展望
这座新建于库尔市最密集的城市开发区域的宏大建筑将创新与现代主义结合了起来。项目具有良好的通达性，靠近各种本地娱乐休闲区，日照充足，享有良好的视野，以上条件与迷人的建筑外形共同使建筑成为了深受欢迎的商住两用大厦。

流畅的过渡与公园
这座八层高的综合建筑具有多重功能。临街的一楼设有宽敞的零售空间。建筑开发的核心是位于二楼的公园。公园被设计成天井的形式，是周边居民楼居民的"邂逅场所"，通过宽大的楼梯与一楼和外界相连。

公园与五层生活空间之间形成了流畅的过渡。公寓通过错落的设计拥抱着公园。公寓既可以通过公园进入，又可以通过宽敞的内部入口进入。

细部与材料
波浪与石灰华
建筑立面由石灰华组成，令人回想起从前的工业建筑，形成了全新的城市美学。它为嵌入式公园茂密的植物提供了背景，同时也推动着该地区未来的城市开发。

公寓的波浪形阳台极富吸引力，它们在公寓和公园之间形成了微妙的联系，赋予了建筑引人注目的外观。

建筑明显脱离于外部环境。阳台被凸显成观景台，享有令人赞叹的视野。

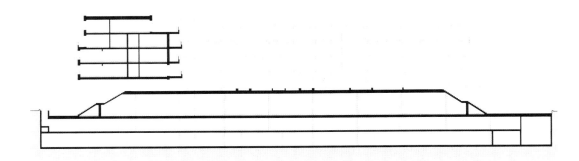

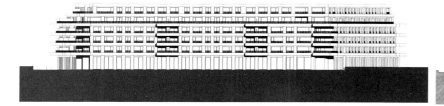

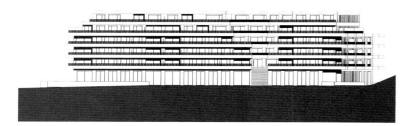
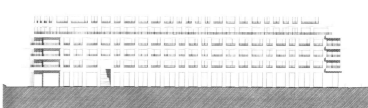

Cross Section Façade
1. Vapour barrier
2. Thermal insulation 200mm
3. Waterproofing
4. Protection mat
5. Soil substrate
6. Gravel strip
7. Metal covering
8. Concrete parapet prefabricated
9. Natural stone slabs bonded
10. Cantilever slab connectors
11. Venetian blinds
12. Window
13. Steel railing laser cut
14. Window sill prefabricated
15. Thermal insulation 200mm
16. Thermal insulation 20mm
17. Impact sound insulation 20mm
18. Separation layer
19. Underlay 60mm
20. Flooring 20mm
21. Thermal insulation 100mm
22. Support pads
23. Natural stone slabs
24. Connection plate

立面横截面
1. 隔汽层
2. 隔热层 200mm
3. 防水层
4. 保护垫
5. 土壤基质
6. 碎石带
7. 金属封盖
8. 预制混凝土护墙
9. 天然石板
10. 悬臂板连接件
11. 活动百叶窗
12. 窗户
13. 激光切割钢栏杆
14. 预制窗台
15. 隔热层 200mm
16. 隔热层 20mm
17. 冲击声隔音板 20mm
18. 分离层
19. 衬垫 60mm
20. 地板 20mm
21. 隔热层 100mm
22. 支撑垫
23. 天然石板
24. 连接板

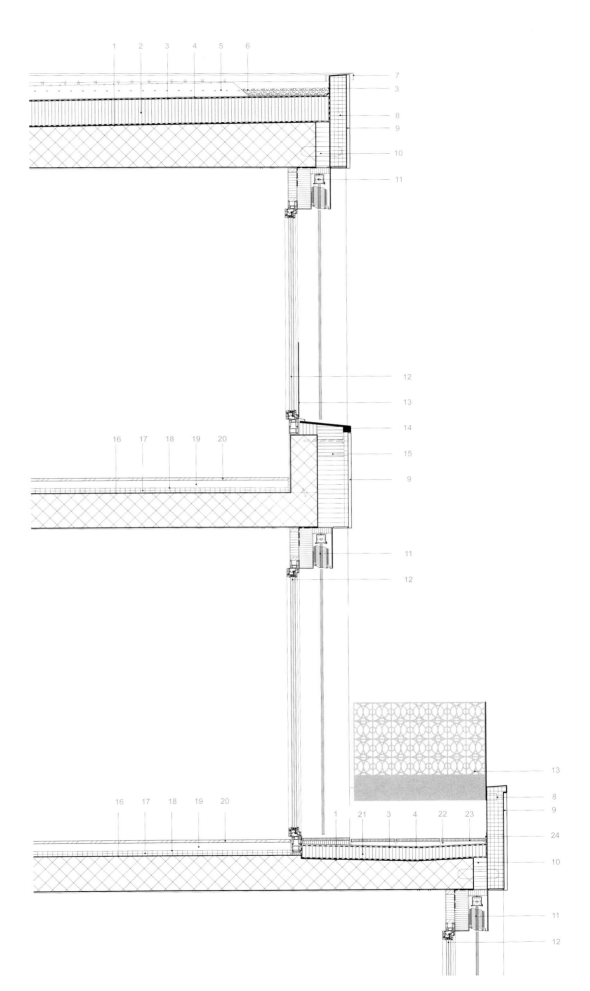

Sheikh Zayed Desert Learning Centre
扎伊德酋长沙漠学习中心

Location/ 地点：Al Ain, UAE/ 阿联酋，阿尔艾因
Architect/ 建筑师：Chalabi Architekten & Partner ZT GmbH
Photos/ 摄影：Antje Hanebeck, Fresh Media FZ LLC, Chalabi Architekten & Partner ZT GmbH
Site area/ 建筑面积：8,396m²
Site area/ 占地面积：13,000m²
Completion date/ 竣工时间：2013

Key materials: Façade – marble stone panels
主要材料：立面——大理石板

Overview
The Sheikh Zayed Desert Learning Centre (SZDLC) is an iconic and prestigious project in the city of Al Ain in the United Arab Emirates. As part of the first phase of the Wildlife Park & Resort, it is a project of national significance to demonstrate the sustainable eco-existence in the region, developed by the Zoo and Aquarium Public Institution in Al Ain.

The client's brief is for a destination indoor attraction that will honour the legacy of Sheikh Zayed (the founder of the modern UAE) and his attitude to the environment, wildlife and conservation, provide a changing look at the natural and cultural history of the Arabian deserts, and as well as deserts worldwide. Its role is to provide an in-depth and scientific overview of the entire zoo experience.

The brief envisaged a continuum of spaces and galleries, adaptive and flexible for future use as well as containing a wide range of exhibits on subjects of desert life: climate, flora, fauna, and how to respect and care for the desert environment. Basic information on the geological formation of the Arabian Peninsula is included, as information on living in the desert context historically.

The SZDLC in function and geometry is an accessible sculpture that affords a multitude of spatial experiences, which create the shape of the building.

Detail and Materials
Rhombic marble stone panels were mounted on a galvanised steel substructure in front of 15cm extruded polystyrol insulation; façade grid and divisions were generated with Grasshopper design tool.

项目概况
扎伊德酋长沙漠学习中心是阿联酋阿尔艾因一座著名的标志性建筑。作为野生动物园与度假区的一期工程，它是当地可持续发展和生态设计的典范，具有全国性意义。项目由阿尔

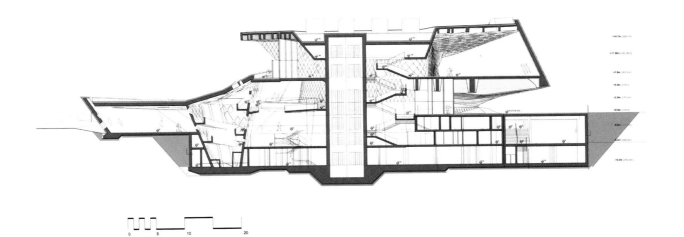

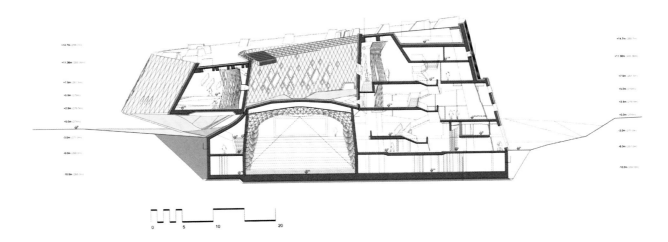

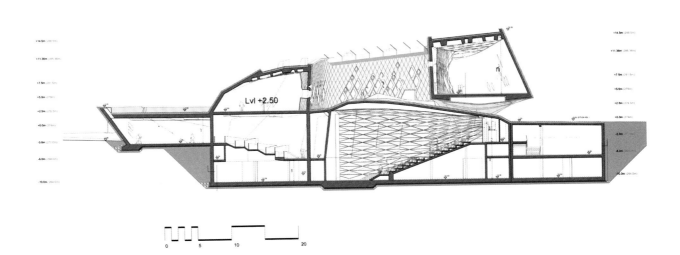

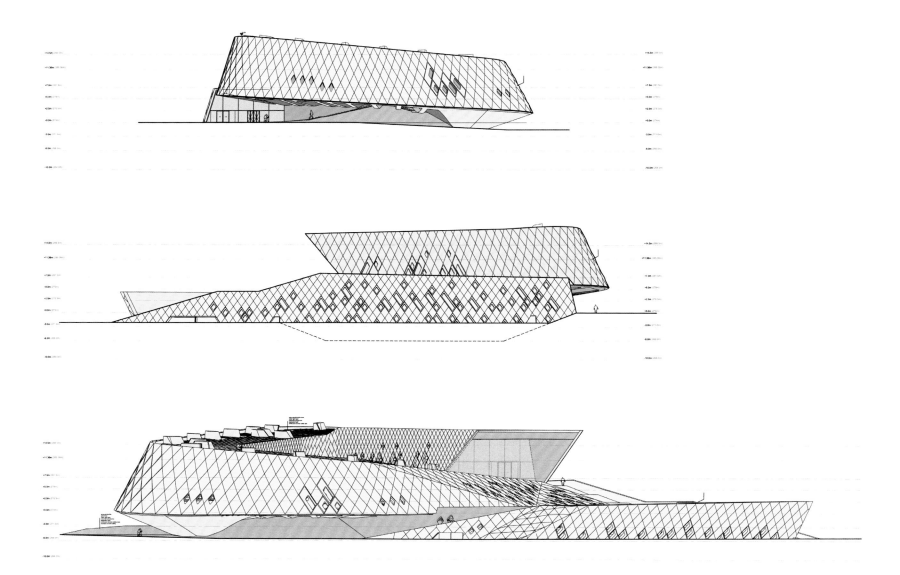

艾因公园及水族馆公共管理局进行开发。

委托人的要求是打造一个室内景点，同时向现代阿联酋的建立者扎伊德酋长及其在环境、野生动植物和历史保护方面的贡献致敬，展示阿拉伯沙漠地区及全球沙漠地区的自然和文化历史。它将为整个野生动物园提供富有深度的科学概况。

项目要求实现空间和展厅的连续性，便于未来的灵活改造，并且对沙漠的气候、动物和植物以及如何保护沙漠环境进行全方面的展览。展览还包含阿拉伯半岛的基本地质构造以及沙漠生活的基本历史。

扎伊德酋长沙漠学习中心在功能和造型上都是一个可进出的雕塑，它为人们提供了多方面的空间体验，塑造了一座独特的建筑。

细部与材料

菱形大理石板被安装在镀锌钢支架上，配有15厘米厚的挤塑聚苯乙烯隔热层。建筑立面的网格和分格都由Grasshopper设计工具生成。

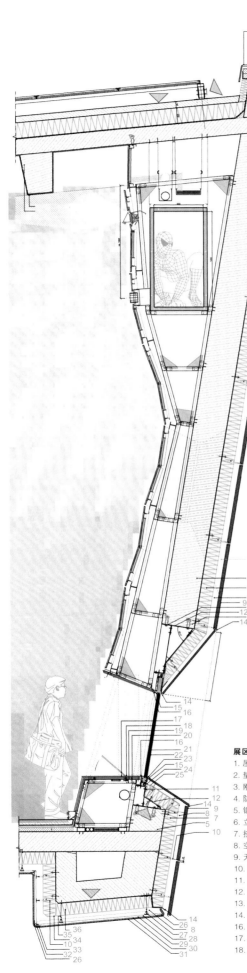
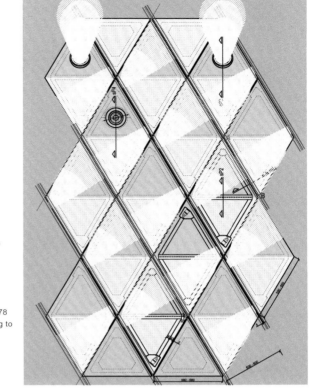

1:20

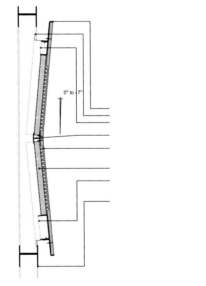

Section through Inner Wall in Exhibition Area
1. Roof specialities parapet sheet metal flashing d=0.2cm
2. Spacer
3. Rigid insulation d=12cm
4. Waterproof membrane
5. Reinforced concrete, variable d=40-50cm
6. Façade flashing
7. Extruded polystyrene (XPS) complying with ASTM C 578 minimum compressive strength requirements according to SPECS
8. Air cavity
9. Natural stone cladding (curtain wall) 2-4 cm
10. Steel Lutz body-anchor or similar
11. Steel frame (nominal depth 12 cm)
12. Ledge forming steel sheet
13. Steel anchor
14. Drip edge
15. Insulated aluminum glazing profile
16. Edge protection
17. Inspection opening
18. 10-15mm sill cover (wood) as a seat
19. 2x12.5mm gypsum boards
20. Knauf UW 55 profile (t=0.6mm) or similar
21. Light diffuser
22. Double glazing
23. Flashing
24. Sheet metal flashing with a drip
25. Sealing
26. Edge forming profile embedded in mortar bed
27. Aluminium runner channel
28. Aluminium furring channel
29. Expanded metal lath
30. Mortar bed
31. Ceramic tiles with partly open joints
32. Safing (allows for expansion and contraction)
33. Framing channel
34. Sealing membrane
35. Steel hanger wires
36. Mineral bound insulation d=15cm

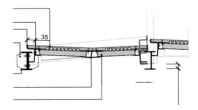

展区内墙剖面
1. 屋顶金属围栏，防水板 d=0.2cm
2. 垫片
3. 刚性隔热 d=12cm
4. 防水膜
5. 钢筋混凝土，d=40~50cm
6. 立面防水板
7. 挤塑聚苯乙烯，遵循美国材料预实验协会 C 578 标准
8. 空气腔
9. 天然石材包层（幕墙）2~4cm
10. 钢锚件
11. 钢架（纵深 12cm）
12. 钢板壁架
13. 滴水檐
14. 隔热铝型材
16. 边缘加固
17. 检查口
18. 10~15mm 窗台板（木制），可作为座椅
19. 2x12.5mm 石膏板
20. Knauf UW 55 型材
21. 灯光扩散器
22. 双层玻璃
23. 防水板
24. 金属防水板，带滴水槽
25. 密封
26. 侧边成形型材，嵌入灰浆层
27. 铝流道
28. 铝贴条
29. 网眼钢皮
30. 灰浆层
31. 瓷砖，半开接缝
32. 保险缝（考虑到膨胀和收缩）
33. 框架槽
34. 密封膜
35. 钢吊线
36. 矿物隔热层 d=15cm

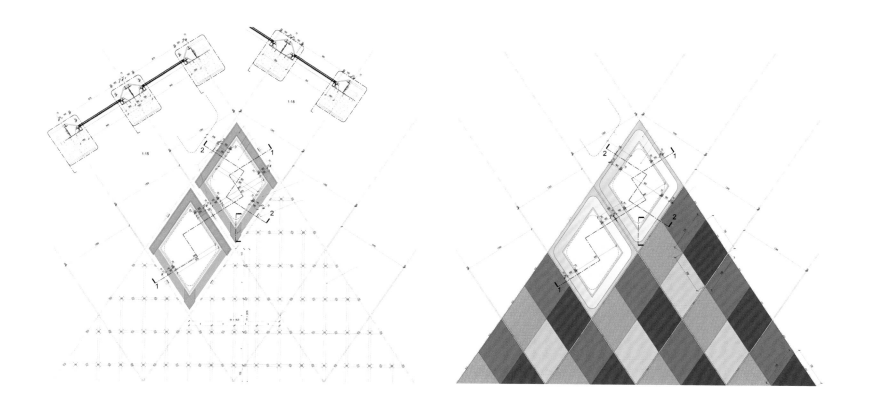

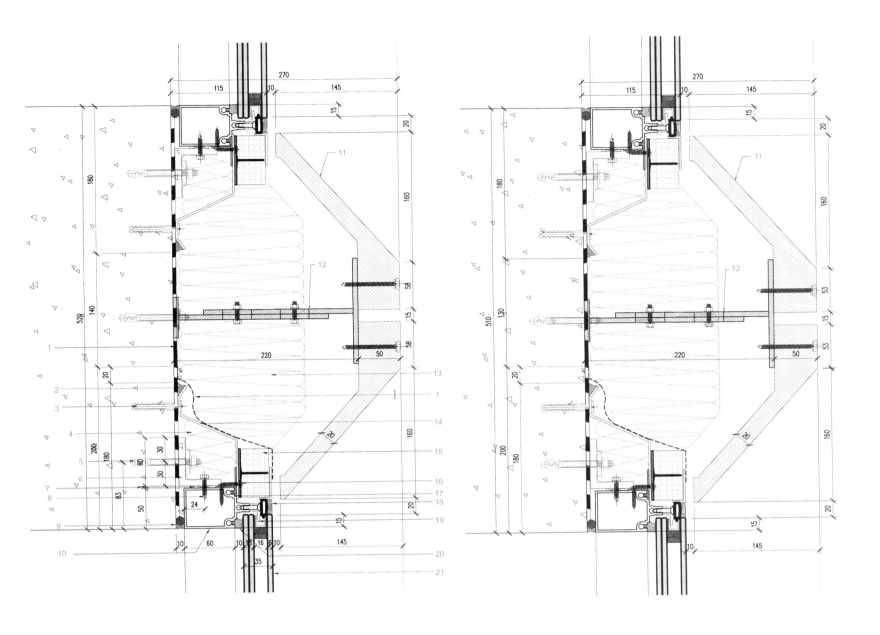

Exterior Stone Cladding
1. Waterproofing membrane
2. Silicone weather seal
3. 4.8×60 pan head screw with flat washer & plastic plug
4. Rock wall insulation
5. 2NOS. M8×75 Hiliti anchor HST-R with lock washer, flat washer & nut
6. Aluminum bracket 60×50×8mm
7. Plastic shims
8. 4NOS. J T2 12 5.5×35 EJOT fixing self taping screw
9. Backer rod with weather sealant
10. Aluminum profile natural anodised finished
11. GRC window reveal
12. GRC bracket
13. Thermal insulation
14. 2mm thick aluminum bended sheet flashing, mill finished
15. Hard foam insulation with aluminum angle 15×40×2mm fixed with silicone
16. 25×25×2mm thick aluminum angle with 4.8×25 pan head screw with flat washer
17. EPDM sponge gasket
18. Silicone gasket backing with weather sealant black
19. Glass toggle
20. Glazing gasket (EPDM)
21. DGU

外墙石材覆层
1. 防水膜
2. 硅胶密封
3. 4.8×60 平头螺丝，配平垫圈和塑料塞
4. 石棉隔热层
5. 2NOS. M8×75 Hiliti 锚件 HST-R，配放松垫圈、平垫圈和螺母
6. 铝支架 60x5x8mm
7. 塑料垫片
8. 4NOS. J T2 12 5.5×35 EJOT 自固定螺丝
9. 泡沫条密封
10. 铝型材，天然阳极氧化面
11. 玻璃纤维增强水泥窗帮
12. 玻璃纤维增强水泥支架
13. 隔热层
14. 2mm 厚铝制曲面防水板，光面
15. 硬泡沫隔热，配铝角 15×40×2mm，硅胶固定
16. 25×25x2mm 厚铝角，4.8×25 平头螺丝，配平垫圈
17. EPDM 海绵封边
18. 硅胶垫圈，黑色密封条
19. 玻璃栓扣
20. 玻璃封边（EPDM）
21. DGU

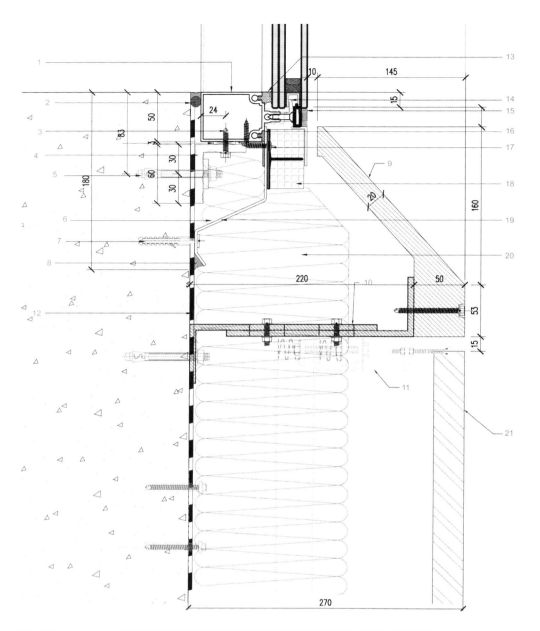

Exterior Stone Cladding
1. Aluminum profile natural anodized finished
2. Backer rod with weather sealant
3. 4NOS. J T2 12 5.5x35 EJOT fixing self taping screw
4. Aluminum bracket 60x50x8 mm
5. 2NOS. M8×75 Hiliti anchor HST-R with lock washer, flat washer & nut
6. Rock wall insulation
7. 4.8×60 pan head screw with flat washer & plastic plug
8. Silicone weather seal
9. GRC window reveal
10. GRC bracket
11. Stone cladding support system
12. Waterproofing membrane
13. Glazing gasket (EPDM)
14. Glass toggle
15. Silicone gasket backing with weather sealant black
16. EPDM sponge gasket
17. 25x25x2mm thick aluminum angle with 4.8x25 pan head screw with flat washer
18. Hard foam insulation with aluminum angle 15x40x2mm fixed with silicone
19. 2mm thick aluminum bended sheet flashing mill finished
20. Thermal insulation
21. Stone cladding
22. Glass

外墙石材覆层
1. 铝型材，天然阳极氧化面
2. 泡沫条密封
3. 4NOS. J T2 12 5.5×35 EJOT 自固定螺丝
4. 铝支架 60x5x8mm
5. 2NOS. M8×75 Hiliti 锚件 HST-R，配放松垫圈、平垫圈和螺母
6. 石棉隔热层
7. 4.8×60 平头螺丝，配平垫圈和塑料塞
8. 硅胶密封
9. 玻璃纤维增强水泥窗帮
10. 玻璃纤维增强水泥支架
11. 石材包层支架系统
12. 防水膜
13. 玻璃封边（EPDM）
14. 玻璃栓扣
15. 硅胶垫圈，黑色密封条
16. EPDM 海绵封边
17. 25x25x2mm 厚铝角，4.8x25 平头螺丝，配平垫圈
18. 硬泡沫隔热，配铝角 15x40x2mm，硅胶固定
19. 2mm 厚铝制曲面防水板，光面
20. 隔热层
21. 石材包层
22. 玻璃

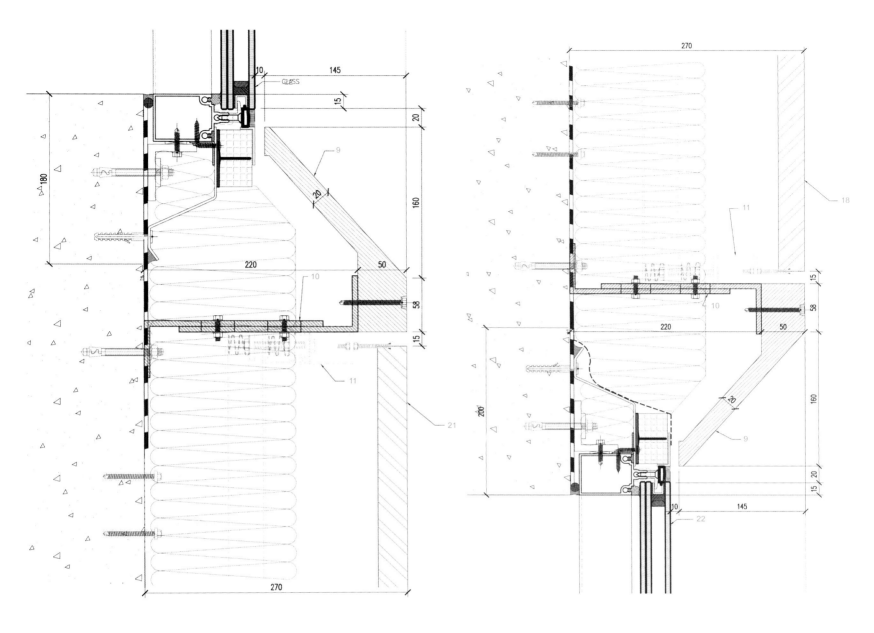

The Samaranch Memorial Museum
萨马兰奇纪念馆

Location/ 地点：Tianjin, China/ 中国，天津
Architect/ 建筑师：Archiland
Photos/ 摄影：Courtesy of Archiland
Collaborators/ 合作设计：HAO architects, Kragh & Berglund landscape, COWI, TADI
Area/ 面积：17,800m^2

Key materials: Façade – sand stone
主要材料：立面——砂岩

Overview
The President of the International Olympic Committee, IOC, Mr Jacques Rogge inaugurated the Samaranch Memorial Museum situated in the coastal town of the new Sport and Health Campus area also bound to host the East Asian Games summer 2013. Even before the inauguration, the 5 Olympic ring building, designed by a team lead by Archiland International was perceived as a symbol and an architectural landmark for the entire area.

"The Samaranch Memorial Museum is in its design a result of the use of the 5 Olympic rings: 2 converted into an infinity shaped looping ramp illustrating the nature of one man's life and further 3 to become sunken courtyards. The building is a landmark, which through its architecture symbolizes the extraordinary achievements of Samaranch with his focus on Legacy and Green Olympics," explains Morten Holm, Partner at Archiland International. He continues: "The museum reflects, both literally and metaphorically, the infinity and looping ramps: legacy, dynamics of time and sports, poetic and beautiful. It appears as a solitary sculpture in interaction with the new recreational park surrounding it." The park has been designed by well reputed Kragh & Berglund Landscape architects (DK) in close relation with the Archiland International. Moreover the project also illustrates the ambition of Archiland International to join multi team partners of HAO (NYC) on concept, COWI (Beijing/DK) on energy concept and many others to reach an integrated design.

Detail and Materials
The significance of the infinity and loop is reflected in the architecture of the museum. The contours of the museum rise as a ramping shape emphasizing the inner circulation ramp mirrored into the roof terrace. The Façade and softly waved windows, clad in sand stone, reflects both the smart design daylight usage and the continuity of the space inside as well as in the story of one man's life.

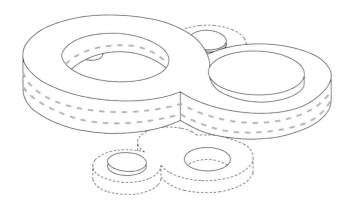

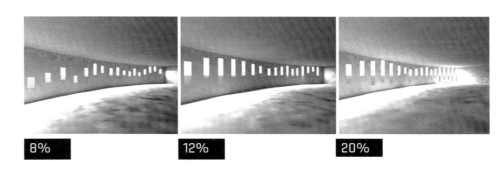

| 8% | 12% | 20% |

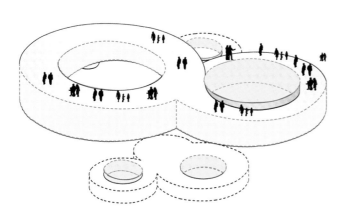

Inside the main area of the museum, the public space has an ever shifting character as they circulates and ramps around the court yards and winter gardens creating a peaceful human scale and contrast to the dynamic exterior of the building. The materials used, raw concrete for the floors and ECO stone for the Façade s, curved lines in panels and ceilings, underline the theme of global human fellowship. Daylight enters the museum Façade through thin, slim, waved line of windows. Advanced daylight modeling has generated systems for courtyard and winter garden skylight and soft north orientated side light creating a distinctive atmosphere in the room.

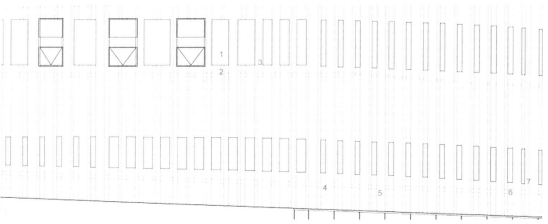

Elevation
1. Electronic open window in gazed façade
2. Window openable in glazed façade
3. Window unopenable
4. Laminated glass façade
5. Stainless steel handle
6. Fixed sun shading lamellas
7. U-profile 100mm (steel)

立面
1. 玻璃幕墙上电动开启窗户
2. 可开启窗户
3. 不可开启窗户
4. 夹层玻璃幕墙
5. 不锈钢扶手
6. 幕墙内衬百叶
7. U 型钢

Façade Detail
1. Stone cladding
2. Connected brackets
3. Aluminium transom
4. Granite cladding façade
5. Aluminium transom
6. Connected brackets
7. Vapor membrane (low resistant)
8. Aluminium transom
9. Rock wool insulation
10. Layer –glass/outside curved
11. Fixed lamella structure vertical façade mollion
12. Mobile aluminum lamella/sunshine protection motor or hand driven
13. Double layer glazing

立面节点
1. 石材饰面
2. 连接件
3. 铝合金横梁
4. 石材幕墙
5. 铝合金横梁
6. 连接件
7. 防水雨布（不透水、透气）
8. 铝合金横梁
9. 保温岩棉
10. 外侧玻璃；单层弧形玻璃
11. 幕墙上固定薄片结构
12. 活动铝薄片 / 遮阳百叶（电动或手动）
13. 内侧双层玻璃

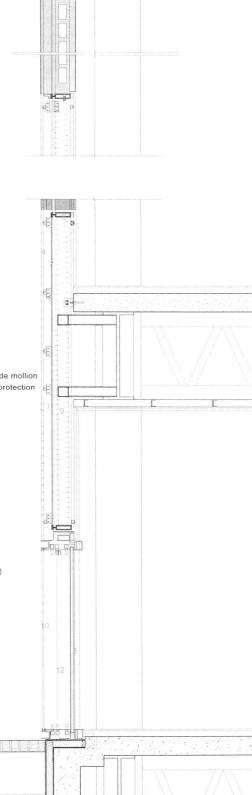

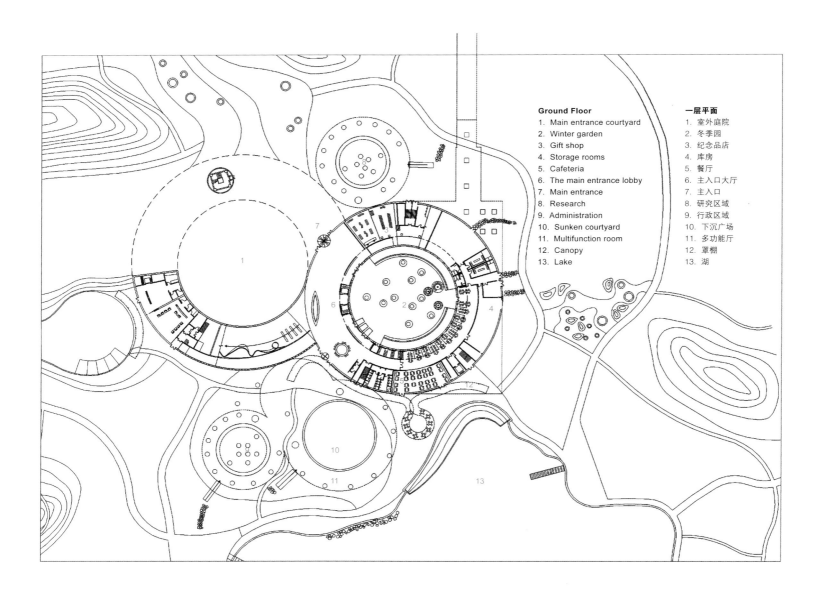

Ground Floor	一层平面
1. Main entrance courtyard	1. 室外庭院
2. Winter garden	2. 冬季园
3. Gift shop	3. 纪念品店
4. Storage rooms	4. 库房
5. Cafeteria	5. 餐厅
6. The main entrance lobby	6. 主入口大厅
7. Main entrance	7. 主入口
8. Research	8. 研究区域
9. Administration	9. 行政区域
10. Sunken courtyard	10. 下沉广场
11. Multifunction room	11. 多功能厅
12. Canopy	12. 罩棚
13. Lake	13. 湖

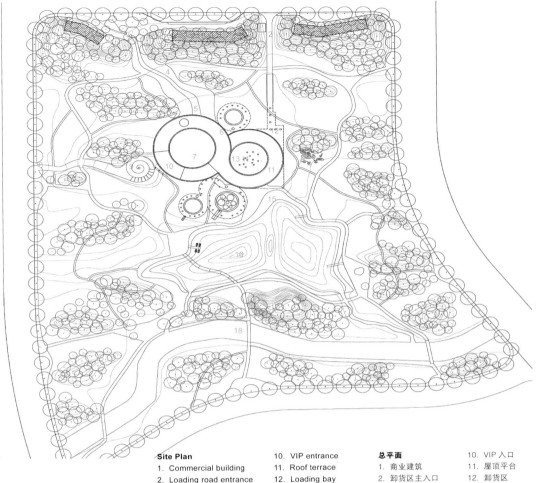

项目概况

国际奥委会主席雅克·罗格先生主持了萨马兰奇纪念馆的开幕仪式。纪念馆坐落于静海县团泊新城西区健康产业园，2013年东亚运动会的举办地。在纪念馆开幕之前，这座由 Archiland 设计的奥运五环大楼就已经是整个区域的建筑地标。

"萨马兰奇纪念馆的设计以奥运五环为基础：两个圆环构成了无穷大符号造型的循环坡道，代表着人类生命的本质，另外三个圆环成为了下沉式庭院。作为一个地标，它在建筑意义上象征着萨马兰奇在绿色奥运、奥运传承方面的卓越成就。"Archiland 建筑事务所的合伙人马坦·豪解释道，"纪念馆在表面和内在都反映了无穷大和循环坡道：传承、时间与体育的动态、诗意而美好。在全新的休闲园区中，它就像一座孤独的雕塑。"健康产业园由著名的 Kragh & Berglund 景观事务所（丹麦）和 Archiland 建筑事务所联合打造。此外，Archiland 还与多个团队共同协作，其中包括 HAO（纽约）（概念设计）、COWI（北京/丹麦）（能源概念）等，实现了完整的设计方案。

细部与材料

无穷大和圆环的意义反映在纪念馆的建筑中。纪念馆的轮廓呈坡形上升，突出了内部交通流线以及屋顶平台的坡道。建筑立面和波浪形窗带包裹在砂岩里，反映了日光照明的巧妙设计以及内在空间和人类生命的连续性。

在纪念馆内部，公共空间围绕着庭院和冬日花园不断变化，营造出平和的人性尺度，与富有动感的建筑外观形成了对比。建筑选择清水混凝土作为地面材料，天然石材作为立面材料，墙面板材和天花板采用曲线造型，突出了全球人类的友谊。日光透过纤细的波浪形窗带进入室内。先进的日光建模通过庭院、冬日花园带来了天光和柔和的北侧光线，为室内营造出独特的氛围。

Site Plan
1. Commercial building
2. Loading road entrance
3. Arrival square
4. Main path
5. Water feature
6. Main entrance
7. Main entrance court
8. VIP entrance road
9. VIP parking
10. VIP entrance
11. Roof terrace
12. Loading bay
13. Winter garden
14. Sunken courtyard
15. Wooden deck
16. Lake
17. Tree clad hill
18. Main canal

总平面
1. 商业建筑
2. 卸货区主入口
3. 入口广场
4. 主入口路
5. 水景观
6. 主入口
7. 主入口庭院
8. VIP 主入口
9. VIP 停车场
10. VIP 入口
11. 屋顶平台
12. 卸货区
13. 冬季花园
14. 下沉庭院
15. 木质甲板
16. 湖
17. 栎树山
18. 主河道

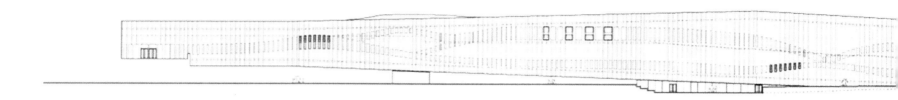

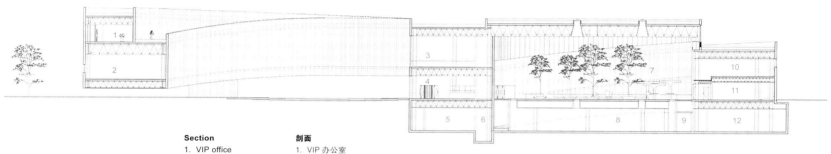

Section 剖面
1. VIP office — 1. VIP 办公室
2. Exhibition — 2. 展厅
3. Cross area — 3. 交换区
4. Foyer — 4. 休息室
5. Tech room — 5. 设备用房
6. Lift — 6. 电梯
7. Winter garden — 7. 冬季园
8. Special exhibition — 8. 特殊展厅
9. Corridor — 9. 走廊
10. Gift shop — 10. 纪念品店
11. Café — 11. 咖啡厅
12. Storage — 12. 储藏室

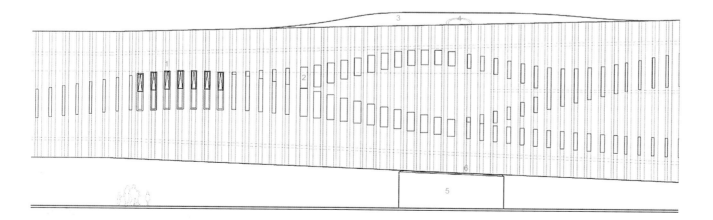

Elevation 立面
1. Window openable in glazed façade — 1. 电动开启窗户
2. Laminated glass behind stone tiles — 2. 磨砂玻璃
3. Green roof above stair — 3. 种植墙面
4. Skylight above stair — 4. 屋顶采光
5. Stainless steel — 5. 不锈钢
6. U-profile 100mm (steel) — 6. U 型钢

萨马兰奇纪念馆 · Samaranch Memorial

Luan Gallery
卢安美术馆

Location/ 地点：Athlone, Ireland/ 爱尔兰，阿斯隆
Architect/ 建筑师：Keith Williams Architects
Area/ 面积：575m²
Completion date/ 竣工时间：2012
Budget/ 预算：€3 million/3,000,000 欧元

Key materials: Façade – limestone, glass
主要材料：立面——石灰岩、玻璃

Overview

The Luan Gallery, designed by Keith Williams Architects, sits on a spectacular site by the Athlone's historic bridge, castle and the church of St Peter & St Paul. Winner of the RIAI Best Cultural Building for 2013, the project involved the adaptation of the historic 1897 Father Matthew Hall into a new gallery, the addition of new build wing to provide temporary white box gallery spaces, and a river gallery overlooking the Shannon, Ireland's largest river.

This project has radically altered the historic Father Matthew Hall. The accretions added over time have been swept away leaving the core form intact, and the elevations modified by the introduction of new large glazed panels opening up the building to the river and the Shannon Bridge. The new wing has provided contemporary gallery space with black out capabilities to enable multi-use gallery, lecture theatre/cinema for film exhibitions, meeting space for literature, music, drama workshops, and digital art exhibitions.

The two galleries are linked by a glazed entrance from the main road, and by the linear river gallery.

Detail and Materials

The palette of materials for the new gallery wing is limited to limestone and zinc. Limestone as a building material has a history of use in public buildings in Ireland, and here has been laid in random cut horizontal strips of varying widths, smooth for the upper gallery and rough cut for the plinth, asserting the contemporary nature of the new wing. Zinc clad roof lanterns have been set back from the parapet wall to centralise daylight penetrating into the gallery.

The Father Matthew Hall has been re-rendered and the roof replaced with a new structure and natural slate tiles.

项目概况

卢安美术馆由 Keith Williams 建筑事务所设计，所在地理位置十分优越，紧邻古桥、城堡以及圣彼得和圣保罗教堂。作

为 2013 年爱尔兰皇家建筑师学会最佳文化建筑奖得主，项目对建于 1897 年的马太神父厅进行了改造，新建的翼楼被作为临时展览空间，还设有一个远眺香农桥的河上画廊。

项目对马太神父厅进行了大规模改造。新添的建筑结构最后只保留了大厅核心空间的完整性，而改造后的玻璃墙面让建筑面向河流和香农桥开放。新建的翼楼提供了现代艺术展览空间，"熄灯"功能实现了多种其他用途：多功能展厅、演讲、电影院、文学/音乐见面会举办地、戏剧工坊、数码艺术展等。

两个展览厅由一个延伸至主路的玻璃入口和一个朝向香农桥的河上画廊连接起来。

细部与材料

新美术馆一楼所选用的材料仅限于石灰岩和锌。在爱尔兰，作为一种建筑材料，石灰岩拥有悠久的应用历史。项目将石灰岩用不同宽度的水平条拼接起来，上半部分墙面呈现为光滑表面，下半部分则为粗糙表面，为新建的翼楼带来了现代气息。由锌板覆盖的灯笼式天窗后撤到护墙后房，让阳光能够照进室内。

马太神父厅进行了重新粉刷，屋顶被全新的结构和天然板岩块所替代。

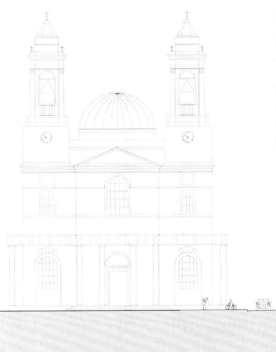

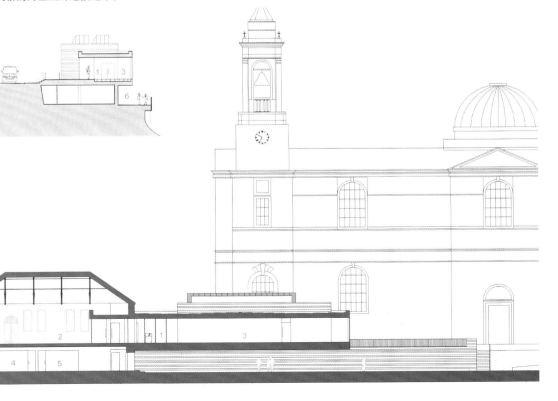

Section through Galleries & Main Entrance
1. Entrance/foyer
2. Main gallery
3. Glazed gallery
4. Administration offices
5. Meeting area
6. Extension to river towpath

展厅及主入口切入剖面
1. 入口/门厅
2. 主展厅
3. 玻璃展厅
4. 行政办公室
5. 会面区
6. 河上纤道延长部分

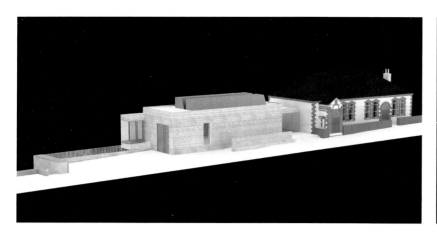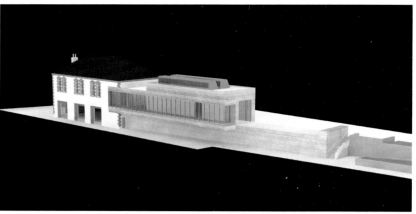

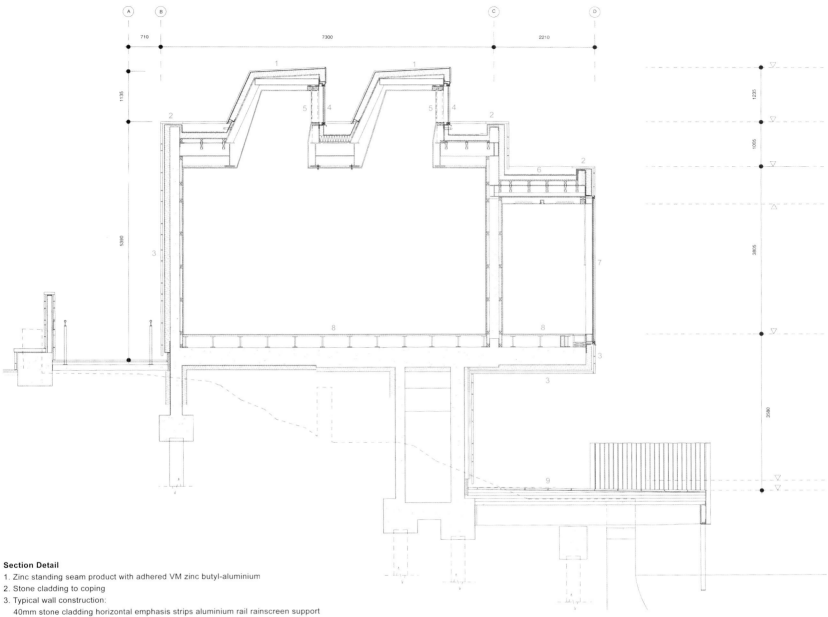

Section Detail
1. Zinc standing seam product with adhered VM zinc butyl-aluminium
2. Stone cladding to coping
3. Typical wall construction:
   40mm stone cladding horizontal emphasis strips aluminium rail rainscreen support system
   120mm rigid insulation with integral vapour control and damp proof layers
   Primary structure substrate
4. Fixed double glazed units
5. Roller blinds for varying light conditions in gallery space
6. Flat roof construction:
   18mm plywood decking with shims to create falls
   120mm rigid insulation
7. River gallery glazing: Schuco FW60 glazing system with horizontal and vertical caps
8. Proprietary raised floor system with pedestals wood veneer on 40mm carrier panel
9. River deck: 40mm stone paving slab

剖面节点
1. 站缝锌板，附有 VM 锌丁基胶合铝
2. 石材覆面顶盖
3. 标准墙面构造：
   40mm 水平石材条覆面，铝轨雨幕支撑系统
   120mm 刚性隔热层，配有水蒸气控制和防潮层
   一级结构底层
4. 固定双层玻璃
5. 遮阳卷帘，配合展览空间的各种照明条件
6. 平屋顶构造：
   18mm 胶合板铺装，用垫片制造下降区
   120mm 刚性隔热层
7. 河上画廊玻璃：Schuco FW60 玻璃装配系统，配水平和垂直顶盖
8. 专利架高式地板系统，木薄板下方是 40mm 支撑板
9. 河上平台：40mm 石铺面板

Tainan Yuwen Library
台南裕文图书馆

Location/ 地点：Tainan, Taiwan, China/ 中国台湾，台南
Architect/ 建筑师：Malone CHANG, Yu-lin CHEN/ MAYU architects+
Photos/ 摄影：Guei-Shiang Ke, Yu-lin Chen
Site area/ 占地面积：2,965m²
Built area/ 建筑面积：1,350m²
Completion date/ 竣工时间：2012

Key materials: Façade – exposed concrete, granite (G682), wood texture aluminium extrusion profiles, low-E glass
Structure – reinforced concrete
主要材料：立面——清水混凝土、花岗岩（G682）、木纹挤制铝型材、低辐射玻璃
结构——钢筋混凝土

Overview
A concrete lower volume is proposed to negotiate urban events on the one hand, and organise internal functions on the other. Large fenestrations articulated with concrete panels and canopies are located at dramatic moments: street corner facing elementary school (children's library), frontal view toward community centre (young-adult area), gaze window viewable from the park (reading room), and finally the horizontal glazing along the boulevard (lobby and new arrival). These symbolic openings convey the public character of the library, allowing citizens' gaze penetrates the library boundaries from all angles. The coexistence of the expansive concrete walls and openings suggests enough aura that lures citizens to explore the knowledge inside.

Sustainability
Sun – On top of the lower concrete base, the architects half-lodge a wood volume which contains collective human knowledge and clad in vertical wood louvers. The harsh direct sunlight is filtered and diffused by these louvers to provide comfortable interior ambiance. The combination of louvres and expansive glazing generates a transparent and universal space.

Air – VRV (Variable Refrigerant Volume) air-condition system is used in this project to achieve higher efficiency and increased controllability. Combined with Total Heat Recovery technology, optimal sustainability can be ensured.

Biological Diversity – Although located in the dense urban setting, the architects strive to protect the local biological system. During construction, every measure is taken to protect the surrounding indigenous Taiwanese flame gold trees. The second floor roof garden, on the level of treetops, becomes biological bridges for local species to travel across the site.

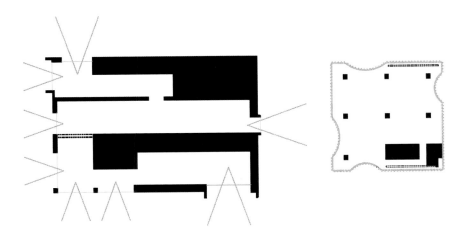
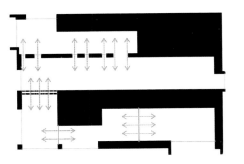
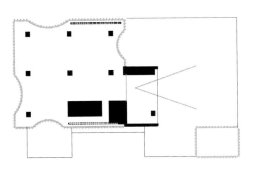

Detail and Materials

In Yuwen Library, the architects deployed two distinct façade strategies. On the upper reading rooms, the main objective was to apply largest transparent surface possible while use the vertical lamella to cut down incident sunrays. The lamella was treated as a second façade, and two alternate profile sizes were used to make it less homogeneous. The size and spacing of individual lamella was determined by careful calculation of sun shading capability required by sustainability building code and aesthetic considerations.

On the lower floors, windows are articulated in a way that they were always perceived as punctured. At important architectural episodes, openings were further enhanced by concrete canopies and frames so that they became "viewing devices". The first floor façade is clad by veneer gold granite as transition from exposed concrete to the wood lamella in terms of tectonic and colour. The subtle depth created by the cladding was purposely kept by flashing and capping detailing so that the perception of solid massing can be preserved.

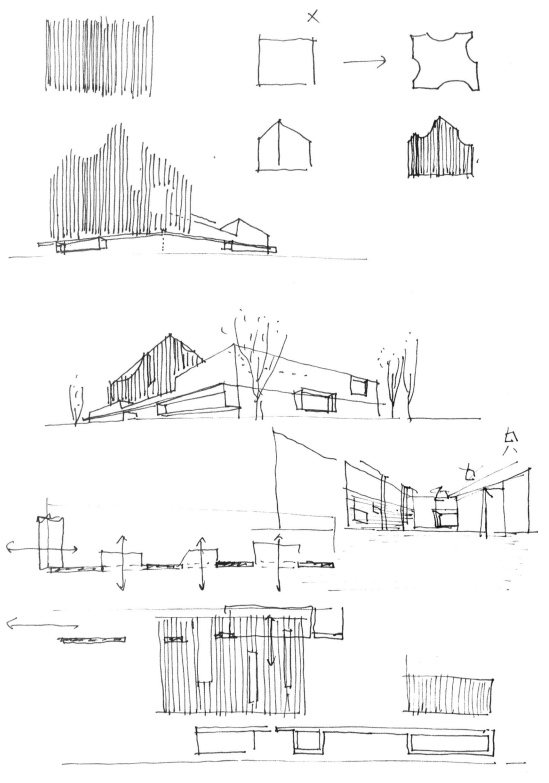

项目概况

混凝土低层建筑结构一方面能与城市活动相互协调,另一方面有利于组织内部功能空间。与混凝土板和雨篷连接在一起的大面积开窗都设在重要的位置:儿童图书馆向小学展示其藏书与活动;青少年图书馆向居民活动中心借景;自修室及社区书房允许公园的市民向内凝视;新书展示区及入口大厅以水平的开口面对林荫道上机动车辆的快速观看。这些象征性的开口传达了公共性的诉求,也使图书馆在大面混凝土墙面及透明性之间,带有些微引诱市民探索的神秘感。

可持续设计

阳光——建筑师在混凝土底座上嵌入一个藏书的木盒:此木盒内藏抽象的人类知识、外在以木格栅披覆。刺目的直射阳光经过这些格栅结构的过滤,营造出舒适的室内环境。格栅和玻璃的组合共同营造出一个透明而统一的空间。

空气——项目应用变制冷剂流量空调系统来实现更高的效率和可控性。它与整体热恢复技术共同保证了项目的可持续性。

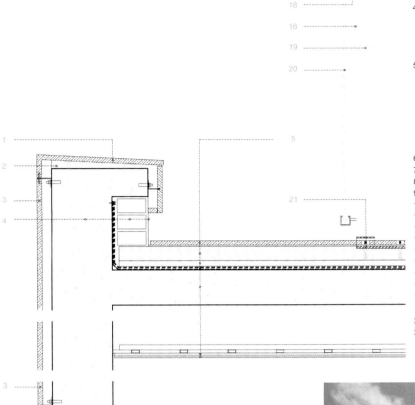

Façade Detail

1. 20mm granite cap w/steel dowel anchors
2. Stone adhesive
3. 20mm granite (G682)
4. 250mm reinforced concrete parapet
   Bituminous roof sealing layer
   3mm heat sealed bituminous sheet membrane
   1/2b brick w/cement mortar coating
   Polyurethane sealing layer
5. 20mm granite slabs (G603)
   60mm reinforced concrete pressure-distribution slab
   25mm rigid thermal insulation
   3mm heat sealed bituminous sheet membrane
   Bituminous roof sealing layer
   Suspended ceiling:
   9mm plasterboard
   12mm plasterboard
6. Polyurethane sealing layer
7. 2.5mm stainless-steel sheet, bent to shape
8. Exposed concrete
9. Aluminum widow w/5+5mm laminated tempered glass
10. 2.5mm fluorocarbon coated aluminum sheet
11. 50x25mm aluminum tube
12. 5mm fluorocarbon coated aluminum panel
13. 24mm maple veneered lumber core plywood
14. Composite maple flooring
15. 1"x1.5" sleepers on sound insulated cushion @20cm
16. 75x19mm coated galvanised steel
17. Glass fixing strip: 40x40x2.8mm coated galvanised steel
18. 25x9mm coated galvanised steel@750mm
19. 50x4.5mm coated galvanised steel
20. 6+6mm laminated tempered glass
21. 75x6mm coated galvanised steel
    Expansion bolt
    75x19mm coated galvanised steel base plate

立面节点

1. 20mm 花岗岩顶盖，配插筋锚件
2. 石材黏合剂
3. 20mm 花岗岩（G682）
4. 250mm 钢筋混凝土护墙
   沥青屋面密封层
   3mm 热封沥青膜
   1/2b 砖，配水泥灰浆涂层
   聚氨酯密封层
5. 20mm 花岗岩板（G603）
   60mm 钢筋混凝土分压板
   25mm 刚性隔热层
   3mm 热封沥青膜
   沥青屋面密封层
   吊顶：
   9mm 石膏板
   12mm 石膏板
6. 聚氨酯密封层
7. 2.5mm 不锈钢板，弯曲成形
8. 清水混凝土
9. 铝窗，配 5+5mm 夹层钢化玻璃
10. 2.5mm 碳氟涂层铝板
11. 50x25mm 铝管
12. 5mm 碳氟涂层铝板
13. 24mm 枫木锯材芯板胶合板
14. 复合枫木地板
15. 1"x1.5" 枕木，隔音衬垫上方，@20cm
16. 75x19mm 涂层镀锌钢
17. 玻璃固定带：40x40x2.8mm 涂层镀锌钢
18. 25x9mm 涂层镀锌钢 @750mm
19. 50x4.5mm 涂层镀锌钢
20. 6+6mm 夹层钢化玻璃
21. 75x6mm 涂层镀锌钢
    膨胀螺栓
    75x19mm 涂层镀锌钢底板

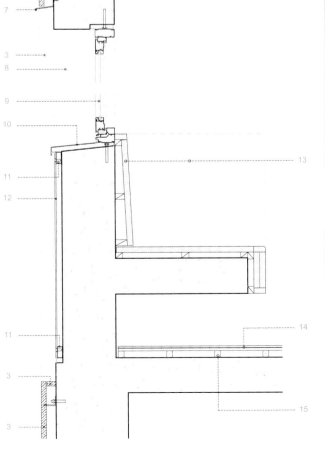

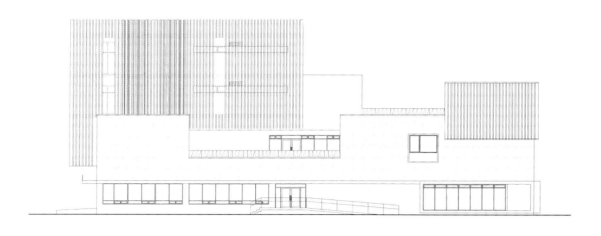

生物多样性——尽管项目处在密集的城市环境中，建筑师仍力求保护本土生态系统。在施工期间，他们竭尽全力、采用各种措施对周围的台湾火焰树进行了保护。三楼的屋顶花园与树冠同高，成为了本土物种穿越建筑场地的生态桥梁。

细部与材料

在裕文图书馆的设计中，建筑师采用了两种独立的立面设计策略。在上半部分的阅览室空间，设计的主要目标是在享受大面积透明表面的同时用垂直木板薄板来减少日光射入。木板包层被处理成第二立面，两种不同尺寸板材的运用避免了建筑的单一感。独立板条的尺寸和间隔均根据可持续建筑标准所要求的遮阳性能计算以及审美考虑联合决定。

下层楼面的窗口呈现为穿孔的形式。混凝土雨篷和窗框让窗口变成了"观景装置"。二楼墙面上包覆着一层金色花岗岩薄板，在构造和色彩上帮助清水混凝土和木格栅形成了过渡。花岗岩覆层的厚度被防水板和顶板巧妙地保留下来，给人以厚重的实体感。

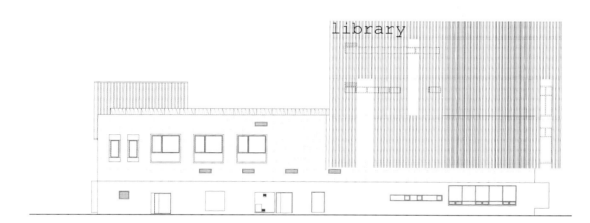

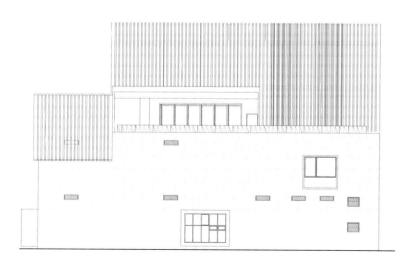

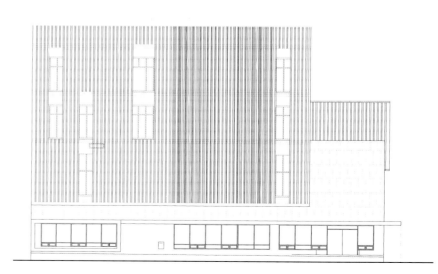

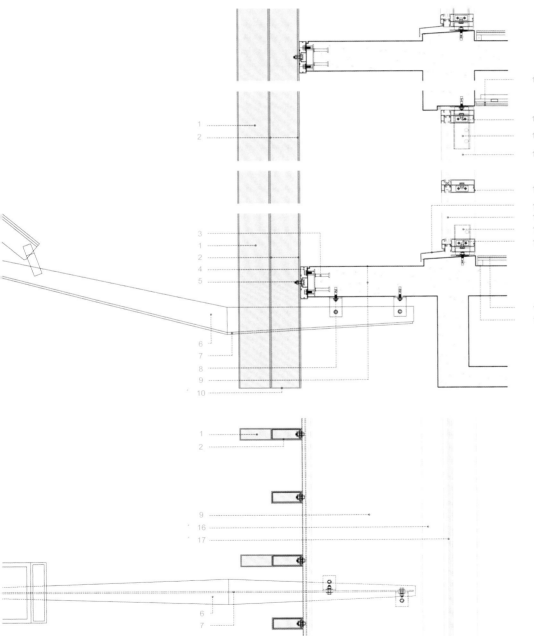

Façade Detail
1. Extruded aluminum vertical louver, fluorocarbon coated in wood texture
2. 140x50x3mm galvanised steel tube
3. Stainless steel anchor embedded in concrete slab
4. Fluorocarbon coated extruded in aluminum cover
5. M12 stainless steel bolt
6. 9mm coated galvanised steel
7. 6mm coated galvanised steel
8. S45C 1x2x4" expansion bolt
9. Polyurethane sealing layer
    150mm concrete slab
10. Aluminum end cover
11. Suspended ceiling:
    9mm plasterboard
    12mm acoustic mineral fibre board
12. Aluminum window frame bracket
13. 4mm galvanised steel socket
14. Vertical aluminum window frame
15. Horizontal aluminum window frame w/anchor
16. 2.5mm fluorocarbon coated aluminum sheet
17. 5+5mm laminated tempered glass
18. Composite maple flooring
19. 1"x1.5" sleepers on sound insulated cushion @20cm

立面节点
1. 挤制铝垂直百叶，木纹，碳氟涂层
2. 140x50x3mm 镀锌钢管
3. 不锈钢锚件，嵌入混凝土板
4. 碳氟涂层挤制铝顶盖
5. M12 不锈钢螺栓
6. 9mm 涂层镀锌钢
7. 6mm 涂层镀锌钢
8. S45C 1x2x4" 膨胀螺栓
9. 聚氨酯密封层
    150mm 混凝土板
10. 铝端盖
11. 吊顶：
    9mm 石膏板
    12mm 隔音矿物纤维板
12. 铝窗框支架
13. 4mm 镀锌钢底座
14. 垂直铝窗框
15. 水平铝窗框，配锚件
16. 2.5mm 碳氟涂层铝板
17. 5+5mm 夹层钢化玻璃
18. 复合枫木地板
19. 1"x1.5" 枕木，隔音衬垫上方，@20cm

County Council of Zamora
萨莫拉郡议会

Location/ 地点：Zamora, Spain/ 西班牙，萨莫拉
Architect/ 建筑师：g+f arquitectos
Photos/ 摄影：Miguel de Guzmán, Joaquín Mosquera
Site area/ 占地面积：636m²
Gross floor area/ 总楼面面积：2,803m²
Completion date/ 竣工时间：2011

Key materials: Façade – sandstone from Zamora
Structure – concrete
主要材料：立面——萨莫拉砂岩
结构——混凝土

Overview
Not only should the new Council offices meet the determined requirements of use, but it also should help define one of the most compromising environments of the city, as Vitiato square – where the building is located – is a compulsory stop in the touristic tour around Zamora. The proposal designs the fourth façade of the square, completing the architectonic complex formed by the Encarnación Hospital to the North and the Ramos Carrion Theatre and the Condes de Alba y Aliste Palace to the South.

The new office building is organised around a patio that allows to:
- Duplicate the rooms with a North-South façade and reduce the blind areas.
- Gain diagonal views to the square and its environment so that they become part of the inner life of the building.
- Generate an external wall image of the building, continuing the aesthetics of the square without losing any light requirements for the interior spaces.

Detail and Materials
The exterior walls, both to the street and of the patio, are covered in sandstone from Zamora, the same material traditionally used in the historic buildings of the city. This stone is used in a trans-ventilated façade that, far from hiding its constructive system, shows its nature and the steel fastenings. For the lower base long pieces were used in order to recover the wall language of the area. The flat roof was also covered in stone becoming another façade.

There are two kinds of window holes. On the one hand, to provide with enough light the interior spaces, big glass surfaces are opened only to the patio. On the other hand, in the walls facing the square and surrounding streets vertical windows are used, so that the wall image is preserved. However, the building avoids enclosing itself to the patio thanks to a deep

Detail
1. Painted round solid steel tube
2. Steel sheet t=8mm
3. Steel sheet 250*150*20mm, welded to base plate 150*150*5mm
4. Sealing between geotextile layer
5. Compression coat t=5cm
6. Hollow core slab floor 25cm
7. Sandstone tile t=10cm
8. Projected polyurethane e=4cm
9. Rigid insulation anchored to structure t=5.5cm
10. Stainless steel adjustable anchoring t=6mm
11. Plasterboard 1.5cm

节点
1. 涂漆脱氧钢圆管
2. 钢板 t=8mm
3. 钢板 250×150×20mm,焊接在 150×150×5mm 底板上
4. 土工布层之间的密封
5. 压缩包层 t=5cm
6. 空心楼板 25cm
7. 砂岩砖 t=10cm
8. 挤塑聚氨酯 e=4cm
9. 刚性隔热层,锚固于结构上 t=5.5cm
10. 不锈钢可调节锚件 t=6mm
11. 石膏板 1.5cm

项目概况

新建的议会楼不仅应满足特定的功能要求,还必须有助于提升并塑造城市环境,因为其所在地韦迪亚多广场是萨莫拉旅游的必游景点。作为广场的第四个面,议会楼将与北面的恩卡纳西翁医院以及南面的拉莫斯卡里翁剧院和阿尔瓦阿里斯特宫隔着广场遥相呼应。

新建的办公楼围绕着天井展开,这种设计有助于:
– 复制南北朝向的房间,减少盲区
– 获得广场及其周边环境的对角线景色,使它们成为建筑内部生活的一部分
– 塑造建筑的外墙形象,延续广场的美学价值,同时又不会影响室内空间的光照

细部与材料

朝向街道和天井的外墙全都覆盖着萨莫拉本地产的砂岩,这种石材被广泛应用于城市的传统历史建筑中。石材所构成的通风立面并没有将构造系统隐藏起来,而是把本质和钢紧固件裸露在外。在下方的底座中,建筑师用长条石板来重现该地区的墙面风格。同样覆盖的平屋顶形成了另一个立面。

hole opened in the main façade that allows views and contact with Viriato square. This area can be accessed from the offices in the first floor and can be used as representative balcony for public activities of the Council.

The patio is designed with sandstone walls, hedges and evergreen climbing plants that will remain green all year long. Dominating the scene stands a populous alba (white poplar). Given its thin high shape, it can be seen from all floors, and, as it is a deciduous tree, its image will change through the seasons, protecting from direct sun in summer but allowing it in wintertime.

Neutral vinyl materials are used in the interior to highlight the contrast between the opaque character of the façade and the opening of the patio and to reinforce the personality of the local stone.

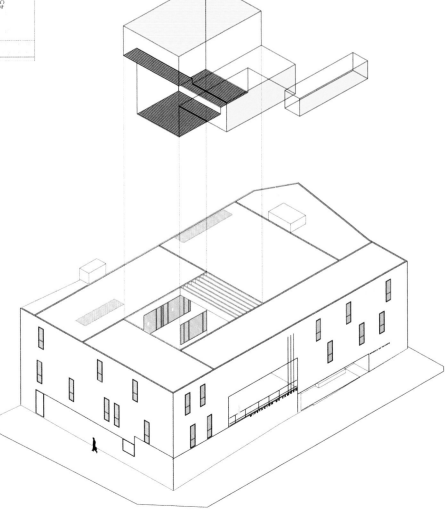

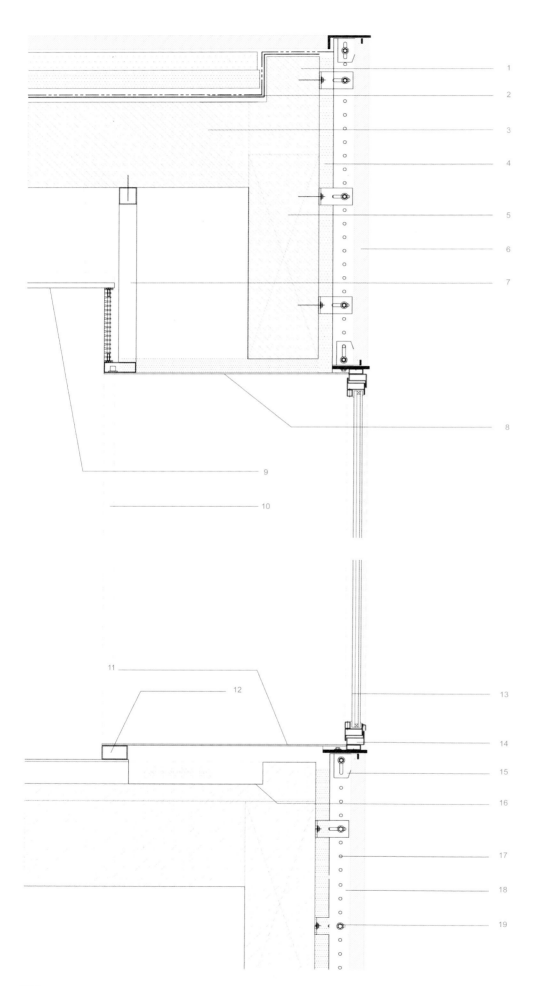

Wall Detail
1. Prefabricated concrete block t=15mm
2. Compression coat t=50mm
3. Prefabricated hollow core slab
4. Projected polyurethane t=4cm
5. Reinforced concrete ring beam
6. Quintanar sandstone t=4cm
7. Supporting truss for perimeter wardrobe doors
8. White lacquered sheet e=2mm
9. Plaster false ceiling
10. Perimeter wardrobe door
11. Grey lacquered sheet t=10mm
12. Steel tube 80.4.2
13. Transparent double glazing
14. Openable steel window, dark grey or brown lacquered
15. Stainless steel anchoring for stone tiles
16. Perimeter duct for system use
17. Φ8.5mm drills for M8 use, distance between centres 40mm
18. Zinc-coated and dark grey or brown lacquered steel plate 100.8, with Φ8.5mm drills each 40mm
19. Galvanised steel adjustable anchoring t=6mm, for post anchoring

墙面节点
1. 预制混凝土块 t=15mm
2. 压缩包层 t=0mm
3. 预制中空板
4. 挤塑聚氨酯 t=4cm
5. 钢筋混凝土环梁
6. Quintanar 砂岩 t=4cm
7. 外围壁柜门支撑桁架
8. 白色涂漆板 e=2mm
9. 石膏假吊顶
10. 灰色涂漆板 t=10mm
11. 透明双层玻璃
14. 可开式钢窗，深灰或棕色喷漆
15. 不锈钢，用于固定石砖
16. 外围系统管道
17. Φ8.5mm 钻孔，中心间距 40mm
18. 深灰色或棕色漆镀锌钢板 100.8，配间距 40mm 的 Φ8.5mm 钻孔
19. 镀锌钢可调节锚件 t=6mm，用于后方固定

建筑有两种窗洞。一方面，为了给室内提供充足的光线，大面积的玻璃表面仅朝向天井一侧。另一方面，在朝向广场和周边街道的墙面上，建筑采用了垂直窗口，从而保护了墙面的统一形象。然而，建筑主立面上一个大型切口为室内空间提供了韦迪亚多广场的景色，避免建筑成为包围天井的围墙。人们可以供二楼办公室进入这一空间，它还可以作为议会举办公共活动的代表台。

天井的设计综合了砂岩墙、树篱和常青藤蔓植物。一棵白杨树处在场景的核心位置。细高的树形让各个楼层都能看到，而作为一棵落叶树，它的形象会随着季节而变化，在夏季保护建筑不受阳光直射，在冬季则让阳光射入室内。

室内运用中性乙烯基材料突出了不透明的外墙与天井方向开窗的对比，并且进一步突出了本地石材的个性。

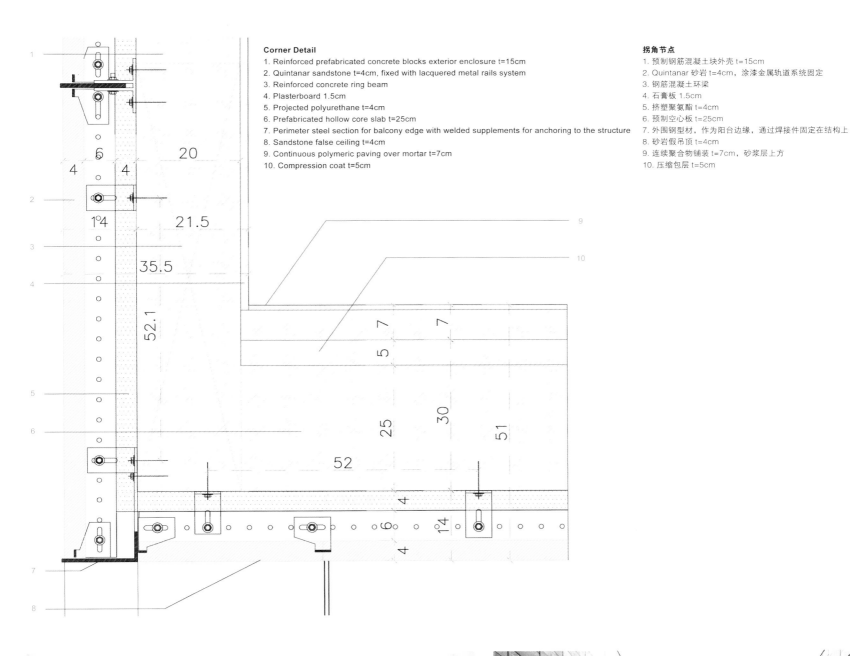

Corner Detail
1. Reinforced prefabricated concrete blocks exterior enclosure t=15cm
2. Quintanar sandstone t=4cm, fixed with lacquered metal rails system
3. Reinforced concrete ring beam
4. Plasterboard 1.5cm
5. Projected polyurethane t=4cm
6. Prefabricated hollow core slab t=25cm
7. Perimeter steel section for balcony edge with welded supplements for anchoring to the structure
8. Sandstone false ceiling t=4cm
9. Continuous polymeric paving over mortar t=7cm
10. Compression coat t=5cm

拐角节点
1. 预制钢筋混凝土块外壳 t=15cm
2. Quintanar 砂岩 t=4cm，涂漆金属轨道系统固定
3. 钢筋混凝土环梁
4. 石膏板 1.5cm
5. 挤塑聚氨酯 t=4cm
6. 预制空心板 t=25cm
7. 外围钢型材，作为阳台边缘，通过焊接件固定在结构上
8. 砂岩假吊顶 t=4cm
9. 连续聚合物铺装 t=7cm，砂浆层上方
10. 压缩包层 t=5cm

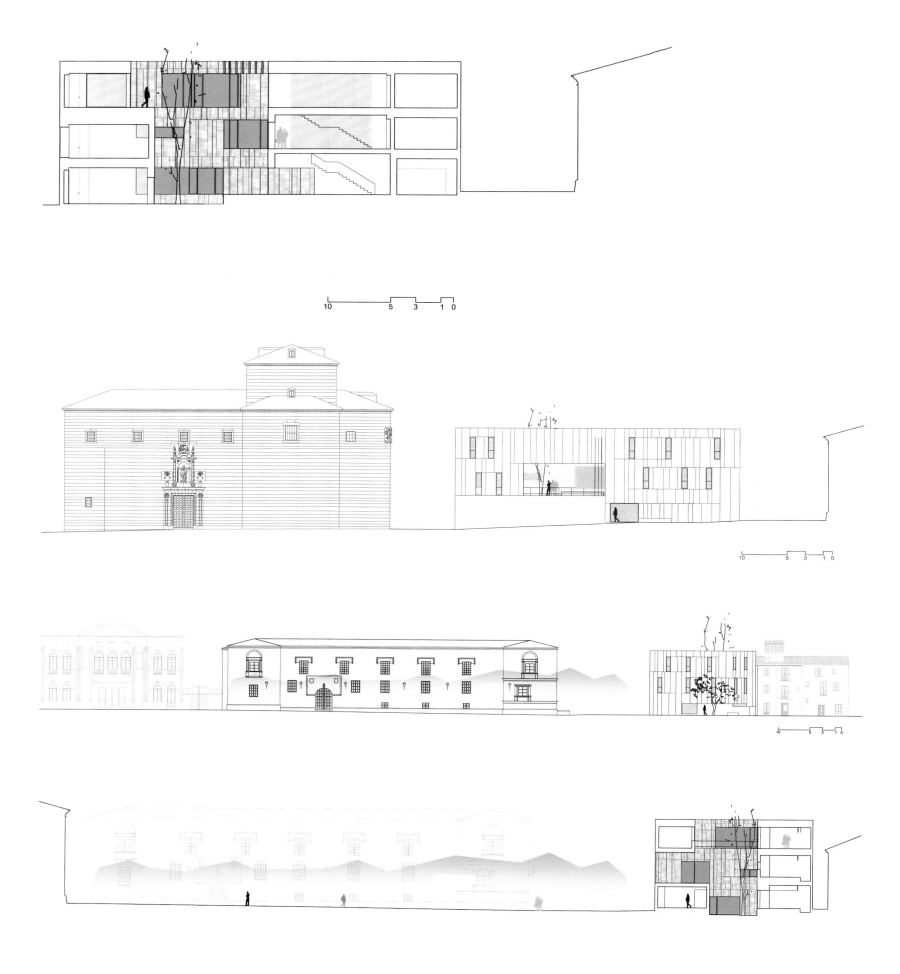

Argul Weave
阿古尔波浪大楼

Location/ 地点：Bursa, Turkey/ 土耳其，布尔萨
Architect/ 建筑师：binaa.co&smart-arch.com
Photos/ 摄影：thomasmayerarchive.de
Area/ 面积：4,500m²
Completion date/ 竣工时间：2014

Key materials: Façade – marble
主要材料：立面——大理石

Façade material producer:
外墙立面材料生产商：
Stone Fabricator - Bayburtlular, Afyon；（石材）
Wood Fabricator - Yapisan, Bursa
（木材）

Overview

Argul Weave is sited in Bursa, Turkey, 100km south of Istanbul in the Asia Minor peninsula. Bursa is home to Turkey's historic and celebrated textile industry which is witnessing a rebirth to position itself as a regional and international leader. Located on the corner of Koklu Cd and Kirkpinar Cd the project site sits at the centre of this manufacturing district and is the first in a planned series of interventions to redevelop the area.

The client is a textile distributor with his existing facility directly adjacent to the East of the project site. From its inception the Argul Weave was planned as a mixed-use development to attract international textile businesses to refocus and rebrand the district as a leading manufacturing textile hub. The Weave is 4,500 sqm distributed over three stories with retail on the ground floor, offices on levels two and three and capped off with a rooftop restaurant.

Detail and Materials

To satisfy the client's vision the design considered the rich tradition of Turkish textiles and their making through the interweaving of individual threads by giant looms. This motif offered a means to integrate the disparate parts of the project into a singular and coherent whole while making a clear statement as to the importance of this district. The material palette was kept to a minimum to emphasise the continuity and plasticity of the design. Patara Marble quarried from Burdur, Turkey and fabricated in Afyon forms the sinuous banding of the façade. Dark red marble from the Turkish Aegean region forms the building's plinth and reddish brown Iroko wood from West Africa clads the inner areas of the weave. As one walks along the façade its undulations create a changing rhythmical pattern that are enhanced by the continuous play of light and shadow evolving throughout the day.

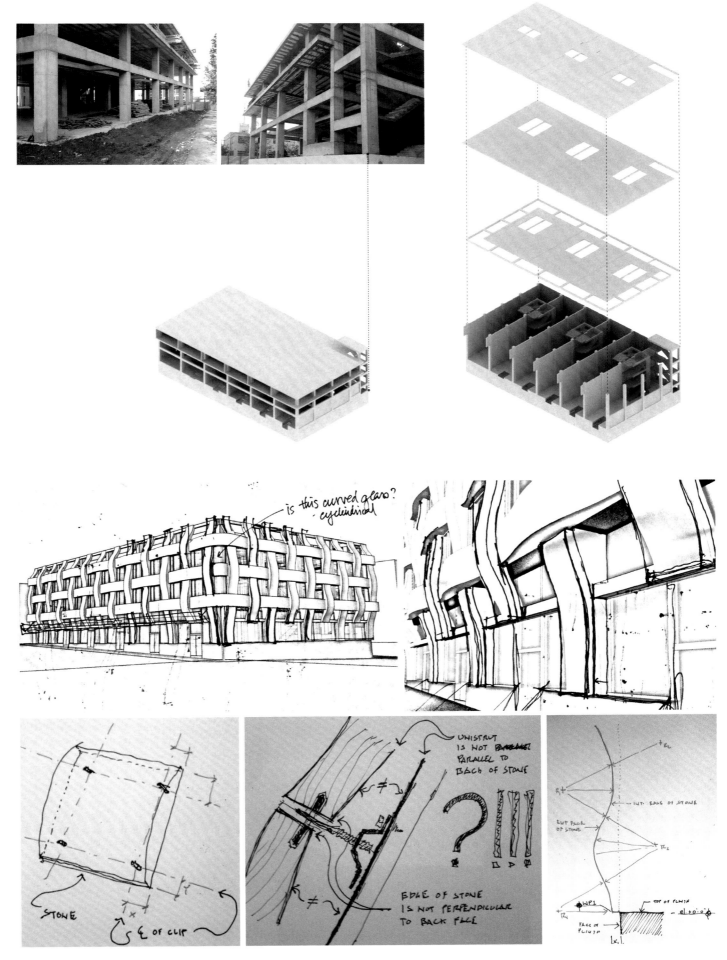

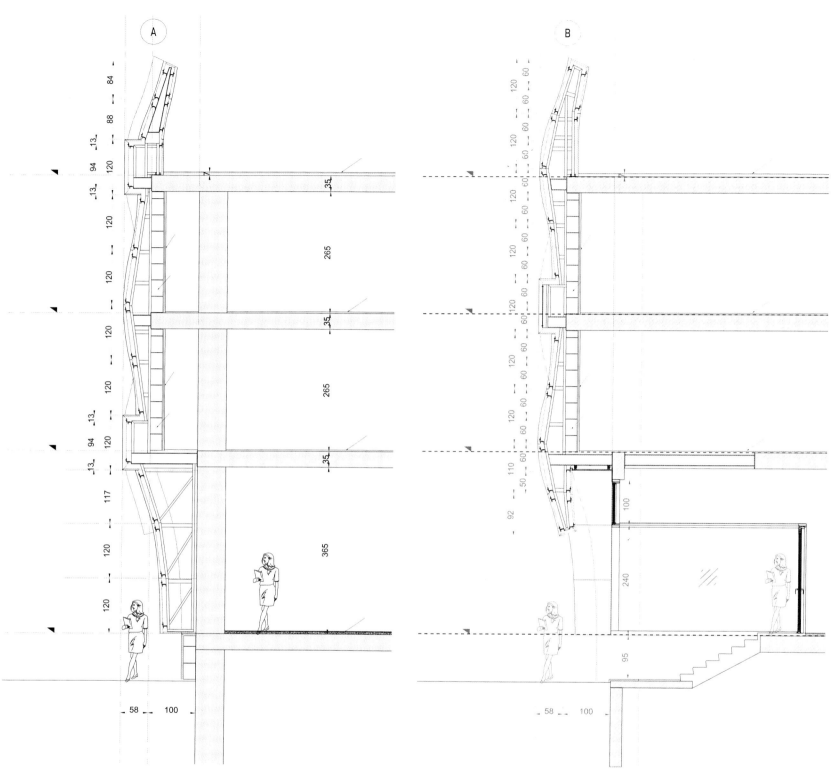

The design's execution relied on a unique and collaborative process that included a diverse group of team members situated in different regions and continents. Digital technologies were tailored to adapt to local practices for fabrication and construction. Master digital models were used to fine tune and control all geometries and construction assemblies and also used as a means to evaluate costs.

Often time's constraints from the fabricators were fed back into the model to further tune the geometries and design as required. As in the design the overall process can be conceived as an interwoven collaboration of actors located in distinct geographies. Research, design and execution fed each other in parallel progressions throughout the project that led to the realisation of the Argul Weave.

项目概况

阿古尔波浪大楼位于土耳其布尔萨，距伊斯坦布尔 100 千米，位于小亚细亚半岛上。布尔萨是土耳其的历史名城，拥有发达的纺织工业，定位是区域及国际领军城市。项目场地位于制造工业区的中心，是该地区重新开发项目的第一个工程。

委托人是一位纺织品经销商，他的现有机构就位于场地的东侧。阿古尔波浪大楼被设计成一个混合功能开发项目，以吸引来自全球各地的纺织贸易商，使该地区重新

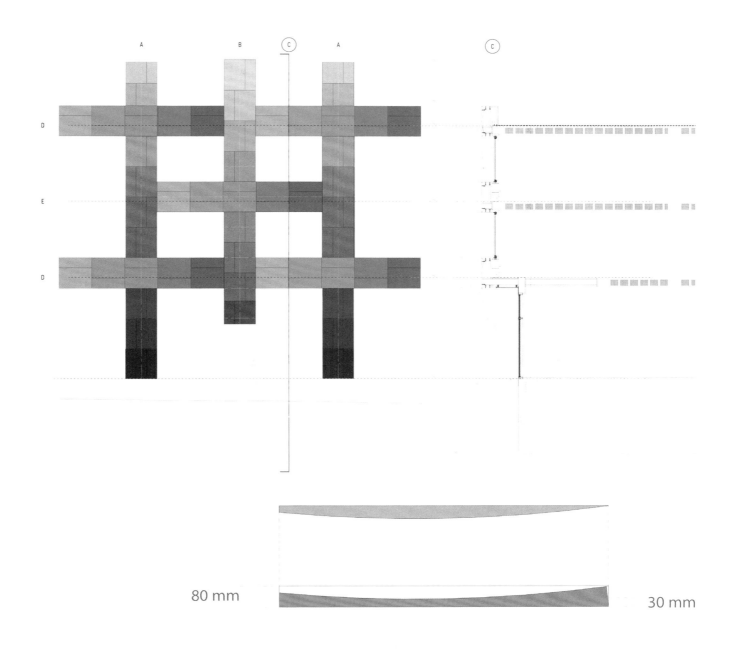

80 mm 30 mm

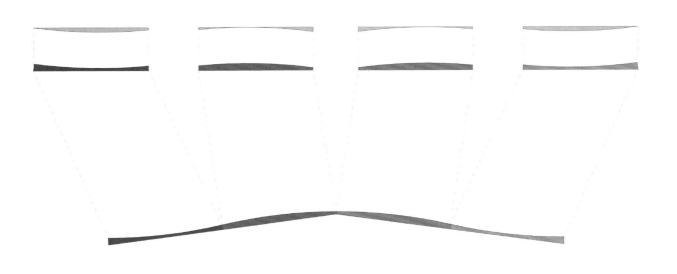

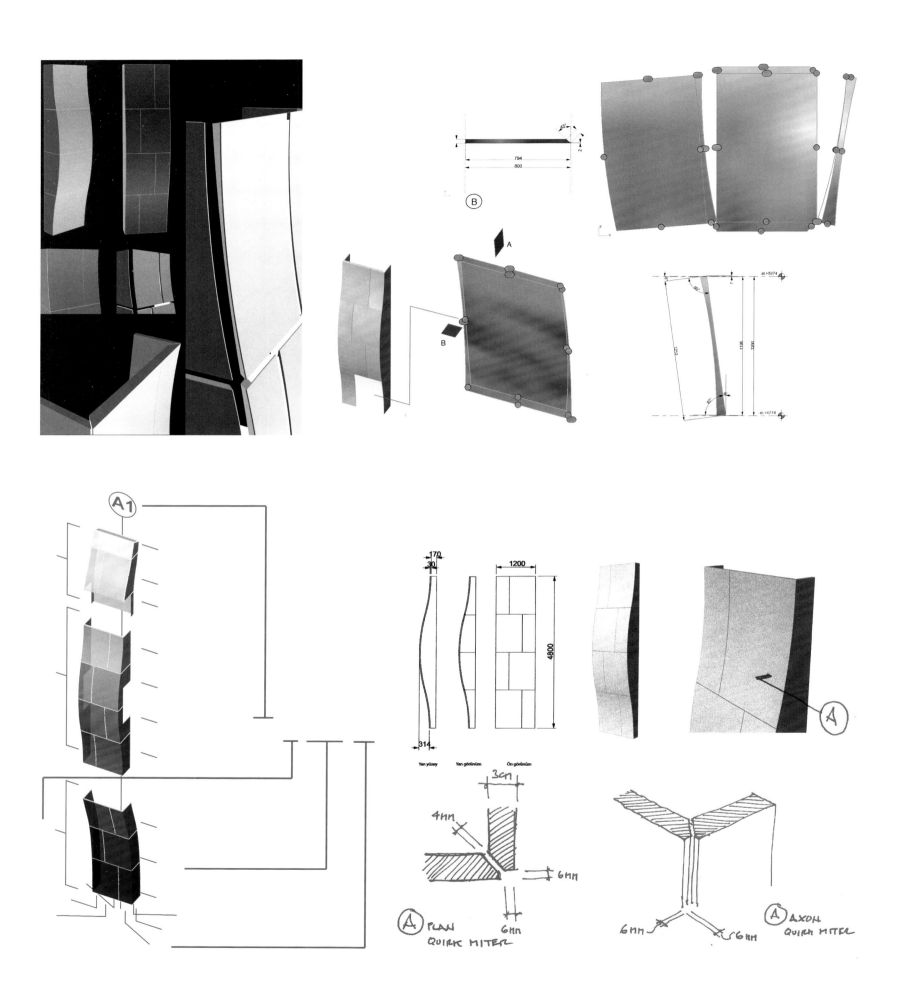

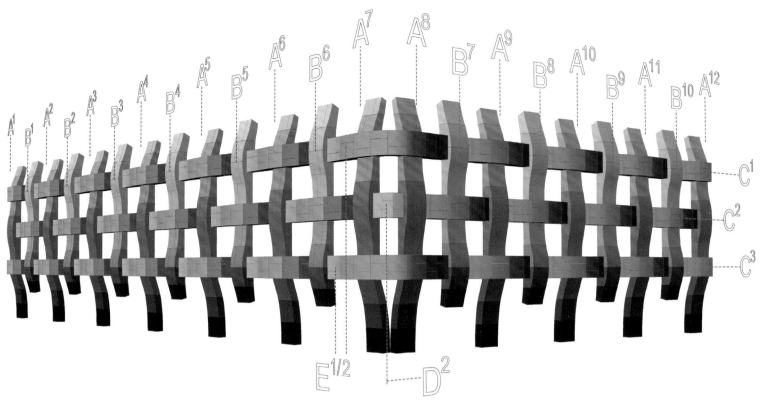

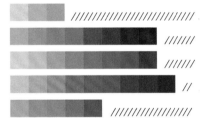
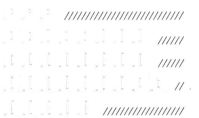

成为领先的纺织制造中心。波浪大楼总面积 4,500 平方米,分为三层:一层是零售空间,二、三层是办公楼,同时屋顶还设有一家餐厅。

细部与材料

为了满足委托人的要求,设计考虑到了土耳其纺织品悠久历史。纺织品的制作是通过巨型织布机上交织的纺线来实现的。这种图案将项目的各个独立的部分结合成一个完整统一的整体,同时也在该地区形成了清晰有力的地标。建筑师尽量将材料的种类缩减至最少,突出了建筑连续性和可塑性。帕塔拉大理石开采自土耳其巴尔杜尔,在阿菲永进行加工,形成了建筑弯曲的立面。来自土耳其爱琴海地区的深红色大理石形成了建筑的底座,西非红棕色绿柄桑木将波浪结构的内部区域包裹起来。沿着建筑立面行走,它的起伏会营造出变换的韵律感,而一天之中变化的光影效果则会进一步提升这种感觉。

设计的成功实施要归功于来自不同地区、不同大陆的团队人员的通力合作。为了适应当地制作和施工,项目特别定制了数字技术,采用主数字模型来微调和控制所有几何造型和施工装配件,同时也有效评估了成本。

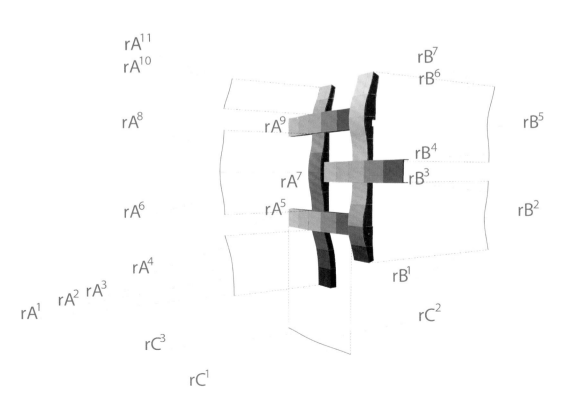

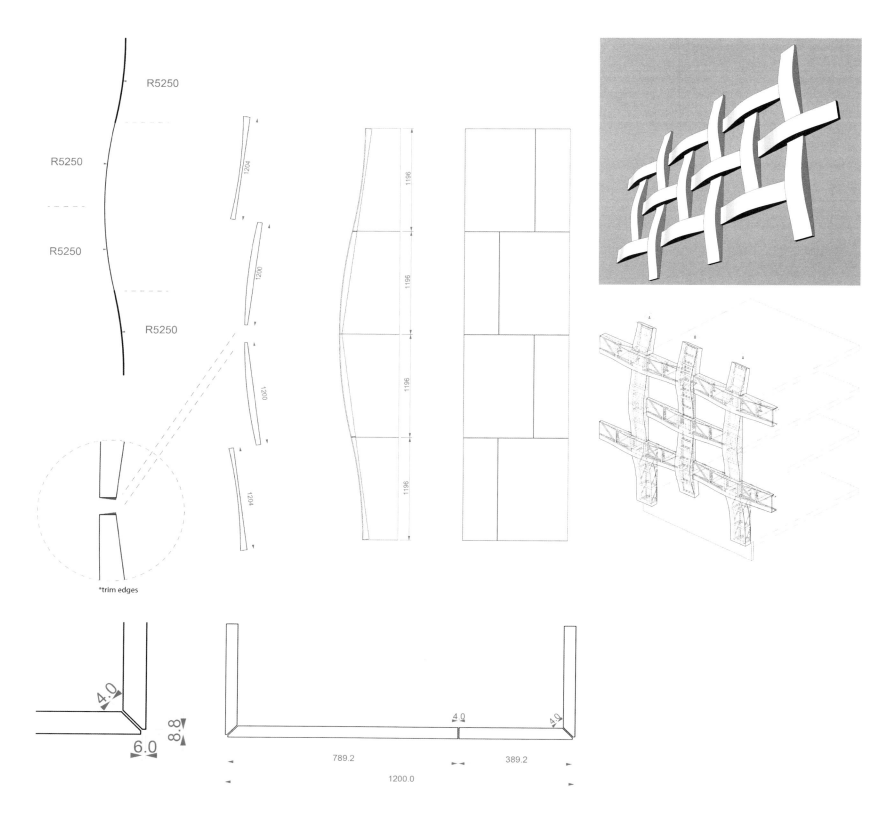

制造商的制约条件被及时反馈到模型上，以便根据设计要求进行进一步的结构调整。在设计中，整个流程可以看作是演员在特定场景下的通力合作。研究、设计和执行互相反馈，最终保证了阿古尔波浪大楼的成功封顶。

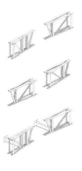
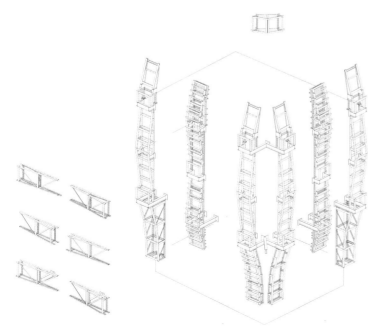

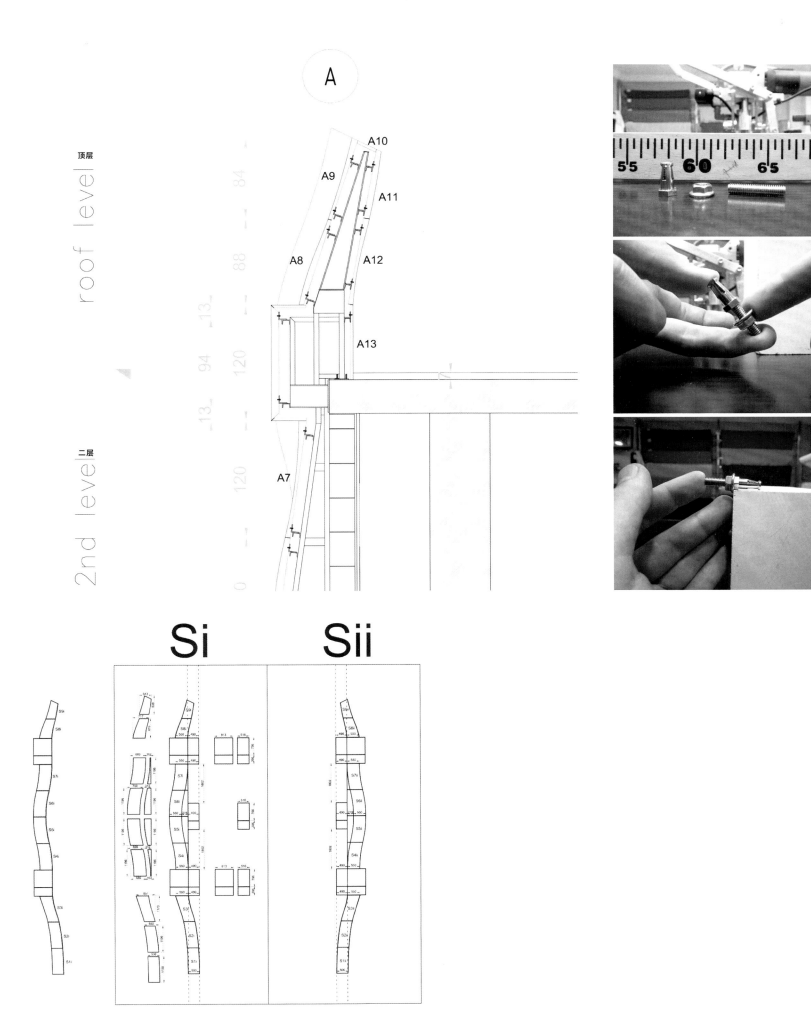

Head Office Archipel Habitat/ Municipalpole of The Quarter

阿希佩尔置业总部办公楼/区域行政中心

Location/ 地点：Rennes, France/ 法国，雷恩
Architect/ 建筑师：Bruno Gaudin Architectes
Photos/ 摄影：Stéphane Chalmeau, Nicolas Borel
Area/ 面积：8,535m²

Key materials: Façade – granite
Structure – concrete
主要材料：立面——花岗岩
结构——混凝土

Overview
The building houses a mixed programme: a district public facility and tertiary spaces for the Archipel Habitat head office. Its mission is to give form to a major public space: the place of the Community of the Metropolis of Rennes. A courtyard penetrates the monolith and allows sun and air within the built volume. The building is irrigated by a clear and generous distribution that enhances the unit. A glass roof that diffuses and spreads natural light into the building crowns the distributive core, in the nerve centre. The reflections roll up around the full metal railing and carve the central void.

A delicate attention is paid to build the space of the office, offering a generous height, an extensive window, and allowing one to understand the materials employed: concrete, wood, lighting, ventilation ducts, suspended radiant panels (heating)…which provide these spaces with their nobility and singularity as well.

Detial and Materials
The façade is in granite and it forms a unit with the ground. As a counterpoint to this mineral material the blades of glass compose a scintillating and coloured rhythm.

项目概况
该项目是一个混合式项目，兼具区域公共设施中心及阿希佩尔置业总部办公楼两种功能。项目的目标是塑造雷恩都市社区的公共空间。深入建筑的庭院为室内带来了充足的阳光和新鲜的空气。建筑布局宽敞清晰，具有优异的环境。玻璃屋顶让自然光线从屋顶射入建筑中央的交通中心。全金属围栏上的倒影极富现代感，切割出一个中空空间。

建筑师十分注重办公空间的塑造，配置了充足的高度、大面积的开窗。混凝土、木材、照明设施、通风管道、悬挂式辐射板（暖气）等建筑材料为空间增添了高端而独特的氛围。

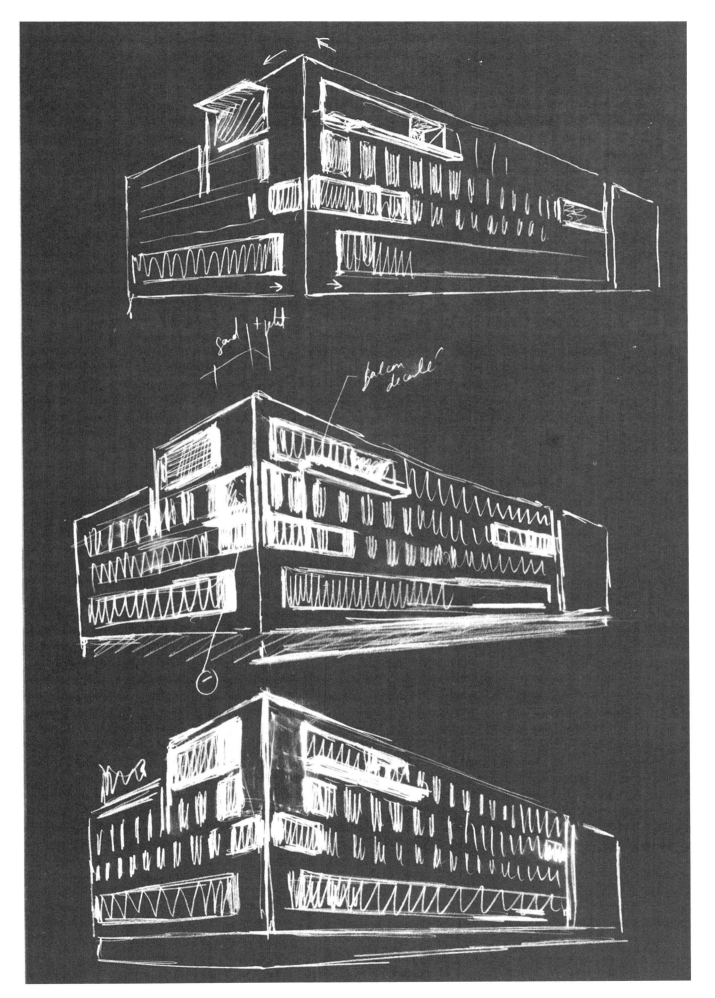

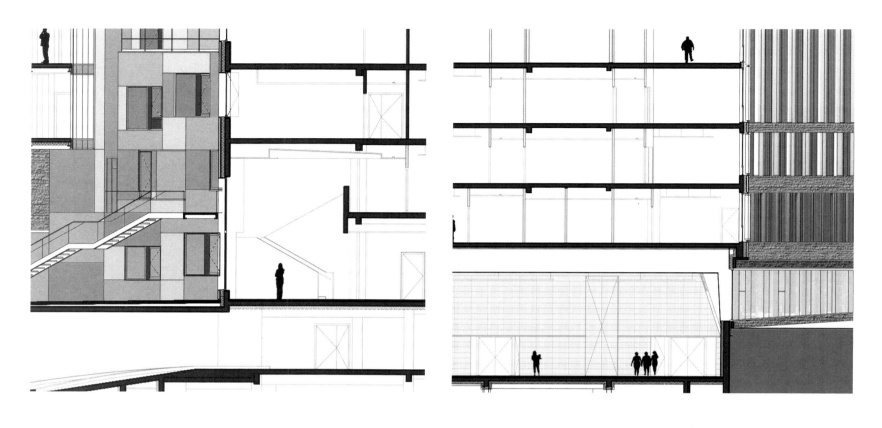

East Elevation Wall Detail

1. Granite entablature
2. Thermal insulation
3. Fixing lug of guardrail
4. Self-supporting granite façade
5. Blade of ventilated air (ventilated air space)
6. Thermal insulation
7. Upturn waterproof membrane
8. Thermal insulation
9. Waterproof coating
10. Continuous flat angle bracket stainless steel, support granite
11. Upper rail aluminum support sun breaker
12. Motorisation sun breaker
13. Side mounting lug of sun breakers, coated galvanised steel
14. Expander of window frame, mixed frame wood aluminum
15. Wood frame opening
16. Opening mixed frame wood aluminum
17. Cross mixed frame wood aluminum
18. Light fixed frame wood and aluminum
19. Granite return in table
20. Pivoting blade of glass sun breaker
21. Side mounting fixing lug of sun breaker, coated galvanised steel
22. Aluminum bearing plate blade of sun breaker
23. Lower rail aluminum support of sun breaker
24. Coated aluminum flap
25. Peripheral expander of window frame, mixed frame wood and aluminum
26. Lug of sash
27. Wood frame opening
28. Way of recessed joint
29. Baseboard
30. Slipping by square stainless steel support granite
31. Upper rail aluminum support of sun breaker
32. Aluminum bearing plate blade of sun breaker
33. Return (cassette) aluminum panel
34. Pivoting blade of glass sun breaker
35. Side mounting fixing lug of sun breaker, coated galvanized steel
36. Aluminum bearing plate blade of sun breaker
37. Lower rail aluminum support of sun breaker, coated aluminum flap
38. Peripheral expander of window frame, mixed frame wood aluminum
39. Self-supporting granite façade
40. Blade of ventilated air (air space)
41. Thermal insulation
42. Concrete shell
43. Interior thermal insulation
44. Stainless steel continuous corner plate support granite
45. Air tight sealed membrane
46. Upper continuous angle fastener sash
47. Curtain wall combined wood and aluminum
48. Profile of finishing batch frontage
49. Opening combined wood and aluminum
50. Curtain wall combined wood and aluminum
51. Glazed spandrel
52. Lower shelf continuous wood
53. Curtain wall combined wood and aluminum
54. Lower continuous flap coated aluminum
55. Continuous granite bench

东立面墙壁节点

1. 花岗岩檐口
2. 隔热层
3. 护栏固定耳
4. 自承式花岗岩立面
5. 通风翅片（通风气腔）
6. 隔热层
7. 向上防水膜
8. 隔热层
9. 防水涂层
10. 连续的不锈钢平角支架，支撑花岗岩
11. 上方铝轨，支撑遮阳板
12. 电动遮阳板
13. 侧安装遮阳板支架，涂漆镀锌钢
14. 窗框扩展，木铝混合框
15. 木框窗口
16. 木铝混合框窗口
17. 木铝交叉混合框
18. 轻质固定木铝框
19. 花岗岩板
20. 玻璃遮阳翅片转轴
21. 侧安装遮阳板支架，涂漆镀锌钢
22. 遮阳板的铝支承板
23. 下方铝轨，支撑遮阳板
24. 涂层铝翻板阀
25. 外围窗框扩展，木铝混合框
26. 窗框支托
27. 木框窗口
28. 凹缝
29. 踢脚板
30. 不锈钢方钢滑道，支撑花岗岩
31. 上方铝轨，支撑遮阳板
32. 遮阳板的铝支承板
33. 铝板箱
34. 遮阳玻璃翅片转轴
35. 侧安装遮阳板支架，涂漆镀锌钢
36. 遮阳板的铝支承板
37. 下方铝轨，支撑遮阳板
38. 外围窗框扩展，木铝混合框
39. 自承式花岗岩立面
40. 通风翅片（通风气腔）
41. 隔热层
42. 混凝土外壳
43. 内部隔热层
44. 不锈钢连续角板，支撑花岗岩
45. 气密膜
46. 上方连续扣件框
47. 幕墙，木铝混合
48. 正面饰面型材
49. 木铝混合窗口
50. 木铝混合幕墙
51. 玻璃拱肩
52. 下方连续木支架
53. 木铝混合幕墙
54. 下方连续涂层铝翻板阀
55. 连续的花岗岩长凳

细部与材料

建筑立面采用花岗岩铺装，与地面形成了一个整体。与这种矿物材料相对比，玻璃翅片为建筑带来了闪耀的彩色韵律。

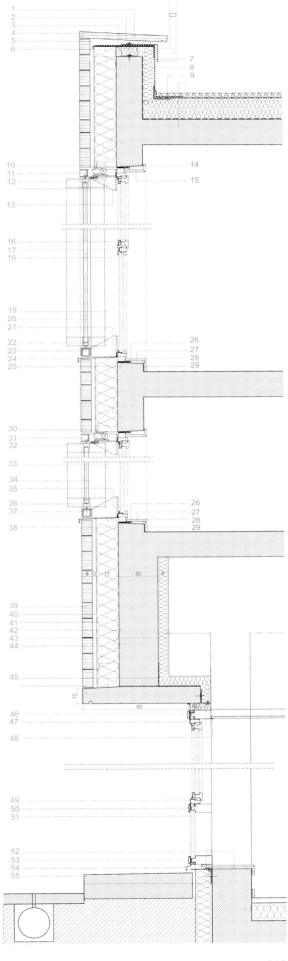

Crescendo Maaswaard Elderly Housing
马斯沃德老年住宅

Location/地点: Venlo, The Netherlands/荷兰, 芬洛
Architect/建筑师: Benthem Crouwel Architects
Photos/摄影: Jannes Linders
Gross floor area/总楼面面积: 24,350m²
Completion date/竣工时间: 2011

Key materials: Façade – cut stone
主要材料: 立面——琢石

Overview
In the Maaswaard area, at the central point between the four cores today constituting the town of Venlo, a new sheltered housing project for the elderly is constructed. This ensemble will be visually defining for Venlo, being high-profile and metropolitan yet sensitive to the surrounding residential areas.

The design proceeds from a sturdy four-storey basement from which rise towers of a further four to six storeys. This composition of towers combined with a nearby existing office tower (Nedinscotoren), takes account of sight lines and ways through in the adjoining neighbourhoods. Two of the project's towers, their fronts wrapped in an "open" screen of glass panels, add a lighter touch, affording some relief from the ensemble's otherwise imposing metropolitan demeanour. A great many different dwelling types are to look out over the river Maas.

Detail and Materials
Cladding basically consists of cut stone and sandwich panels in a range of colours from light to dark. Balconies are set back inside the blocks as loggias where they can be developed into sunrooms. A semi-public, sun-filled courtyard in the south spills over into the park on that side. The block overlooking the courtyard seems to float due to its 2.5 metre tall glass plinth which contains health care and other facilities. There is also a prominent south-facing restaurant cum grand café with terrace. The ambience of this outdoor area reflects the human touch, giving residents both a sense of seclusion and the fullest experience of social life there.

项目概况
在荷兰芬洛市的玛斯沃德区中心，新建了一个老年住宅项目。这个住宅综合体将成为芬洛市的新地标，既高端大气，又能很好地融入周边的住宅区。

设计以四层高的底座空间为基础，上方建造四至六层塔楼。

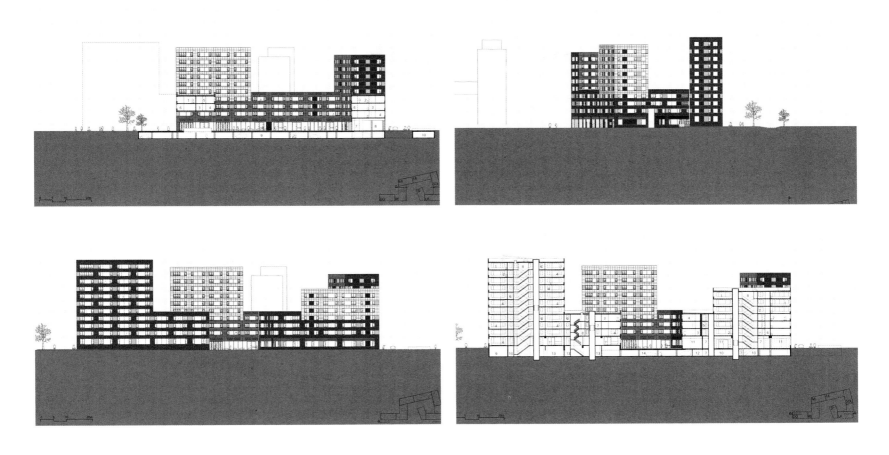

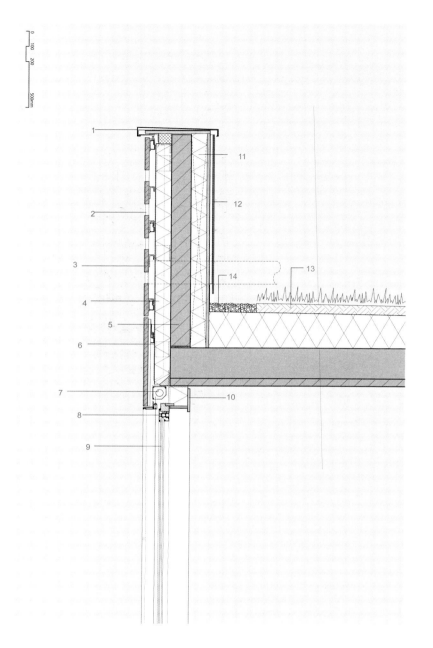
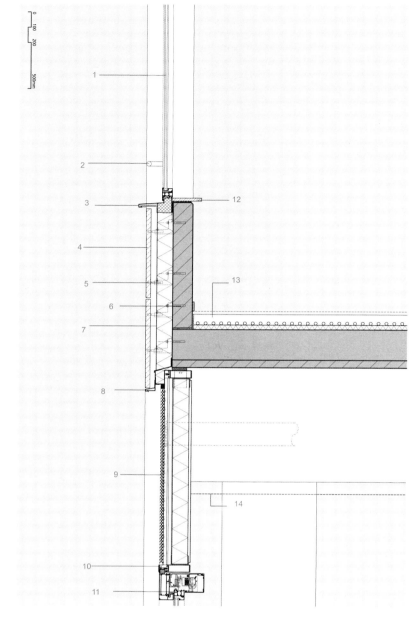

Detail 1
1. Aluminum cover
2. Slots in concrete composite I.V.M. underlying system
3. Concrete composite panels
4. Insulation, 100mm thick
5. Precast concrete wall panel
6. Waterproof layer
7. Blinds
8. Aluminum frame
9. Insulated glazing
10. Detachable valance
11. 100mm insulation
12. 8mm Trespa cladding
13. Sedum roofing
14. Gravel strip

节点 1
1. 铝顶盖
2. 混凝土复合 I.V.M. 下层系统插槽
3. 混凝土复合板
4. 隔热层，100mm 厚
5. 预制混凝土墙板
6. 防水层
7. 遮阳帘
8. 铝框
9. 隔热玻璃
10. 可拆卸挂帘
11. 100mm 隔热层
12. 8mm 千丝板包层
13. 景天属植物屋顶
14. 碎石带

Detail 2
1. Insulated glazing
2. Fall prevention
3. Aluminum water hammer
4. Basaltic lava panels
5. Insulation, 100mm thick
6. Precast concrete wall panel
7. Waterproof layer
8. Aluminium edge finish
9. Vent in wall system
10. Aluminium curtain wall
11. Aluminium sliding door system
12. Stone windowsill
13. Anhydrite with underfloor heating
14. Ceiling line

节点 2
1. 隔热玻璃
2. 防跌保护
3. 铝防水板
4. 玄武质熔岩板
5. 隔热层，100mm 厚
6. 预制混凝土墙板
7. 防水层
8. 铝边饰
9. 墙面系统通风口
10. 铝幕墙
11. 铝拉门系统
12. 石窗台
13. 硬石膏，地热供暖系统
14. 天花线

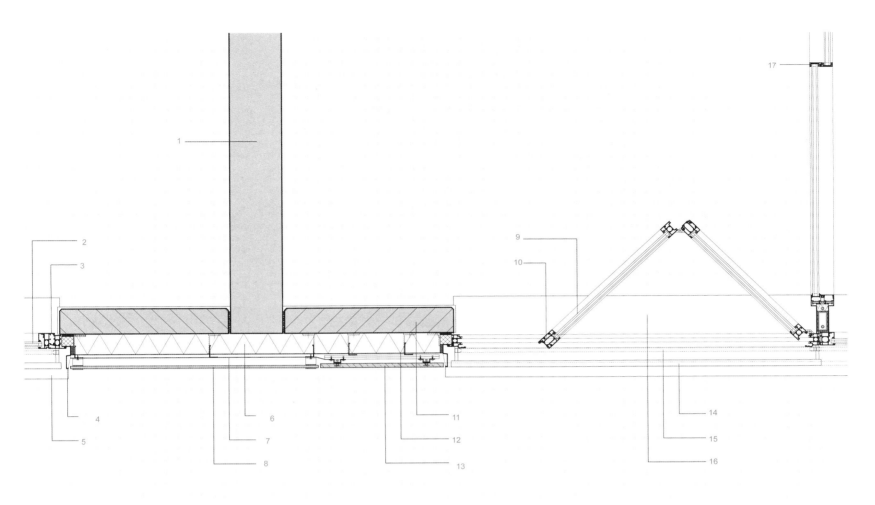

Detail 3
1. In situ concrete wall
2. Insulation glazing
3. Aluminium hinged window
4. Aluminium edge finish
5. Aluminum water hammer
6. Insulation, 100mm thick
7. Coloured foil
8. Patterned glass panels
9. Insulated glazing
10. Aluminum folding frame
11. Precast concrete wall panel
12. Waterproof layer
13. Concrete composite panels
14. Fall prevention
15. Blinds
16. Stone windowsill
17. Aluminum sliding door

节点 3
1. 现浇筑混凝土墙
2. 隔热玻璃
3. 铝铰链窗
4. 铝边饰
5. 铝防水板
6. 隔热层，100mm 厚
7. 彩色膜
8. 压花玻璃板
9. 隔热玻璃
10. 铝折叠框
11. 预制混凝土墙板
12. 防水层
13. 混凝土复合板
14. 防跌保护
15. 遮阳帘
16. 石窗台
17. 铝拉门

这种塔楼的组合与附近的办公楼相互结合，充分考虑了周边地区的视线和视野。项目中，两座塔楼正面都采用玻璃幕墙覆盖，给人以轻盈的感觉，减轻了建筑整体的厚重感。各种不同类型的住宅单元都能远眺马斯河的美景。

细部与材料
建筑覆面基本由琢石和夹心板构成，深浅色兼具。阳台向内凹进形成了游廊，可以被开发成日光浴室。南侧的半公共日光庭院一直延伸到后方的公园。俯瞰庭院的楼体看起来好像悬浮在 2.5 米高的玻璃底座（内设健康医疗及其他便民设施）上。户外空间的氛围十分人性化，让居民既有归隐之感，又能充分体验社交生活。

247

Police Station and Civil Protection Headquarters
警察局和民众防护总部

Location/ 地点：Lugo, Spain/ 西班牙，卢戈
Architect firm/ 建筑事务所：Mestura architects
Architects/ 建筑师：Humbert Costas Tordera, Manuel Gómez Triviño, Jaime Blanco Granado, Carlos Durán Bellas, Josep M. Estapé i de Roselló
Photos/ 摄影：Andrés Fraga
Gross area/ 总面积：5,293.83m²
Compleiton date/ 竣工时间：2012

Key materials: Façade – basalt stone
Structure – brick
主要材料：立面——玄武岩
结构——砖

Overview

The volumetry of this building is defined by a U-shaped structure around a courtyard. This courtyard articulates and prioritises different uses, providing lighting inside the main routes.

This implementation allows to differentiate the location of the offices of the local police and civil protection which requires a completely separate operation. This is possible thanks to a clear segmentation of the ground and first floors of circulations both indoors and volumetric configuration of the building, giving an image capable of conferring identity to each of the services and strengthening its presence in the two main façades shown from the access.

Detail and Materials
Why Choose Stone?
The general appearance is that of a compact building with controlled openings in the exterior walls and airy interiors in protected façades facing the courtyard and in the areas of access and citizen services. The position of the three vertical communication cores, it means generating a mass fragmentation pursued built to accommodate the scale of the building to the parcel.

The architects have tried the prefabrication and industrialisation in various systems of the construction process in order to facilitate the works on site, economise in times of placing and give the building a high level of technology incorporating new technologies in construction and environmental and seeking optimisation between quality and enforceability.

The building is conceived as a large mass of basaltic rock. The technology used for this stone

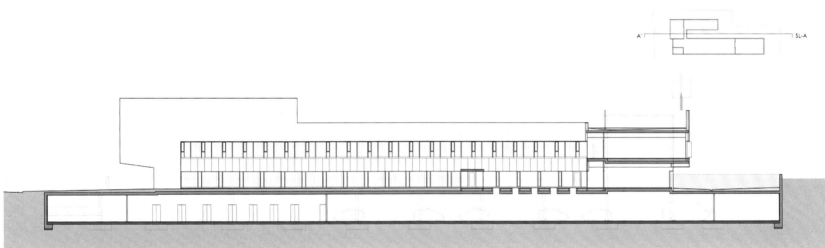

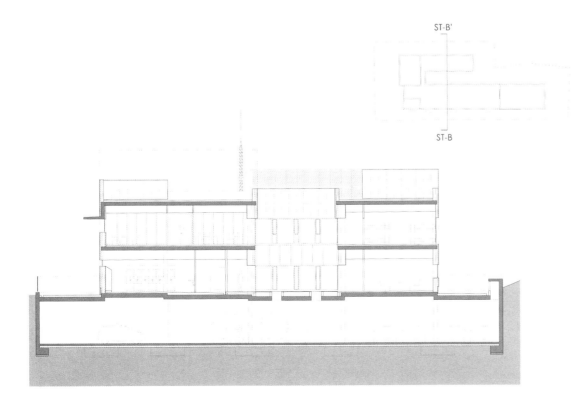

façade is a ventilated system by aluminium profiles hanging basalt pieces.

Features of the Constructional Material Products
- Pieces of large format (up to 2.00 x 0.60 metres) basalt stone, 3 centimetres thick.

- Hanging stone façade through a vertical main substructure of aluminium and aluminium auxiliary profiles in all horizontal joints.

Technology of This Material
The composition chosen for this façade, with large formats, requires a high degree of mechanical strength of the aluminum substructure, were carried load and wind tests resistance for sizing these profiles. Also the exterior ceilings are covered with the same type of stone and the same substructure system.

It is ventilated façades and roofs, there is a ventilated air space between the stone and the walls of the building and that the stone is always hung. This air chamber allows for better thermal performance of the building.

To prevent dampness, all the walls were coated with waterproof mortar before placing the thermal insulation.

项目概况

建筑呈 U 形结构，中间环绕着一个庭院。这个庭院连接并划分出不同的功能区，并且为建筑内的主通道提供了照明。

这种配置有助于区分本地警察办公室和民众防护办公室，方便二者的独立运行。一楼和二楼交通的清晰分割让不同的服务区各司其职，突出了两个建筑主立面的存在感。

细部与材料

为什么选择石材？
建筑的外观显得十分紧凑，外墙上仅有少量的门窗开口。但是朝向庭院的空间、出入空间以及民众服务空间都显得轻快通透。三个垂直交通核将建筑分散成不同的结构，使其与当地其他建筑的规模相匹配。

建筑师力求在施工过程中实现不同系统的预制作和工业化，从而减少装配施工时间，赋予建筑更多的高新技术含量，优化建筑品质与可执行性的比率。

建筑被设计成一块巨大的玄武岩。石材立面所采用的特殊技术体现在由铝型材悬挂玄武岩板所构成的通风系统。

建筑材料产品的特性
– 大尺寸玄武岩板（最大尺寸可达 2.00x0.60 米），3 厘米厚
– 悬挂式立面系统，由铝制垂直下层结构和铝制水平接头辅助型材支撑

材料的技术应用

构成立面的大尺寸玄武岩板要求铝制下层结构必须具有极高的机械强度，经过了承重和风力测试。同时，外部天花板上也采用了同种类型的石材和同样的支撑结构系统。

建筑采用通风立面和通风屋顶，在石材和建筑墙面之间设有通风空气腔。这层空气腔保证了建筑拥有更好的热性能。

为了防潮，所有墙面在安装隔热层之前都涂有防水灰浆。

Detail
1. Stone basalt brushed and flamed
2. Chamber air ventilated façade
3. Substructure aluminum set to brick
4. Thermal insulation Styrofoam extruded
5. Brick perforated wall
6. Indoor air chamber
7. Wall of double hollow brick
8. Galvanised steel lintel
9. Asphalt sheet glued to the wall
10. Top of galvanised steel lintel
11. Top of stone basalt

节点
1. 玄武岩，经过刷洗煅烧
2. 通风气腔立面
3. 固定在砖墙上的铝制支撑结构
4. 挤塑泡沫聚苯乙烯隔热层
5. 穿孔砖墙
6. 内部气腔
7. 双层空心砖墙
8. 镀锌钢过梁
9. 沥青板，黏合在墙面上
10. 镀锌钢过梁顶盖
11. 花岗岩顶盖

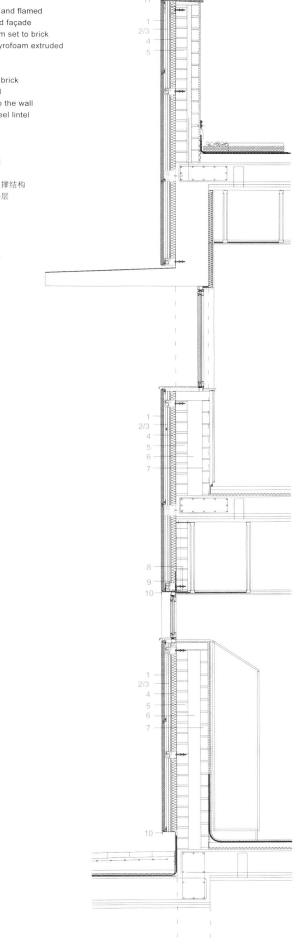

Social Housing for Mine-Workers
矿工社会福利住宅

Location/ 地点：Asturias, Spain/ 西班牙，阿斯图里亚斯
Architect/ 建筑师：Nacho Ruiz Allén &José Antonio Ruiz Esquiroz/ZON-E ARCHITECTS
Photos/ 摄影：Ignacio Martinez, Jose Antonio Ruiz
Gross internal floor area/ 室内总楼面面积：2,385m²
Total cost/ 总成本：1,212,784 €/1,212,784 欧元

Key materials: Façade – slate
Structure – brick
主要材料——立面：板岩
结构——砖

Façade material producer:
外墙立面材料生产商：
cufa

Overview
This project comes up from a tendering process to build state subsidised housing in Cerredo (Asturias), a mining town located in the very heart of the Cantabrian Mountains where no residential construction had been made for over 25 years.

The project has two stages that materialise in two perpendicular buildings forming an L. In the first stage the architects undertake the biggest building, which faces the road that crosses the town.

The project's nature as object is emphasised by the way the groud floor is approached: this has been set back along its perimeter, reinforcing the idea of a "floating body."

Detail and Materials
The volumetric proposed has an angular shape. It is a geometry crystallised from some elemmentary laws that are given by the town-planning regulations. The formal result is something halfway between a petrified object, a mountain's shape and a disturbing organism floating over the mountainside.

This "crystallographic" object has the same dark colour as the local slate. Like a piece of coal, it absorbs almost all the light it gets and reflects a small amount of it, calmly showing us its rich geometry.

The building's unity contrasts with the individuality of each of the 15 apartments that show through some galleries in the façade. These are cubes which drill the volume using a herringbone pattern and work as heat and light exchangers.

Each of the apartments is different, both in size and in its floor plan distribution, in the location of its gallery and in its roof's configuration. However, all of them enjoy cross ventilation and breathtaking views of Asturias' craggy landscape.

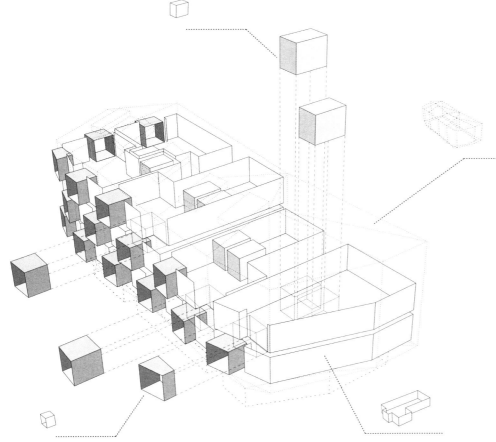

项目概况

项目是一个位于西班牙阿斯图里亚斯省喜拉多市的国家补贴住房项目。这是一座位于坎塔布连山脉中心地带的矿业城镇，已经25年都没有新建住宅项目了。

项目分为两个阶段完成，有两座相互垂直的建筑构成L形。在第一阶段，建筑师建造了朝向横跨城市的主干道的大型住宅楼。

项目的特色凸显在一楼空间的设计方式上：它从建筑外围向后撤，为上方的建筑结构提供了一种"飘浮感"。

细部与材料

建筑结构呈现为棱角分明的造型。这一造型的形成受到了一些城镇规划的基本法规的限制。最终的建筑造型融合了石化体、山体以及浮动在山腰的有机体的综合形象。这个"结晶体"呈现出与本地板岩相同的深色色彩。就像一块煤块，它能吸收大部分光并反射少量光，平静地展示出丰富的几何形态。

建筑的整体感与15套公寓的独立感形成了对比，并且在建筑立面的窗口上展现出来。这些窗口呈人字形纹理穿过建筑体块，作为热量和光线的交换器。

每套公寓在尺寸、平面布局、窗口的位置以及屋顶的构造上都不尽相同。然而，它们都享有交叉通风和阿斯图里亚斯独特的山色美景。

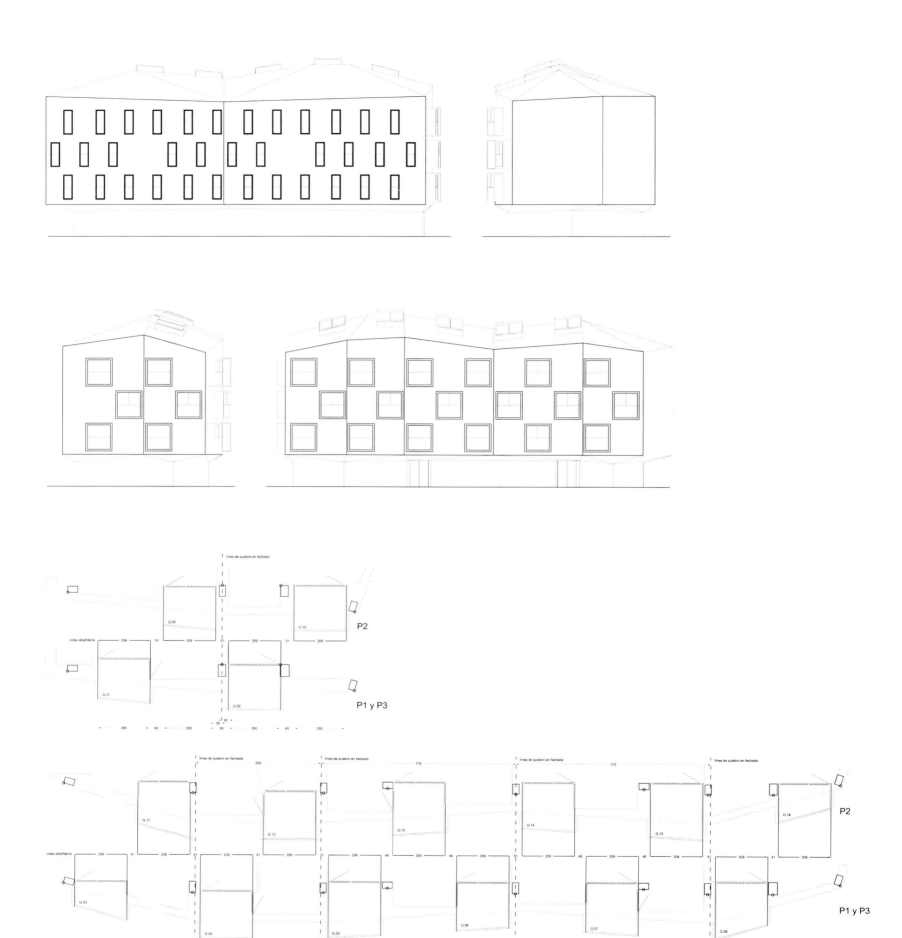

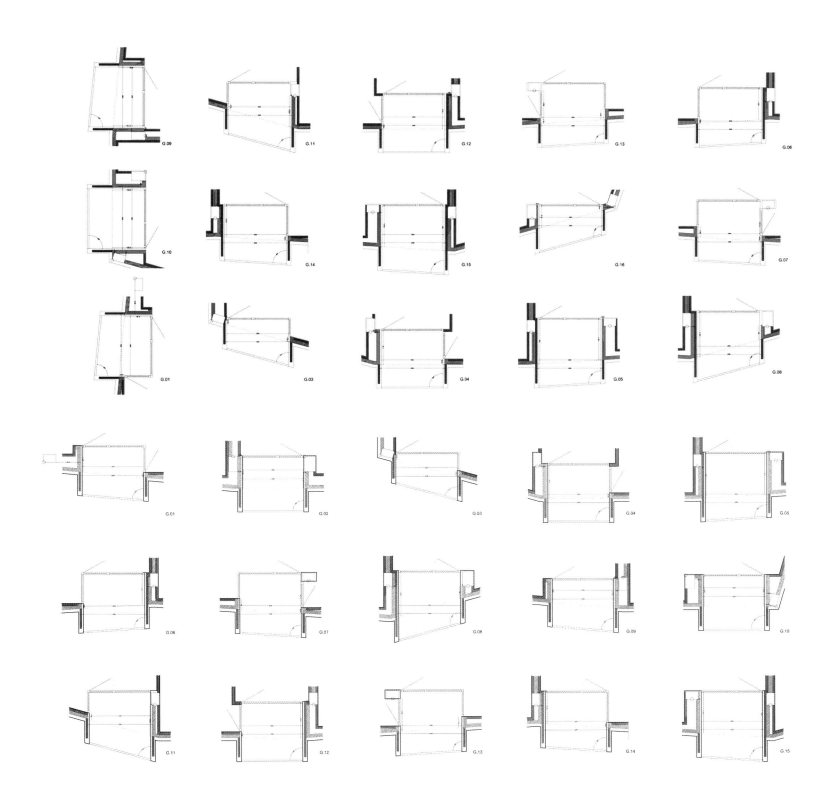

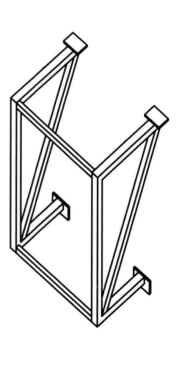

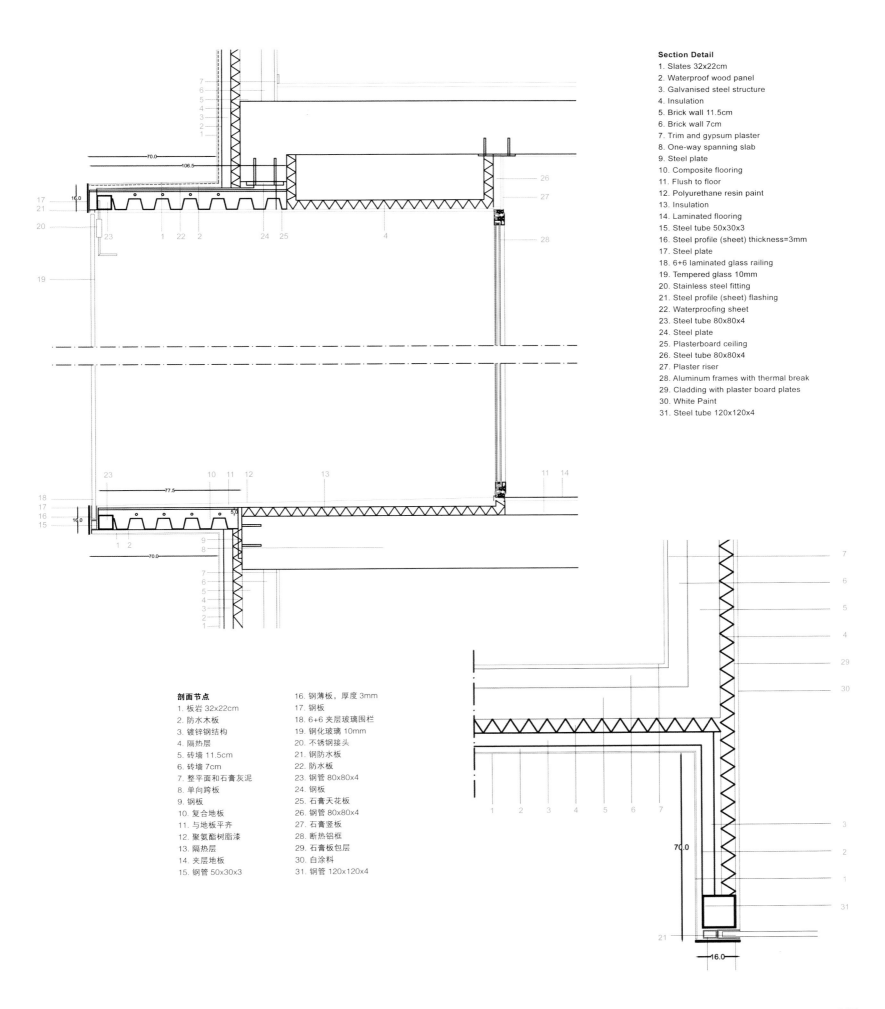

CULT
文化中心

Location/ 地点: Gournay-en-Bray, France/ 法国，古尔奈昂布莱
Architect/ 建筑师: L'ATELIER
Completion date/ 竣工时间: 2013
Cost/ 成本: 5,790,000 €/5,790,000 欧元

Key materials: Façade – slate, copper
主要材料: 立面——板岩，铜

Overview
In the town centre, close to the Place d'Armes, the town of Gournay-en-Bray has just completed a cultural centre which houses a multimedia library and cultural activities. This new centre is built on the site formerly occupied by the Damau factory.

The "Workshop" (a reference to the old Damau factory workshops) meets two major challenges:
- building a contemporary structure at the back of the plot by redefining the relationship with its natural environment (gardens, alleys, neighbouring stone walls) and surrounding structures (noble styles, traditional shelters, buildings with character);
- ensuring the new structure is readable in relation to existing buildings by framing the views from the main and secondary entrances.

Detail and Materials
Dark Purple Slate
A homogenous material and a natural stone has been suggested for the cladding of the archetypes: dark purple slate from Canada. This is implemented in a traditional manner by the local crafts and trades guild. Its use is often reserved to prestigious buildings, but here it is implemented in streamlined design, masking the gutters, with no overhang on the roof. The whole (façades and roof) creates strong visual continuity. The envelope of scales created by the slates and the reflections of the schist, make the dark purple hue more vegetable than mineral. Thus, the poetry of the gardens and courtyards of the old Damau factory is preserved.

The Copper Grid
The extensive glass façades are covered, in places, by a copper grid to dress the large volumes and transparent bays, creating intimacy or protection from the sun. This grid appears intermittently on the ground floor and extends over the entire height of the South truss to shield the large meeting room from outside

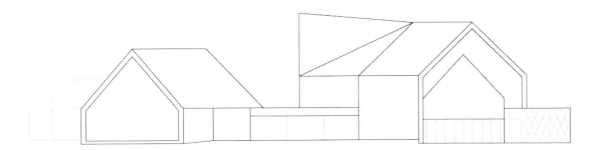

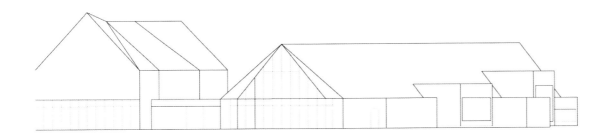

eyes. Furthermore, it enters into conversation with the sign-building at the entrance. The diamond shaped corners of the copper boxes lend them a hard edge which brings them to life. The external gates are made from solid copper.

Larch
The "protruding boxes" which house the reading rooms, the children's library and the storytelling corner are made from a larch cladding of varying thickness. The wide openings are fitted with external textile blinds to protect them from the sun.

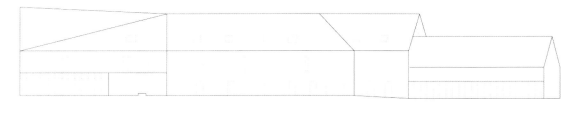

项目概况

古尔奈昂布莱镇在镇中心靠近兵器广场的位置上新建了一座文化中心，内设多媒体图书馆和文化活动空间。这个新中心所在场地的前身是达矛工厂。

"文化工坊"（命名参考了达矛工厂的工坊）的设计面临着两个主要挑战：
— 在场地后方建造一座现代化结构，重新定于与自然环境（花园、小巷、附近的石墙）及周边建筑结构（高端风格、传统小屋、特色建筑）的关系
— 保证新建结构与已有建筑之间的关系，在主入口和次入口处提供必要的视野条件

细部与材料

深紫色板岩

建筑的覆面设计选择了一种均质的天然石材：加拿大深紫色板岩。当地工匠用传统的方式对石材进行了安装。这种板岩通常用于宏伟的建筑，但是在这里它采用流线型设计，将沟槽掩藏起来，使屋顶没有任何突出的部分。石材的整体外观（立面和屋顶）营造出强烈的视觉延续性。板岩构成的鳞片式外壳以及片岩的反射让深紫色调更具植物感而不是矿物感，这样一来，达矛工厂花园和庭院的诗意就被传承了下来。

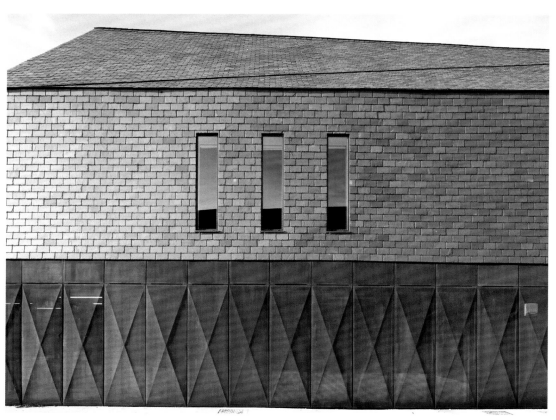

Façade Detail
1. Rigid insulation
2. Drainage strip
3. Ventilation grid of the air gap
4. Cleat insulating panel holder
5. Peripheral zinc bib
6. Lacquered aluminum frame in steel subframe
7. Exterior insulation rock wool
8. Airspace
9. Chevron 50x50 metal tabs on staggered
10. Sheathing
11. Purple slate cladding 400x250
12. Ventilation grid
13. Chaîneau recessed zinc
14. Wooden support for Chaineau
15. Roofing purple slate 400x250
16. Purlin HEA 120
17. Half-style dressing + false ceiling BA13
18. Skylight
19. Ridge in Lignolet
20. Blowing grills
21. Sheaths of VMC
22. Handrail scale
23. Pendant light, helix formed by fluorescent strip light
24. Alveolar slab
25. Tie beam
26. Solid foundation

立面节点
1. 刚性隔热层
2. 排水带
3. 空气层通风格栅
4. 楔形隔热板支架
5. 外围锌围边
6. 涂漆铝框，钢架支撑
7. 外部隔热石棉
8. 空气层
9. Chevron 50x50 金属连接，错列式
10. 包板
11. 紫色板岩包层 400x250
12. 通风格栅
13. Chaîneau 嵌入式锌板
14. Chaineau 木支架
15. 屋顶紫色板岩 400x250
16. 桁条 HEA 120
17. 半装式假吊顶 BA13
18. 天窗
19. Lignolet 屋脊
20. 鼓风格栅
21. VMC 包板
22. 扶手
23. 吊灯，荧光灯构成
24. 齿槽板
25. 系梁
26. 实心地基

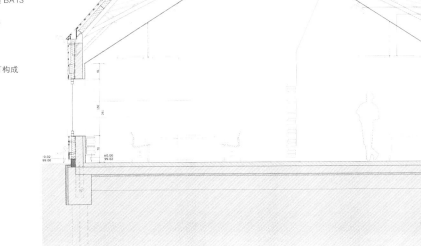

铜格栅

大面积的玻璃幕墙上覆盖着一层铜格栅，实现了隐私保护和遮阳的作用。这层格栅在一楼空间断断续续，向上则一直延伸到南侧桁架的顶部，将整个会议室都包裹起来。此外，格栅还与入口处的标志楼结合起来。铜网的菱形开口让他们看起来坚毅而充满活力。文化中心的大门采用实心铜板制成。

落叶松木

阅览室、儿童图书馆以及故事角所在的突出结构外覆盖着不同厚度的落叶松木板。宽敞的窗口配置着外部遮阳帘来实现遮阳。

Health Centre
健康中心

Location/ 地点: Monterroso, Lugo, Spain/ 西班牙，卢戈，蒙特罗索
Architect firm/ 建筑公司: abalo alonso arquitectos
Architects/ 建筑师: Elizabeth Abalo, Gonzalo Alonso
Photos/ 摄影: Santos Díez / bisimages
Completion date/ 竣工时间: 2013

Key materials: Façade: slate, galvanised steel
主要材料：立面——板岩、镀锌钢

Overview

Monterroso is a small village in the interior of the province of Lugo, famous for its annual Expo, more than five hundred years old. As often happens in these cases, the service of the health centre covers one larger area than the own core of the City Council, and with a significantly ageing population. It acquires, therefore, significant importance the Bus Station, located in the extreme northwest and perhaps not so much possible relationship with the nearby municipal headquarters.

The property occupies the whole of the plot. Both the existing slope, and the location of the station suggest access, level, by the southwest corner, so that the building collapses slightly and favours not only the sunlight of upper Street, but read as "tapia", more chord in a matter of scale with the rural environment in which we find ourselves. This access floor includes, towards the South, paediatrics, north ward, consultations and reception with ancillary services back.

Two volumes emerge from the compact construction: the private medical area, with a gap above the main entrance, and a skylight in the West elevation. Four courtyards moderate contact with the outside. In the plant basement are areas of women, physiotherapy, dentistry and facilities, lighted and ventilated by these patios to favour the intimacy of views.

Detail and Materials

A continuous slate skin covers façades and roofs, with steel galvanised in access, patios and major gaps, strategic breaks and contrasts with the warm interior lined in oak.

The architects have explained why they chose it to cover the façade as follows: "We choose this façade material because it is a local material. There are several types of slate in this area of Galicia, Spain. We like to use local material in our designs to integrate them. We used a habitual technology as you can see in the detailWe had used similar techniques before with another type of stone."

项目概况

蒙特罗索是位于西班牙卢戈省中部的一个小城镇,以拥有 500 多年历史的年度博览会而著称。与许多同类的城镇相同,健康中心的服务范围覆盖了市议会核心区及其外围的大面积地区,并且面临着大幅增长的老龄化人口。因此,健康中心与汽车站紧密联系起来,位于城镇的西北面,与附近的市政中心并没有过多的联系。

项目占据了整个地块。缓坡和汽车站的位置都突出了西南角的入口通道,使得建筑缓缓地下行,不仅享有上面街道的温暖阳光,还在乡村环境中形成了独特的和弦,让我们找回自己。入口层朝南,设置着儿科、北侧病房、咨询诊室和接待处,后部设有辅助服务设施。紧凑的建筑结构由两个空间构成:私人医疗区和西立面的天窗。建筑外面的四个庭院温和地联系起来。地下室内设有妇科、物理治疗区、牙科等设施,通过天井实现了采光和通风,同时也拥有良好的私密性。

细部与材料

建筑的立面和屋顶被连续的板岩表皮所覆盖,其中通道、天井和主要开口处由镀锌钢板所取代。它们与以橡木为主的室内装饰形成了鲜明的对比。

建筑师是这样阐述他们对立面材料的选择的:"我们选择这种立面材料的原因是因为它是本土材料。西班牙加利西亚地区盛产若干种不同的板岩。我们喜欢在设计中使用本土材料,以便使其融入环境。我们有一套惯用的技术系统,这点可以在细部图中看到。我们也曾将这种技术应用在其他类型的石材中。"

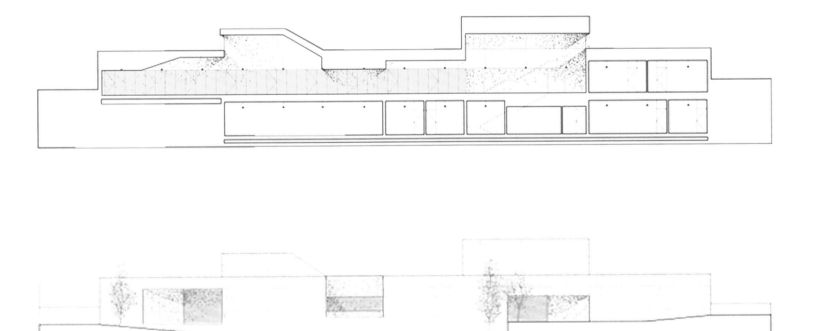

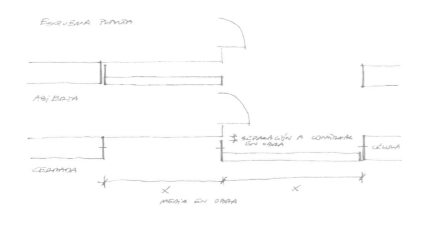
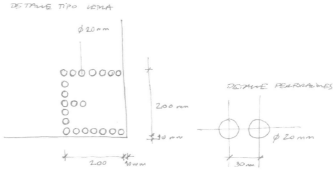
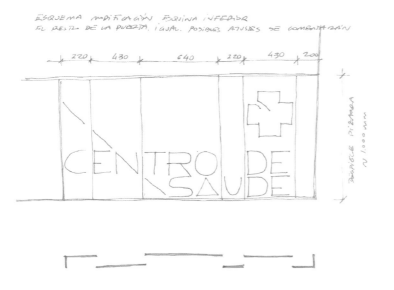
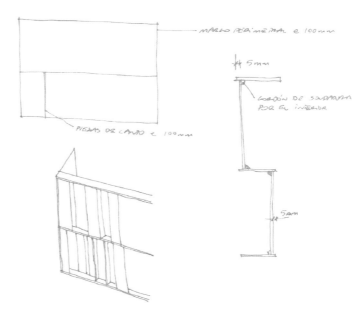

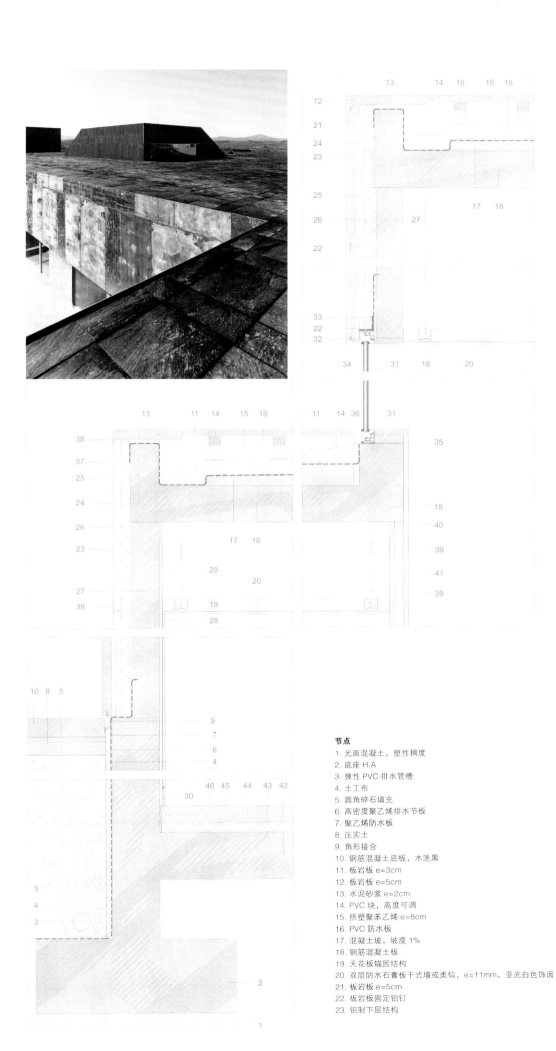

Detail

1. Concrete cleaning. Plastic consistency
2. Brake shoe H. A
3. Flexible PVC drainage pipe grooving
4. Geotextile sheet
5. Gravel fill rounded
6. Sheet draining nodules of high density polyethylene
7. Waterproof polyethylene sheet
8. Compacted earth
9. Angular attachment
10. Reinforced concrete floor, washing black
11. Slab of slate e=3cm
12. Slab of slate e=5cm
13. Cement mortar e=2cm
14. Plots of PVC adjustable in height
15. Extruded polystyrene e=8cm
16. PVC waterproofing sheet
17. Formation of concrete slope. Slope of 1%
18. Reinforced concrete slab
19. Anchoring structure of ceiling
20. Double plate type waterproof plasterboard drywall or similar e=11mm, matt white finish
21. Slab of slate e=5cm
22. Pin for fastening aluminum slabs of slate
23. Aluminum substructure
24. Air chamber e=5cm
25. E=5cm taken with extruded polystyrene plastic rosettes
26. Plastering mortar waterproof e=1cm
27. Fireproof block size 30*19*14cm
28. Plasterboard panelling semi-direct type e=13mm
29. Profile of galvanised steel for fastening the backing semi-direct plasterboard
30. Baseboard of DMH to paint e=10mm. Height=100mm
31. DMH e=30mm
32. For fixed glass metal structural stadip 5+5
33. Angular anchor profile L 80.8 for carpentry
34. Goteron chamber in stainless steel sheet folded e=0.5mm
35. Stud
36. Flashing the carpentry
37. Smooth metallic plate e=5mm
38. Galvanised steel pin for the fixing of steel metal
39. Double plate type plasterboard e=13mm per plate
40. Substructure: profile plasterboard sheet steel galvanised cold rolled. e=0.6mm. Total thickness of 76mm, anchored to concrete slab with steel self drilling screws and studs at 600mm
41. Wool insulation mechanically fastened touch
42. Health fan type forged caviti c 15+5
43. Rigid extruded polystyrene insulation e=50mm
44. Lightweight concrete mortar
45. Leveling mortar e=20mm
46. Linoleum e=2.5mm

节点

1. 光面混凝土，塑性稠度
2. 底座 H.A
3. 弹性 PVC 排水管槽
4. 土工布
5. 圆角碎石填充
6. 高密度聚乙烯排水节板
7. 聚乙烯防水板
8. 压实土
9. 角形接合
10. 钢筋混凝土底板，水洗黑
11. 板岩板 e=3cm
12. 板岩板 e=5cm
13. 水泥砂浆 e=2cm
14. PVC 块，高度可调
15. 挤塑聚苯乙烯 e=8cm
16. PVC 防水板
17. 混凝土坡，坡度 1%
18. 钢筋混凝土板
19. 天花板锚固结构
20. 双层防水石膏板干式墙或类似，e=11mm，亚光白色饰面
21. 板岩板 e=5cm
22. 板岩板固定铝钉
23. 铝制下层结构
24. 空气腔 e=5cm
25. e=5cm，挤塑聚苯乙烯塑形花状装饰
26. 石膏灰泥防水 e=1cm
27. 防火块 30×19×14cm
28. 石膏板镶板，半直接型 e=13mm
29. 镀锌钢型材，用于固定半直接石膏板背面
30. DMH 底板 e=10mm，高度 100mm
31. DMH e30mm
32. 用于固定玻璃金属结构 stadip 5+5
33. 角形锚固型材 L90.8
34. Goteron 折叠不锈钢板盒 e=0.5mm
35. 螺柱
36. 窗框防水板
37. 光面金属板 e=5mm
38. 镀锌钢钉，固定金属钢
39. 双层石膏板，每块 e=13mm
40. 下层结构：石膏板，冷轧镀锌钢，e=0.6mm，总厚度 76mm，固定在混凝土板上，配有 600mm 钻尾自攻螺丝和嵌钉
41. 石棉隔热层，机械固定
42. 健康扇，forged caviti c 15+5 型
43. 刚性挤塑聚苯乙烯隔热层 e=50mm
44. 轻质混凝土砂浆
45. 找平砂浆层 e=20mm
46. 油毡 e=2.5mm

Churchyard Offices and Staff Housing in Gufunes Cemetery
古芬斯墓园办公楼和员工宿舍

Location/ 地点：Reykjavik, Iceland/ 冰岛，雷克雅未克
Architect/ 建筑师：ARKIBULLAN architects
Photos/ 摄影：Auja - Audur Thorhallsdottir
Site area/ 占地面积：30,300m²
Built area/ 建筑面积：808m²

Key materials: Façade – basaltic stone
Structure – concrete
主要材料： 立面——玄武岩
结构——混凝土

Overview
The building for the Churchyard Offices and Staff Housing in Gufunes Cemetery is the first building of several in Reykjavik's main cemetery complex. The phases yet come to include a church, a chapel and a crematorium.

This two-storey building is situated on a basaltic hill south of the cemetery grounds. The project approach is based on the site's premises, working in different scales, looking at the space between the stones with the aim of translating it to architectural spaces. The upper floor includes a reception and offices, while the lower floor contains technical and working areas.

Detail and Materials
The building defines the entrance to the whole area of the Gufunes Cemetary, standing as the eastern wall of the structures that are yet to come. The structure constitutes of two masses. White concrete walls in contrast to walls made of local basaltic stone sets its mark on the building's appearances.

项目概况
古芬斯墓园办公楼和员工宿舍是雷克雅未克近年来的首座墓园综合建筑。未来，墓园还将新建一座教堂、一座礼拜堂和一间火葬场。

这座两层高的建筑位于墓园南侧的玄武岩山丘之上。项目的设计以场地地形为基础，综合改革方面，将岩石之间的空间转变为建筑空间。建筑上层是接待厅和办公室，下层则是技术和工作区。

细部与材料
建筑标志着整个古芬斯墓园区域的入口，代表着整个墓园的东墙。建筑结构由两部分组成。白色混凝土墙壁与本地玄武岩建成的墙壁形成了鲜明对比，为建筑带来了醒目的外观。

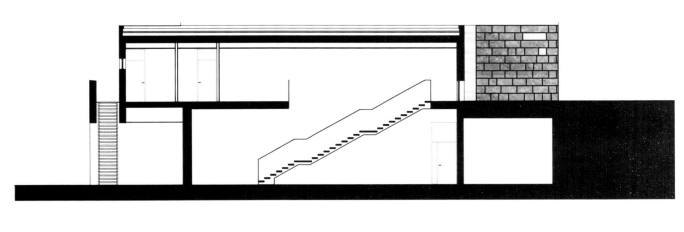
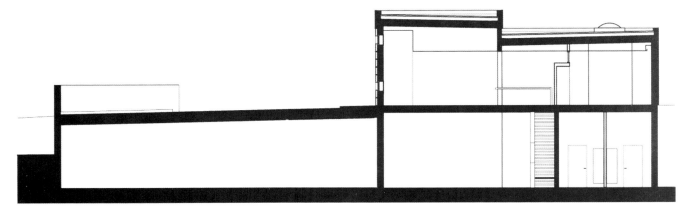

267

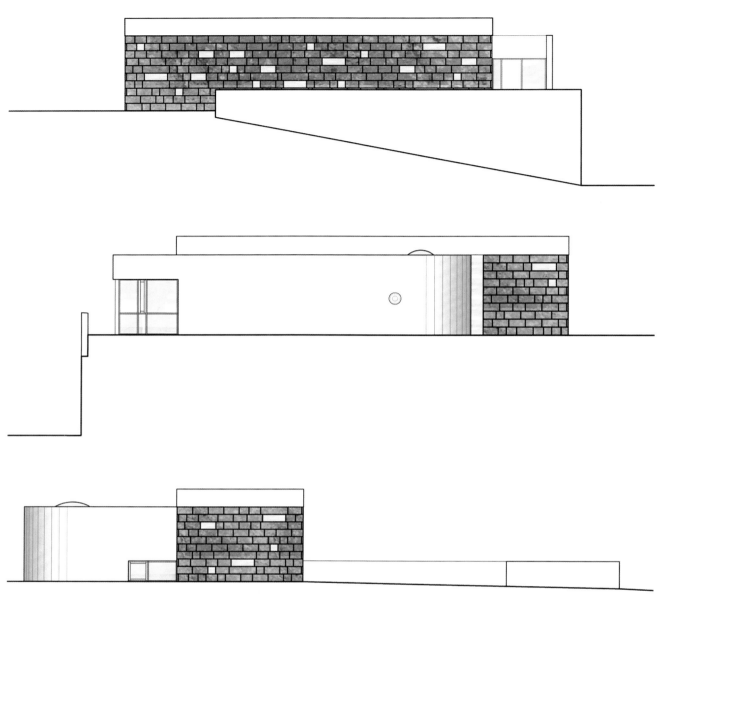
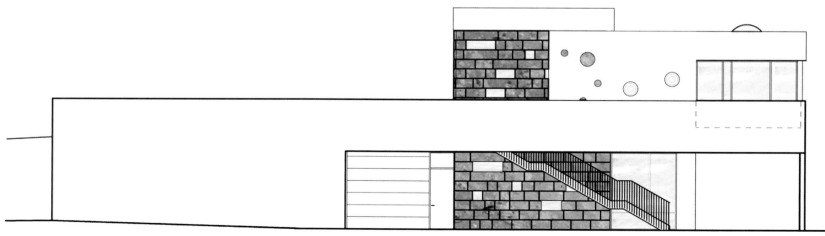

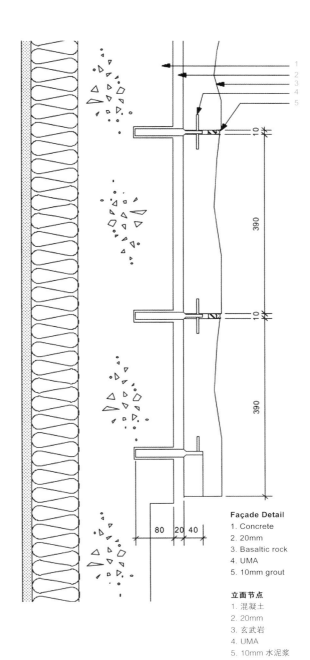

Façade Detail
1. Concrete
2. 20mm
3. Basaltic rock
4. UMA
5. 10mm grout

立面节点
1. 混凝土
2. 20mm
3. 玄武岩
4. UMA
5. 10mm 水泥浆

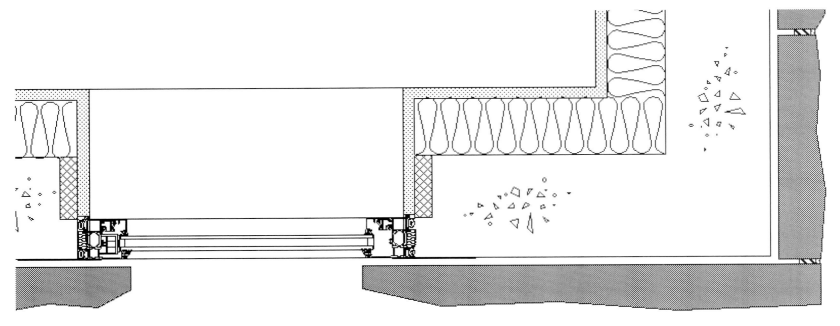

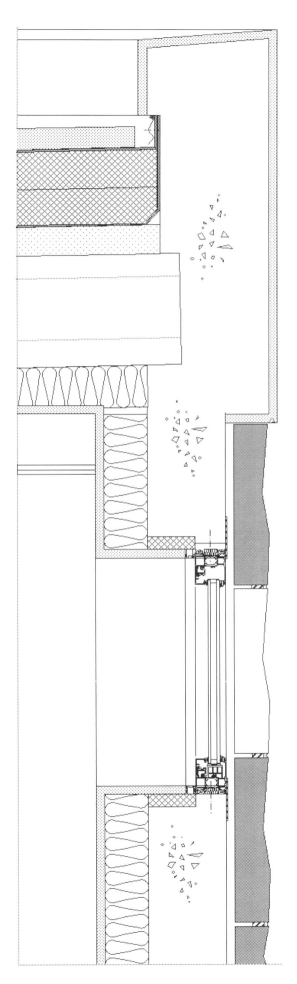

School Centre Paredes
帕雷德斯学校中心

Location/ 地点：Alenquer, Portugal/ 葡萄牙，阿连卡
Architect/ 建筑师：André Espinho / André Espinho – Arquitectura
Photos/ 摄影：FG + SG Fernando Guerra
Area/ 面积：6,700m²

Key materials: Façade – slate, marble stone
Structure – concrete
主要材料：立面——板岩、大理石
结构——混凝土

Overview
The School Centre has been designed to accommodate around 600 children, from the ages of three to nine. Within the school, the children are divided by the 1st cycle, kindergarten and ATL (Leisure Activities). The first floor consists of administration areas, service and a reception for parents whilst the ground floor contains a gymnasium and the majority of the school classrooms, with a direct link to the playgrounds (both covered and uncovered).

School Centre Paredes is composed of a white volume resting on four black volumes, thus marking the separation between floors. The majority of the project works around the creation of three patios/playgrounds and the relationship of the building with the slope of the existing ground. Contact with the outside was key to this project, with the organisation and shape of the interior space allowing all circulations to enjoy a large amount of natural light. By including a number of covered outdoor spaces, the building now provides excellent leisure facilities for children in all seasons. Several wall paintings in the playgrounds and atriums were carried out by artists invited by the designer in an attempt to enrich the interior space.

Detail and Materials
The building structure is in concrete and the exterior walls are made with the ventilated façade system. The black panels are slate stone (from Portugal) and white panels are "Estremoz" marble stone (from Portugal). The rest of the exterior walls are covered with white cement covered with micro marble stones. The roof is covered with ventilated white thermal plates made of cement and Roof Mate. In the interiors, the designers applied hydraulic mosaic floor in the corridors and vinyl flooring in classrooms and school offices. The ceilings are in plasterboard with acoustic correction. The wall paintings are done in vinyl from "vescom". The school is equipped with solar panels.

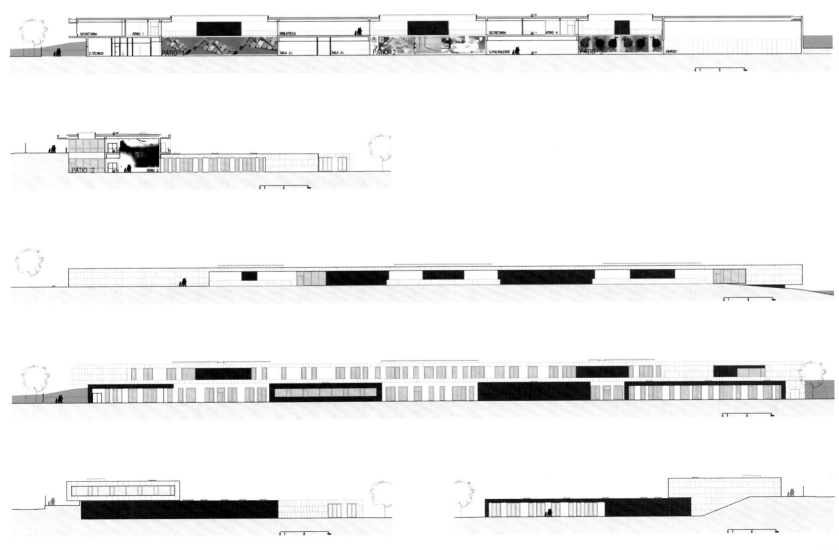

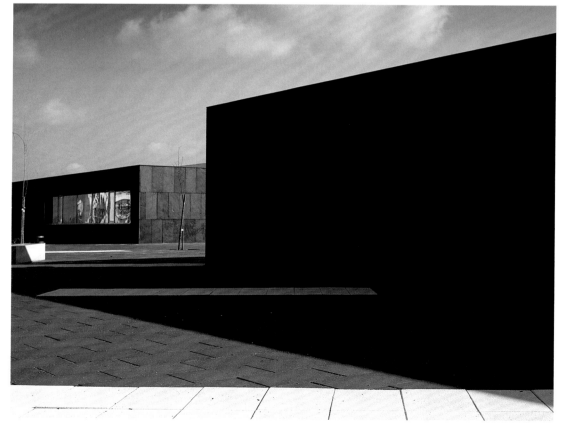

项目概况

学校中心将容纳 600 名三至九岁的学生。在学校里，学生将通过第一阶段课程、幼儿园和休闲活动分开。学校二楼是行政区、服务区和家长接待室；一楼是体育馆和学校的大部分教室，与操场（分为塑胶操场和普通操场两种）直接相连。

帕雷德斯学校中心由一个白色结构坐落在四个黑色结构上组成，楼层之间的区分十分明显。项目的大部分结构都围绕着三个天井/操场以及建筑与地面坡势之间的关系展开。与外界的联系对项目来说十分重要，室内空间的组织和造型让所有内部交通流线都能享受大量的自然采光。一系列带顶户外空间让学生一年四季都能享受优质的休闲设施。设计师邀请艺术家在操场和中庭墙上绘制了壁画，进一步丰富了室内空间。

细部与材料

建筑采用混凝土结构，外墙配有通风立面系统。黑色板材是板岩（葡萄牙本国产），白色板材是艾斯特雷莫斯大理石（葡萄牙本国产）。外墙的其他部分由白水泥覆盖，上面布满了大理石微粒。屋顶上覆盖着由水泥和防水涂料制成通风白色隔热板。在室内设计中，走廊应用了液压马赛克地面，教室和办公室则应用了乙烯基地面。天花板采用具有音效处理功能的石膏板。壁画由乙烯基涂料 vescom 绘制而成。学校还配有太阳能电池板。

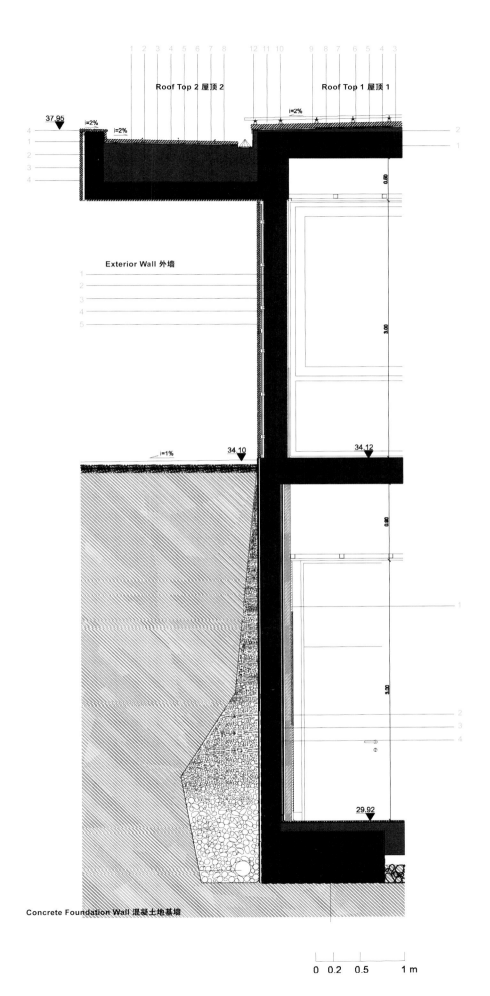

Roof Top 1
1. Concrete slab
2. Form layer of concrete lightened with expanded clay aggregates
3. Primary bitumen emulsion Imperkote F
4. Polyster 40
5. Polyster 40T
6. Aluminium insulation Boltherm 4mm
7. Polyster impersep 250
8. Concrete with water repellent additives armed with steel 50mm
9. Primary bitumen emulsion imperkote F
10. PVC support for concrete slabs
11. Air space circulation 0.5cm
12. Concrete white slabs 600x400x50mm

Roof Top 2
1. Concrete slab
2. Form layer of concrete lightened with expanded clay aggregates
3. Primary bitumen emulsion Imperkote F
4. Polyster 40
5. Polyster 40T
6. Aluminium insulation Boltherm 4mm
7. Polyster impersep 250
8. Concrete white slabs 600x400x50mm

Exterior Wall
1. Designed plaster and painting 20mm
2. Brick wall 0.25cm
3. Insulation Boltherm 4mm
4. Air space circulation 0.3cm
5. Black slate stone cleaved 0.3 cm fixed with halfen system

Concrete Foundation Wall
1. Concrete wall
2. Primary bitumen emulsion imperkote F
3. Air space circulation 0.3 cm
4. White marble stone "estremoz" 0.3 cm fixed with halfen system

屋顶 1
1. 混凝土板
2. 混凝土层，以膨胀黏土骨料减重
3. 初级沥青乳剂 Imperkote F
4. 聚酯纤维 40
5. 聚酯纤维 40T
6. 铝隔层 Boltherm 4mm
7. 聚酯纤维 250
8. 钢筋混凝土，防水处理，50mm
9. 初级沥青乳剂 Imperkote F
10. 混凝土板的 PVC 支撑
11. 空气流通层 0.5cm
12. 混凝土白板 600x400x50mm

屋顶 2
1. 混凝土板
2. 混凝土层，以膨胀黏土骨料减重
3. 初级沥青乳剂 Imperkote F
4. 聚酯纤维 40
5. 聚酯纤维 40T
6. 铝隔层 Boltherm 4mm
7. 聚酯纤维 250
8. 钢筋混凝土，防水处理，50mm

外墙
1. 石膏和涂料 20mm
2. 砖墙 0.25cm
3. Boltherm 隔层 4mm
4. 空气流通层 0.3cm
5. 黑色石板 0.3cm，固定在哈芬系统上

混凝土地基墙
1. 混凝土墙
2. 初级沥青乳剂 Imperkote F
3. 空气流通层 0.5cm
4. Estremoz 白色大理石板 0.3cm，固定在哈芬系统上

"La Grajera" Institutional Winery
格拉赫拉酒庄

Location/ 地点：La Rioja, Spain/ 西班牙，拉里奥哈
Architect/ 建筑师：VIRAI ARQUITECTOS
Completion date/ 竣工时间：2011

Key materials: Façade – ceramic, sandstone (ARENISCAS DE LOS PINARES http://www.areniscas.com)
Structure – prefabricated concrete
主要材料：立面——瓷砖、砂岩（ARENISCAS DE LOS PINARES http://www.areniscas.com）

Façade material producer:
外墙立面材料生产商：
Sandstone- ARENISCAS DE LOS PINARES http://www.areniscas.com
（砂岩）

Overview

The project seeks a balance between its representative character and the desire to merge with the landscape. Volumes follow the forms of the land and are interrupted, as they approach the forest, remaining close to its border and thus respecting the existing vegetation. The aim is to surprise.

A large sandstone basement located partially below ground, which houses the production area of the winery, bends and rises becoming an element of the landscape. The glass surfaces of the southern side of the institutional pavilion are protected by ceramic shutters and porches which shield against the sun in summer, and let it go through in winter. The massiveness of the sandstone basement helps to increase the thermal inertia of the building. The orientation and section of the winery allows natural ventilation of the building reducing the need of mechanical ventilation for the wine making process.

The existence of two entrances at different levels allows the bodega to be underground, taking advantage of the natural slope of the land and helping the movement of the must and the wine using gravity, according to Riojan tradition, reducing the need for pumps. The winery uses 100% geothermal heat pumps for the air conditioning and for the winemaking process.

Detail and Materials

The use of the natural stone is of great importance in the project. In the exterior, the local sandstone basement helps to merge the winery with the landscape. The stonework used is like those from the old and modest retaining and countryside walls from the small vineyards in the Riojan landscape. In the interior, the use of black honed slate stones in walls and floorings helps to create a special itinerary and an atmosphere.

Both stones come from local quarries. The sandstone comes from a local quarry 100 km away

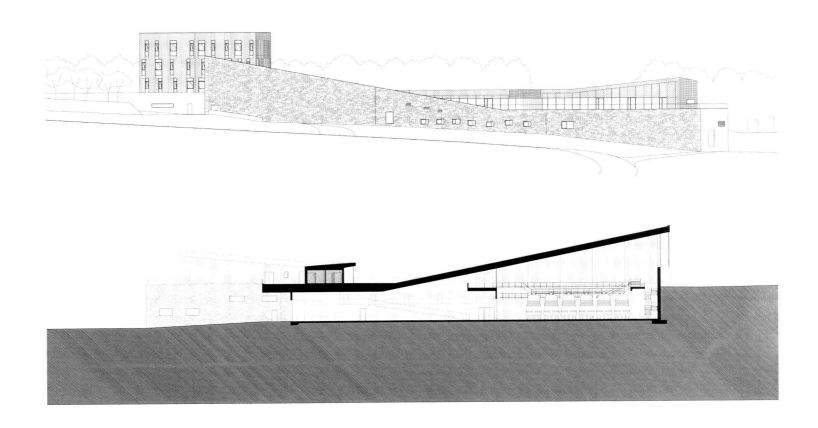

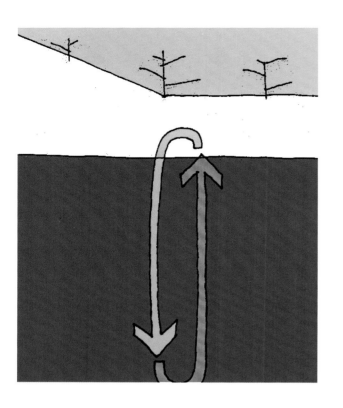
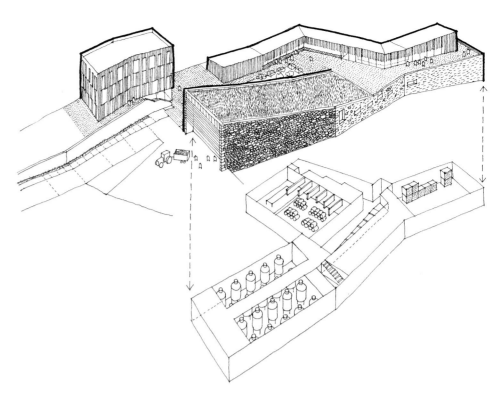
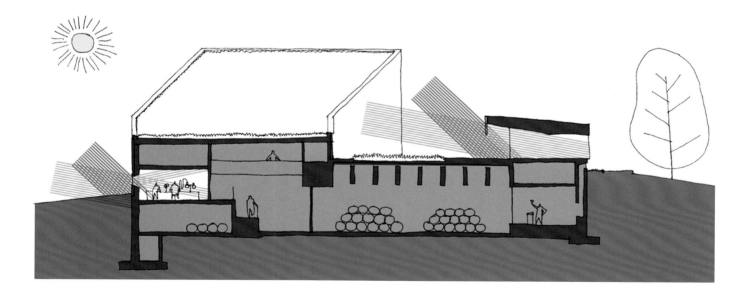

from the building site (www.areniscas.com). The slate comes also from a spanish quarry, Bernardos in Segovia (www.naturpiedra.com).

A ceramic cladding element has been specifically designed for the façades of the event pavilion, with jutting parts that keep the same rhythm and dimension of shutters, allowing to create visual uniformity between the opaque areas of the façade and the transparent ones with shutters.

项目概况

项目试图在自身代表性特征与景观的融合性之间找到一个平衡点。建筑随着地势起伏，并且在靠近森林的地方被打断，充分保护了原有的植被，形成了令人惊喜的效果。巨型砂岩底座的一部分嵌入地下，内设酒庄的生产区。它随着地势起伏，形成了景观元素的一部分。建筑南侧的玻璃墙面被陶瓷遮阳板和门廊保护起来，起到了冬暖夏凉的恒温效果。

砂岩底座的体块有助于增加建筑的热惯性。酒庄的朝向和截面保证了建筑的自然通风，减少了在酿酒过程中机械通风的需求。

将入口设在不同的楼层保证了酒窖在地下，可以利用自然坡势和重力来实现酿酒过程中必需品和酒品的运送，从而减少对电动泵的需求。酒庄采用100%地源热泵进行空气调节和酿酒。

细部与材料

天然石材的运用在项目中起到了重要的作用。外部的本

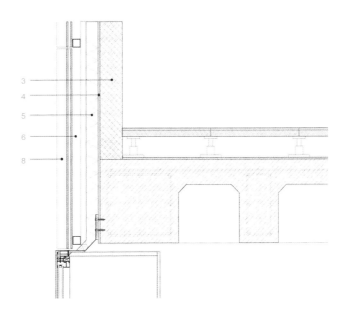
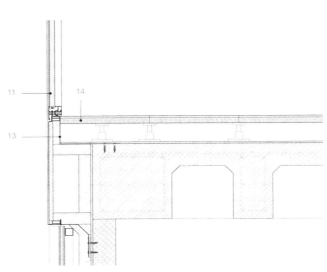

Detail Façade
1. Setting coat of plaster
2. Hollow brick wall, t=7cm
3. Solid brick wall, t=11cm
4. Cement render
5. Thermal insulation of mineral wool
6. Ventilated air chamber
7. Aluminum profile
8. Special extruded ceramic piece designed by Faveton and Virai arquitectos
9. Lacquered MDF
10. Galvanised steel folded sheet preframe
11. Anodised aluminum window frame with thermal break (4+4/10/5)
12. Anodised aluminum angle trim
13. Folded sheet
14. Raised floor

立面节点
1. 石膏抹灰层
2. 中空砖墙 t=7cm
3. 实心砖墙 t=11cm
4. 水泥抹面
5. 隔热矿物棉
6. 通风气腔
7. 铝型材
8. 特制挤制瓷砖，Faveton and Virai arquitectos 设计
9. 涂漆中密度纤维板
10. 镀锌钢折叠板预装配
11. 断热阳极氧化铝窗框（4+4/10/5）
12. 阳极氧化铝角饰
13. 折叠板
14. 活动地板

地砂岩底座有助于酒庄融入景观。项目的石砌结构类似于里奥哈地区传统葡萄园古老而低调的挡土墙。内部墙面和地面上的黑色亚光板岩帮助酒庄塑造了一段特殊的旅程和奇妙的氛围。

两种石材均来自本土采石场。砂岩来自于距离建筑场地仅有 100 千米距离的本地采石场（www.areniscas.com）。板岩同样来自于一家西班牙采石场，塞戈维亚的 Bernardos（www.naturpiedra.com）。

陶瓷覆面元件是特别为酒庄的活动馆所设计的，突出的瓷砖与遮阳百叶保持了相同的韵律和尺寸，在立面上透明和不透明的区域之间形成了一种视觉统一感。

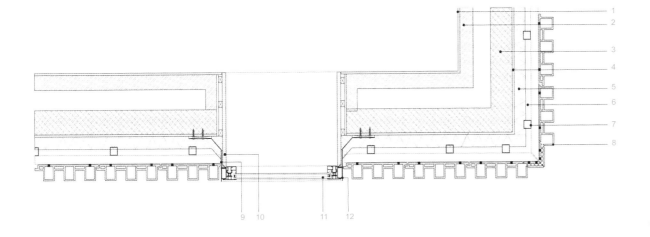

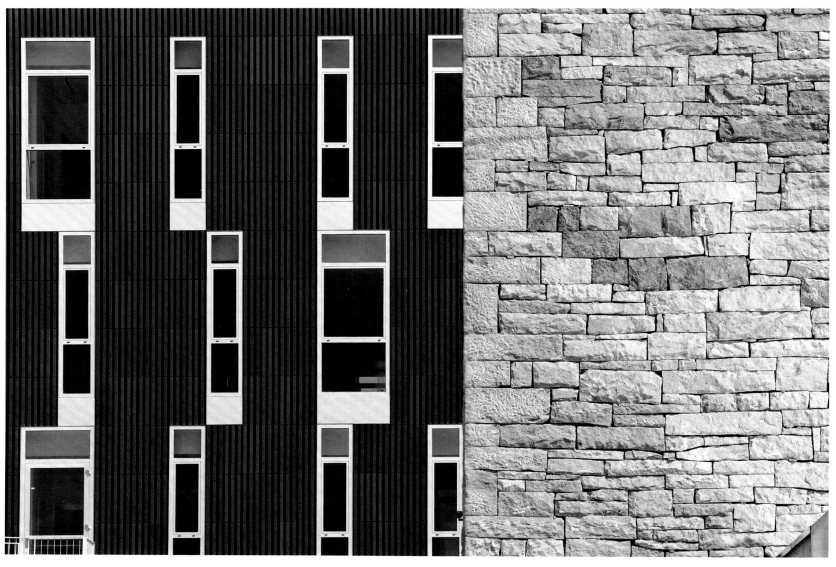

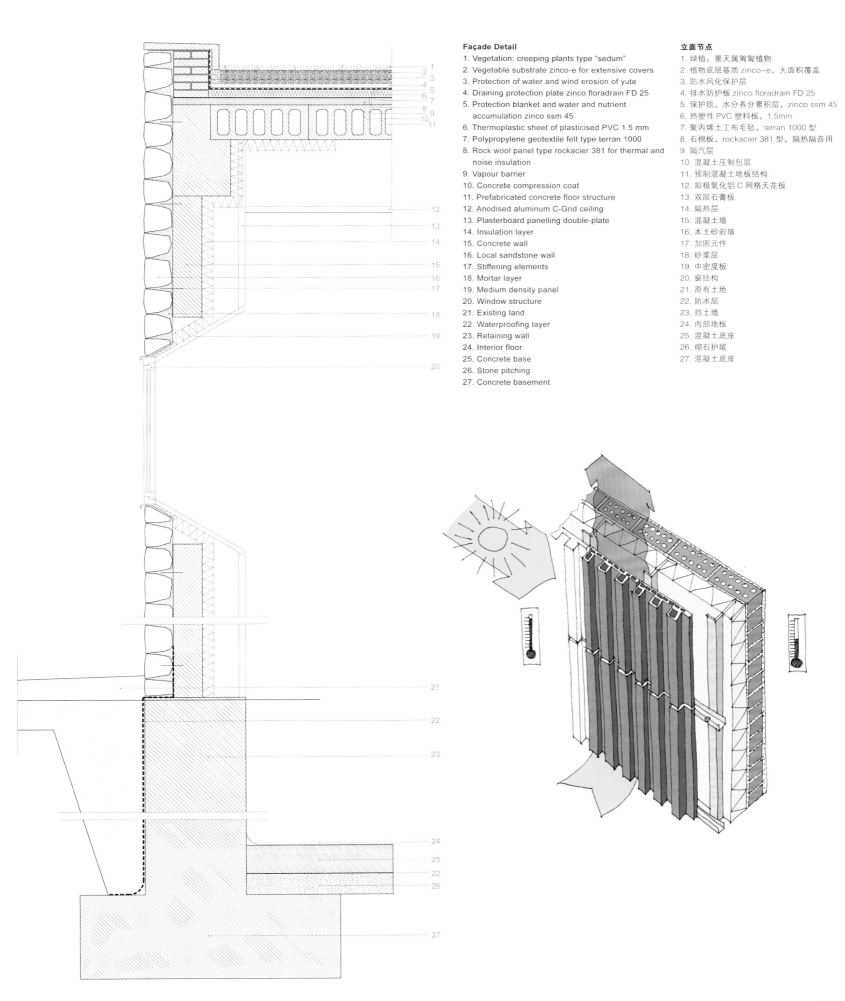

Façade Detail

1. Vegetation: creeping plants type "sedum"
2. Vegetable substrate zinco-e for extensive covers
3. Protection of water and wind erosion of yute
4. Draining protection plate zinco floradrain FD 25
5. Protection blanket and water and nutrient accumulation zinco ssm 45
6. Thermoplastic sheet of plasticised PVC 1.5 mm
7. Polypropylene geotextile felt type terran 1000
8. Rock wool panel type rockacier 381 for thermal and noise insulation
9. Vapour barrier
10. Concrete compression coat
11. Prefabricated concrete floor structure
12. Anodised aluminum C-Grid ceiling
13. Plasterboard panelling double-plate
14. Insulation layer
15. Concrete wall
16. Local sandstone wall
17. Stiffening elements
18. Mortar layer
19. Medium density panel
20. Window structure
21. Existing land
22. Waterproofing layer
23. Retaining wall
24. Interior floor
25. Concrete base
26. Stone pitching
27. Concrete basement

立面节点

1. 绿植：景天属匍匐植物
2. 植物底层基质 zinco-e，大面积覆盖
3. 防水风化保护层
4. 排水防护板 zinco floradrain FD 25
5. 保护毯，水分养分累积层，zinco ssm 45
6. 热塑性 PVC 塑料板，1.5mm
7. 聚丙烯土工布毛毡，terran 1000 型
8. 石棉板，rockacier 381 型，隔热隔音用
9. 隔汽层
10. 混凝土压制包层
11. 预制混凝土地板结构
12. 阳极氧化铝 C 网格天花板
13. 双层石膏板
14. 隔热层
15. 混凝土墙
16. 本土砂岩墙
17. 加固元件
18. 砂浆层
19. 中密度板
20. 窗结构
21. 原有土地
22. 防水层
23. 挡土墙
24. 内部地板
25. 混凝土底座
26. 砌石护坡
27. 混凝土底座

Hotel Hospes Palma
帕尔马霍斯佩斯酒店

Location/ 地点：Calvià (Mallorca), Spain/ 西班牙，卡尔维亚（马略卡岛）
Architect/ 建筑师：EQUIP Xavier Claramunt
Photos/ 摄影：Adrià Goula
Site area/ 占地面积：4,065m²
Built area/ 建筑面积：5,500m²

Key materials: Façade – limestone
主要材料：立面——石灰岩

Overview
It is a project that refurbishes and extends the Maricel Hotel. Maricel Hotel was built in 1948 as one of the first hotels specially designed for tourists. Thanks to his privileged situation, it was easy to extend its facilities towards the sea with terraces as a giant's stair to the water. The building first opened itself over areas more related to these terraces and the sea, using a series of arcades to enlarge the basement and focus it on the rocky seashore.

The extension is to be built on two plots placed just in front of the original building. The main issues are how to connect, across the public street, and how to deal with the urban surroundings. In that direction, the extension seeks to stress the importance of the original building as main entrance and to establish an access to the new areas capable of generating an alternative context to the existing urban development. The new situation is rearranged as a valley that makes its way recovering the technique of the so call marjades, the terraces used on traditional agriculture activities in Majorca. Creating these new marjades, the valley moves ahead, connecting the new areas to the main building. Dry stonewalls deal with the soil on how to settle on the new areas. Sometimes, they both agree simply with slopes, sometimes, likewise the terraces that the main building uses as a solarium on its way to the sea, the valley sculpts the soil with marjades. Solid and vernacular dry stonewalls that give a desirable environment, detached from the constructions of the neighbouring extensions.

Detail and Materials
The building façade is made by limestone typical from the region. Following the image of the dry stonewalls of the island, the main walls of the building are coated with limestone, regaining the atmosphere of a more natural and agreeable environment. The stone pieces have around 15-20cm thick and they are fixed with mortar directly to the wall. The mortar is

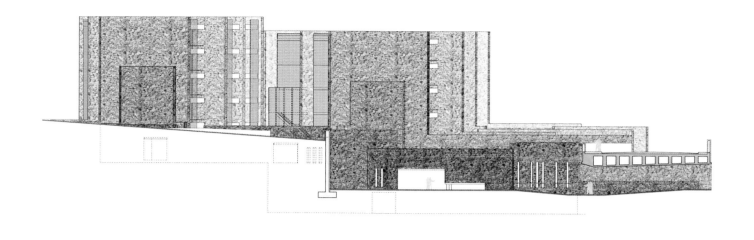

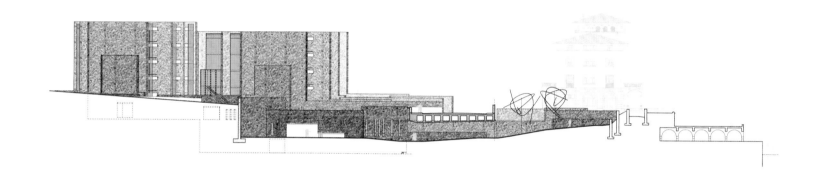

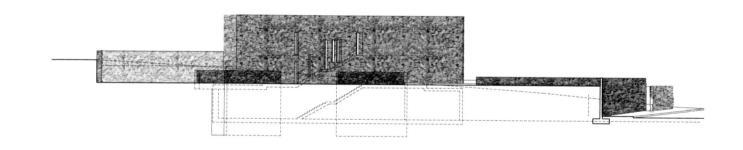

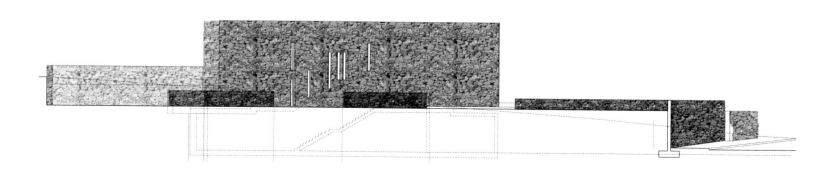

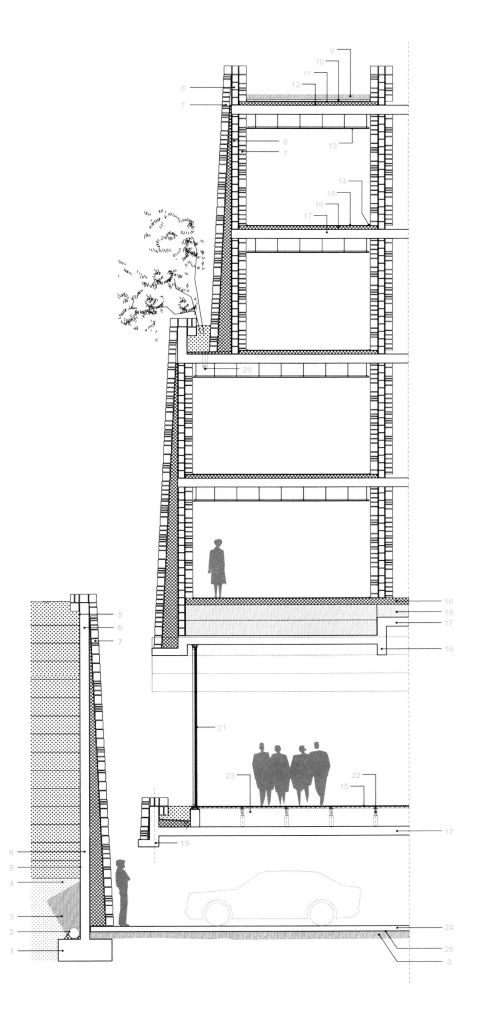

Detail	节点
1. Concrete footing	1. 混凝土底脚
2. Drain tile	2. 排水瓦管
3. Gravels for basement	3. 地下室碎石铺面
4. Terrain	4. 地势
5. Geotextile 100g/m²	5. 土工布 100g/m²
6. Concrete retaining wall for basement	6. 地下室混凝土挡土墙
7. Limestone from the place	7. 石灰岩
8. Concrete brick	8. 混凝土砖
9. Gravels for roof	9. 屋顶碎石铺面
10. Rigid extruded polystyrene 20mm 20kg/m3	10. 刚性挤塑聚苯乙烯 20mm 20kg/m³
11. SBS elastomeric membrane	11. SBS 人造橡胶膜
12. Concrete slopes for water drain	12. 混凝土坡，用于排水
13. Gypsum board 150mm	13. 石膏板 150mm
14. Wood baseboard	14. 木底板
15. Wood pavement	15. 木铺面
16. Under wood pavement	16. 下层木铺面
17. Concrete slab h=25cm	17. 混凝土板 h=25cm
18. Under technical pavement	18. 下层技术铺面
19. Concrete beam w=25cm	19. 混凝土梁 w=25cm
20. Drain pipe	20. 排水管
21. Wood window frames	21. 木窗框
22. Technical pavement system	22. 技术铺面系统
23. Facilities: air, electrical, water	23. 基础设施：排气、电力、排水
24. Concrete ground slab 15cm (parking)	24. 混凝土地面板 15cm（停车场）
25. Polystyrene 0.5mm	25. 聚苯乙烯 0.5mm

not protruding from the joints between the stones, in order to get the image of the typical dry stonewall. The texture of the stones themselves and its thickness give to the building a very good isolation system, without the need of an artificial material to isolate it. The wall top covering is made by mortar, with a rounded shape so that the rain water falls down on its laterals, without affecting the central part of the wall.

项目概况

本项目是对前玛丽塞尔酒店的翻修和扩建。玛丽塞尔酒店建于1948年，是第一批专为游客所设计的酒店之一。酒店享有优越的地理位置，很容易将配套设施延伸到海边，通过酒店露台的巨大阶梯可直达水面。改造工程让建筑的多个区域与这些露台和海洋的联系更加紧密，通过一系列游廊拓展了地下室空间，使其与岩石林立的海岸线联系起来。

扩建工程将建在原有建筑的前方。设计的主要挑战是如何连接并跨越公共街道以及如何处理与周边城市环境的关系。以此为设计方向，扩建工程力求突出原有建筑的重要性，以其为主入口，在新旧建筑之间建立起一条通道，从而与城市环境建立起紧密的联系。

酒店的新址被改造成山谷的形式，复兴了马略卡岛的传统农业形式——梯田技术。山谷通过这些梯田向前推进，将新区域与主楼连接起来。干石墙与土壤的配合树立起了新区域的形象。有时，它们一起顺着坡势而下；有时，主楼的露台则作为日光浴室，沐浴在阳光海风之中，形成了梯田。实心的本土化干石墙营造出令人愉悦的环境，使酒店与周边的扩建施工工程隔开。

细部与材料

建筑立面由当地常见的石灰岩构成。建筑的主墙面参考了岛上干石墙的形象，由石灰岩覆盖，唤回了更自然、更清新的生活环境。石块的厚度约为15~20厘米，通过砂浆直接固定在墙面上。砂浆并没有凸出于石板间的接缝，塑造出传统干石墙的感觉。石头的纹理和厚度为建筑提供了良好的隔热系统，完全没有必要添加人造材料进行隔热保温。墙壁的顶部覆盖着一层砂浆，圆角造型让雨水从侧面留下，丝毫不会影响墙壁的中央部分。

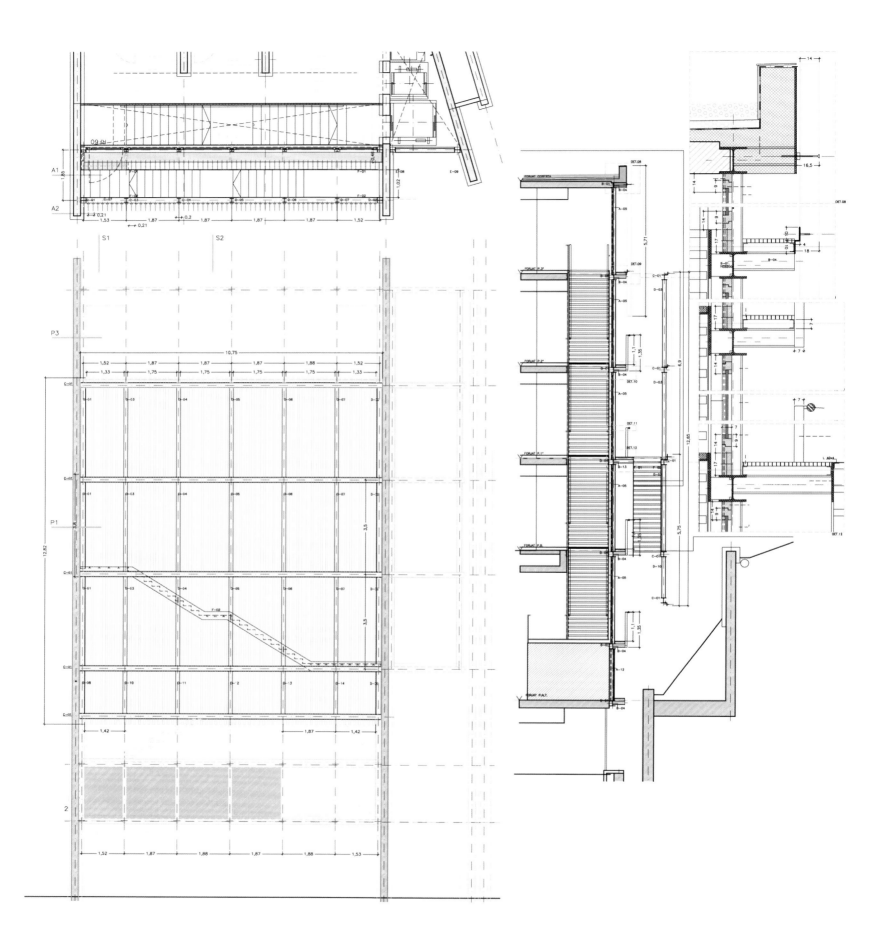

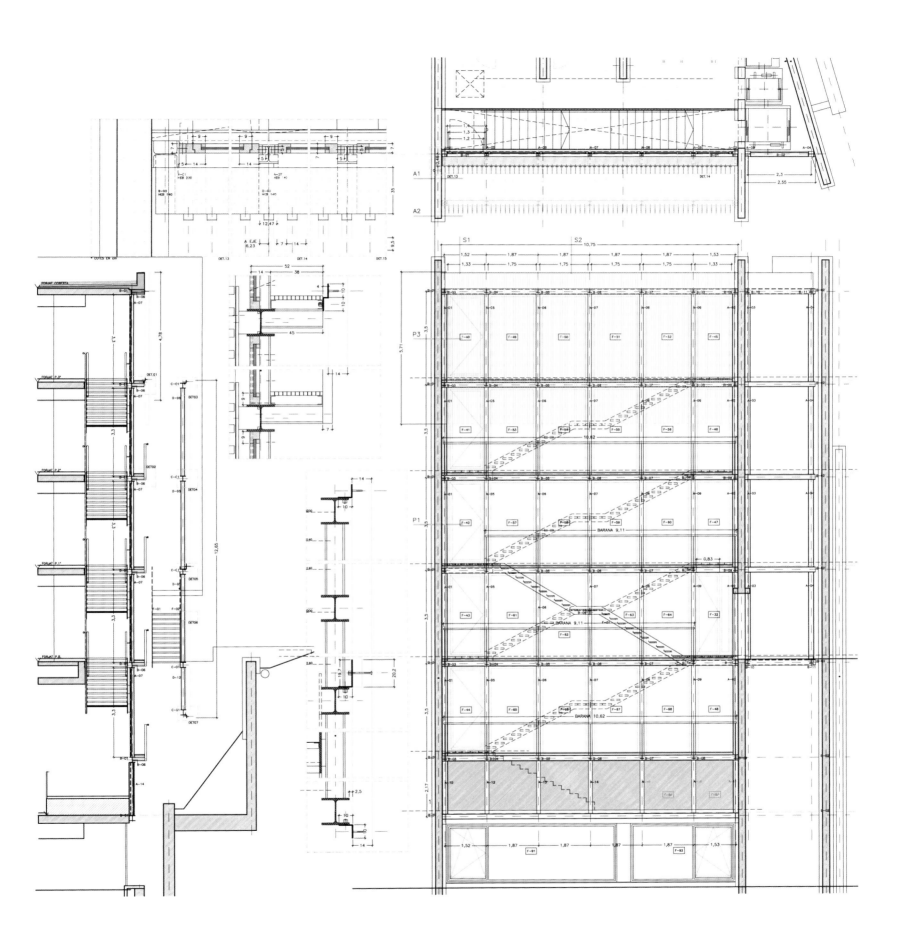

Apartment No.1
一号公寓

Location/ 地点：Mahallat, Iran/ 伊朗，马哈拉特
Architect/ 建筑师：Ramin Mehdizadeh
(ARCHITECTURE by COLLECTIVE TERRAIN, AbCT)
Photos/ 摄影：Omid Khodapanahi
Site area/ 占地面积：420m²
Surface area/ 楼面面积：1,590m²

Key materials: Façade – stone
Structure – stone
(Total recycled stone used: 1,800 square metres)
主要材料：立面——石材
结构——石材
（所使用的回收石材总量：1,800 平方米）

Overview
The project has a contemporary form carved from a heavy mass with sharp edges. The project as a whole resembles a big rock in the quarry which is carved with sharp edges in an artistic way. It has a complex geometry with a texture coming from a context which makes it properly fit to the site. By using the locally recycled rough material, this project comes to negotiate itself with the context of Mahallat and blends well with the neighbourhood in spite of its unique design.

Form is a response to a limited inside space due to the irregular shape of the footprints. In all angels, the project is very proportionate and all the parts are well combined to create a holistic ensemble.

By adding triangular forms to the geometry, the architect creates a more proportional space inside the room. Also, it helps the light to penetrate the space in a more subtle manner.

Detail and Materials
The façade consists of a heavy mass carved with sharp angels which resembling a big rock in the quarry. The triangular prisms protruding from the façade produces dynamic shades on the façade. Windows are covered by wooden shutters which help control the light and heat inside the units, and provide privacy for the residents. The rest of the façade is covered by recycled stones collected from local plants. Windows which are not covered by shutters are small in size, consistent with traditional characteristics of Mahallat's architecture.

The main material used in the project is left-over stones collected from local stone cutting plants which are recycled and used in various forms throughout the project. According to the Mayor's office, one thousand tons of left-over stones are produced in stone cutting factories in Mahallat every day. The main characteristic of the left over stones is having the same thickness and flat surfaces. The architect notes that this

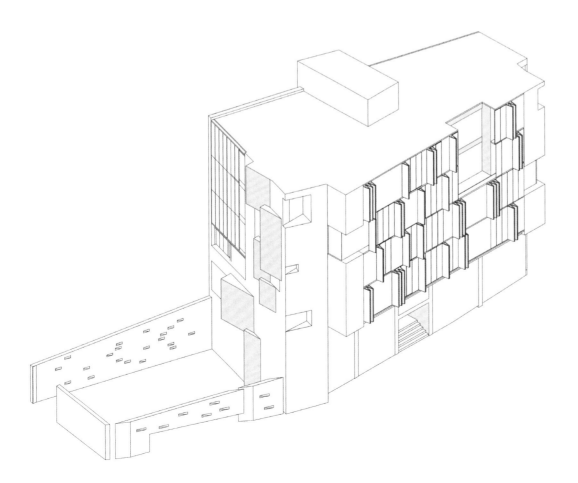

unique characteristics coming from the way stones are cut in the local plants make them to be applicable in different formats: Putting them next to each other in a row results in having a row with a consistent height; Repeating these rows on top of each other creates a coherent yet diverse texture from the horizontal rows of rough-edged leftover stones. This texture covers the whole exterior walls and is used in parts of the interior spaces. Also, the architect has used the left over stones in the mosaic form for the finishing of the parking area. The combination of left-over stones with sands creates an interesting protection and finishing for the water-proofing membrane in the roof. Also, these rooftop mosaics give the residence the option of using the roof top area during seasons with moderate climates to enjoy the view overlooking Mahallat.

At the beginning, investors and potential buyers were highly skeptical of recycling the leftover stones. For them, recycling meant using obsolete material, which adds no value to the project. Also, they believed that using recycled stones would make the project look cheap, hence, decreasing the value of the final product. However, after

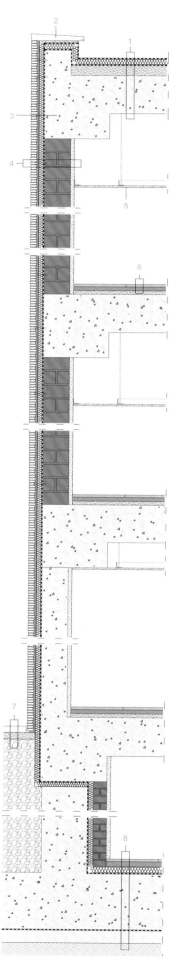
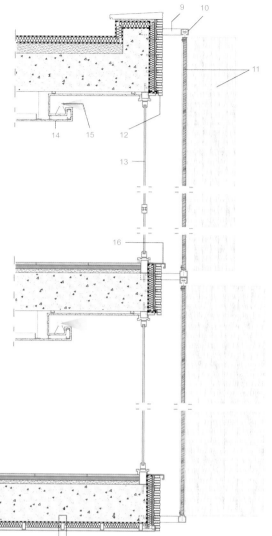

Construction System – Vertical Section

1. Roof formed by waterproofing membrane, 2"(50mm) board insulation, vapour barrier, 2 7/8"(75mm) screed, 11 7/8"(300mm) reinforced concrete slab
2. Stone coping on 2% slope
3. Reinforced concrete beam
4. Façade infill in reclaimed local stone, layer of mortar and steel tiles, 1 3/8"(35mm) board insulation, waterproofing membrane, wall in 7 7/8"x4x2" (200x100x50 mm) brick, 1 1/4"(30mm)render
5. False ceiling in 1/2" (12.5mm) gysum board on steel L-profiles and tie rods suspended from slab
6. Floor in 15 3/4 x 15x3/4 x3/4" (400x400x20mm) travertine blocks, 1 1/4" (30mm) screed, 1 1/4"(30mm) leveling screed.
7. 2"(50mm) stone pavers, 2" (50mm) screed, gravel layer
8. Floor in 15 3/4 x 15x3/4 x 3/4" (400x400x20mm) travertine blocks, 1 5/8" (40mm) screed, 2"(50mm) board insulation, waterproofing membrane, 15 3/4" (400mm) reinforced concrete, 4" (100mm) concrete base, earth
9. Steel box profile painted black securing sun shading to supporting structure
10. Steel track
11. Sun shading consisting of side-hung timber panels on frame of 1 5/8 x 1 5/8" (40x40mm) steel profiles
12. 1 3/8x 1 3/8"(35x35mm) steel L-profile edging
13. Floor-to-ceiling casement window with 3/16 – 1/4 – 3/16" (5/6/5mm) aluminium double glazing unit
14. Gypsum board soffit on steel C-profiles
15. Recessed lighting
16. Aluminium flashing
17. Finish in sheet aluminium, 2 3/8x1 1/4" (60x30mm) box profile support with insulation interlayer, waterproofing membrane
18. Full-height continuous glazed façade with 3/16 – 1/4 – 3/16"(5/6/5mm) aluminium double glazing unit

结构系统——垂直剖面

1. 屋顶：防水膜、50mm 隔热板、隔汽层、75mm 砂浆层、300mm 钢筋混凝土板
2. 石顶盖，坡度2%
3. 钢筋混凝土梁
4. 立面填充：回收利用的本地石材、灰浆层、钢瓦；35mm 隔热板；防水膜；200x100x50mm 砖砌墙、30mm 抹面
5. 假吊顶：12.5mm 石膏板，L 形钢和横拉杆支撑
6. 地面：400x400x20mm 石灰华块、30mm 砂浆层、30mm 找平砂浆
7. 50mm 石材铺装、50mm 砂浆层、碎石层
8. 地面：400x400x20mm 石灰华块、40mm 砂浆层、50mm 隔热板、防水膜、400mm 钢筋混凝土、100mm 混凝土底、土
9. 箱式钢材，涂黑漆，为支承结构遮阳
10. 钢轨
11. 遮阳装置：侧挂木板、40x40mm 钢架
12. 5x35mm L 形钢边
13. 落地平开窗：5/6/5mm 铝框 + 双层玻璃
14. 石膏板拱肩，C 形钢支撑
15. 嵌灯
16. 铝防水板
17. 铝板饰面，60x30mm 箱式型材支撑，配隔热层、防水膜
18. 全高的连续玻璃幕墙：5/6/5mm 铝框 + 双层玻璃

the project was completed, not only the perspectives of locals have changed, but also the techniques used in this project have been emulated in other projects. Recycling also turned out to be an especially attractive option for builders during tough economic situation in Iran in the midst of sanctions.

All the windows are covered with shutters custom-made by local craftsmen with simple local techniques. The combination of hard woods and rough stones texture is unique yet very attractive. The rest of materials have been chosen from Iranian manufacturers and have been used in a coherent manner.

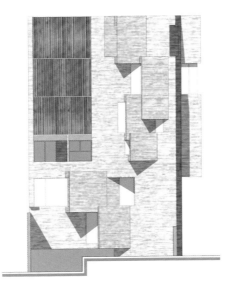
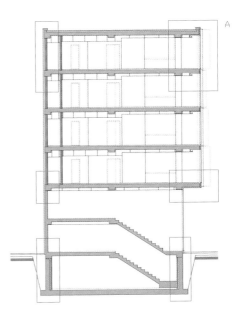

项目概况

项目拥有尖锐的现代外形，就像采石场中的一大块巨石经过艺术切割而成。复杂的几何造型和与环境相适应的纹理让它更好地融入场地。本地回收原料的使用让项目即使设计独特，也能很好地融入马哈拉特的周边环境。

建筑的外形是由内部空间和外部场地的不规则形状所共同决定的。从各个角度来看，项目都十分匀称得体，各个部分共同融合成一个和谐的整体。

三角造型的建筑让建筑师打造了更合适的室内空间。此外，它有助于光线以更微妙的方式进入室内。

细部与材料

建筑立面被切割出尖锐的棱角，就像是采石场中的巨型岩石。凸出于墙面的三棱柱造型为建筑立面带来了动感的阴影。窗户外安装着木制百叶窗，既能够控制室内的光热摄取，又能保护居民的隐私。立面的其他部分由收集自当地工厂的回收石材所覆盖。没有百叶窗的窗户尺寸都较小，与马哈拉特的传统建筑风格相一致。

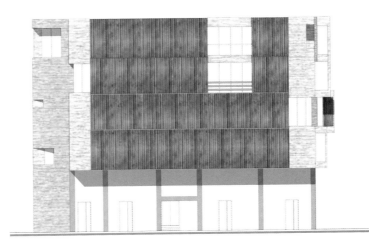
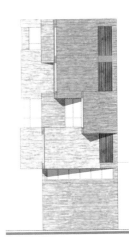

项目所使用的主要材料是收集自当地石材切割厂的边角余料。它们经过回收，以各种不同的形式应用在项目中。据马哈拉特市长办公室统计，马哈拉特的石材切割厂每天能产生1,000吨的边角料。这些边角料的主要特征是拥有相同的厚度和平整的表面。建筑师注意到了这一特征，以不同的形式对其进行了应用：将它们排列成行能形成统一高度的一排石材；而将这些排石材相互叠加就能形成连贯而多样的纹理。这种纹理将整个外墙覆盖起来，并且被应用到了室内空间的某些部分。此外，建筑师还将马赛克形式的石材边角料应用在了停车场的铺装上。石材边角料与沙子的结合形成了屋顶防水膜的保护层。这些屋顶马赛克铺装让人们可以在温和的气候中使用屋顶，享受马哈拉特的城市美景。

Administrative Building of the Croatian Bishops' Conference
克罗地亚主教会议行政楼

Location/ 地点：Zagreb, Croatia/ 克罗地亚，萨格勒布
Architect/ 建筑师：University of Zagreb (The Faculty of Architecture, Institute for Architectural Design)
Photos/ 摄影：Miro Martinić
Site area/ 占地面积：11,500m²
Built area/ 建筑面积：11,900m²
Completion date/ 竣工时间：2011

Key materials: Façade – composite panel of glass and onyx; structure – reinforced concrete
主要材料：立面——玻璃缟玛瑙复合板
结构——钢筋混凝土

Overview
The Croatian Bishops' Conference Building is an administrative and cultural centre and a residence of the top hierarchical structure of Catholic Church in Croatia. The building is located in an attractive northern area of the city with gardens and green public areas. It is positioned on a gentle slope with a park and two individual buildings: the Bishop's Ordinariate building and the Papal Nuncio's residence.

Sustainable Feature
Almost all roof terraces are designed as "green" roofs for the purpose of thermal stability in summer months. The exposed walls are massive (made of reinforced concrete or blocks of brick) with a thermal insulation layer and a ventilated cladding.

The protection from western and eastern (low) insulation was made by outdoor protective elements: vertical brise-soleils, a curtain wall made of composite glass – onyx panels and the outdoor rotating panels in window openings. The horizontal brise-soleils, projections and overhangs act as protective devices against the southern sun exposure.

The installation of geo-thermal heating (deep ground probing) made possible low-temperature floor heating of both the residential and the administrative parts of the building.

Detail and Materials
Structure
The concept of a combined concrete and steel structure with continuous straight lines was a challenging task regarding the construction of the structural elements. The basic structure is reinforced concrete with reinforced concrete

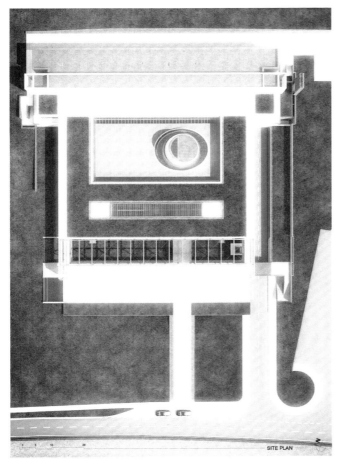

SITE PLAN

slabs and beams supported by reinforced concrete columns and walls on a raft foundation. The front atrium and the access platform are roofed by a steel structure acting as a protection against sunlight and the elements.

In view of the configuration of the terrain, the basement and the ground-floor level were entirely buried along the eastern boundary of the plot. Two main structural requirements were crucial at this point:
1. Protection of the building pit: the construction of the reinforced concrete pylons in a three-row cascade with a shotcrete and anchored reinforced concrete membrane. A retaining wall was built above the last pylon row.
2. Protection and drainage of rainwater: installation corridors were built around the building which accumulate rainwater and channel it outside the building.

Façades
The front of the building faces the main road with a slip road leading to the building. It also functions as the external side of the atrium roofing, the interior courtyard and the ground floor level. It is clad in composite panels of glass and white onyx interpreted as shutters which act as protection against the glare and heat of the western and eastern sun exposure. It produces special effect in counter light or by night owing to its light absorption and emission properties. The decision to use this material was made on the basis of its aesthetic properties and the symbolism of the building in terms of light and purity.

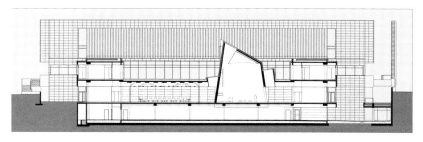
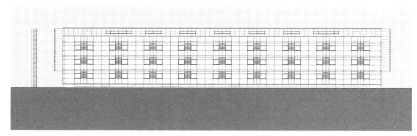
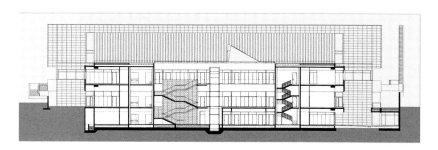
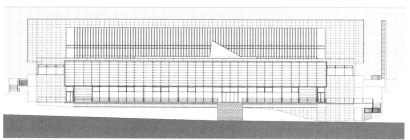
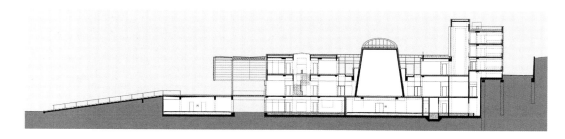

项目概况

克罗地亚主教会议楼是一座行政与文化中心，同时也是克罗地亚天主教会最顶层人员的住所。建筑位于城市北部一处风景优美的区域，周边遍布花园和公共绿地。建筑坐落在一个缓坡上，旁边是公园和两座独立建筑：主教教区楼和教廷大使的住所。

可持续特征

几乎所有的屋顶平台都被设计成了绿色屋顶，以保证建筑在夏季的热稳定性。清水墙面由钢筋混凝土或砖块砌筑而成，配有隔热层和通风包层。

东、西两面的隔热防护由户外防护元件构成，包括：垂直遮阳百叶、复合玻璃幕墙（缟玛瑙板）和窗口的户外旋转遮阳板。水平遮阳百叶、突出结构以及悬臂结构则被用于南侧立面的遮阳防护。

地热能供暖装置为建筑的住宅和行政空间都提供了低温地热供暖。

细部与材料

结构

对结构元件的建造来说，混凝土与钢材混合结构的直线线条设计是一项不小的挑战。项目的基本结构是钢筋混凝土：由钢筋混凝土柱和墙支撑钢筋混凝土板和梁，下方是筏式地基。前方中庭和入口平台的屋顶是一个钢结构，起到了遮阳保护的作用。

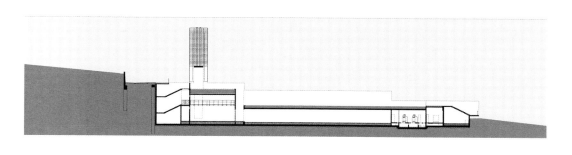
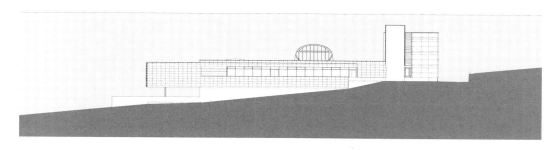
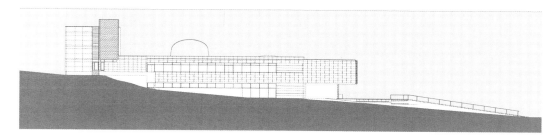

在地形配置上，地下室和一楼空间整个被埋在场地的东侧边缘。这一设计的关键在于两个结构要求：
1. 建筑地基坑的保护：钢筋混凝土塔架的建造采用三排层叠结构，由喷浆混凝土和锚固钢筋混凝土膜构成。挡土墙建造在最后一排塔架的上方。
2. 雨水防护和排水：环绕建筑建造了多条外部走廊，能够收集雨水并将其导入建筑之外。

立面
建筑正面朝向主路，通过一条岔道相连。建筑正面同时还是中庭屋顶、内部庭院和一楼空间的外侧。它的外面覆盖着玻璃和白色缟玛瑙的复合板材，板材作为遮阳结构，保护东、西两面不受眩光和高温困扰。由于具有光线吸收和发射性能，在逆光或夜晚条件下，板材营造出独特的视觉效果。这种材料不仅具有美学价值，而且与建筑纯洁光明的象征意义相符。

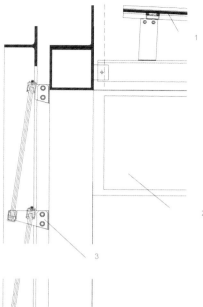
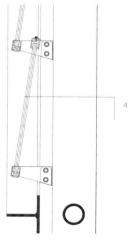
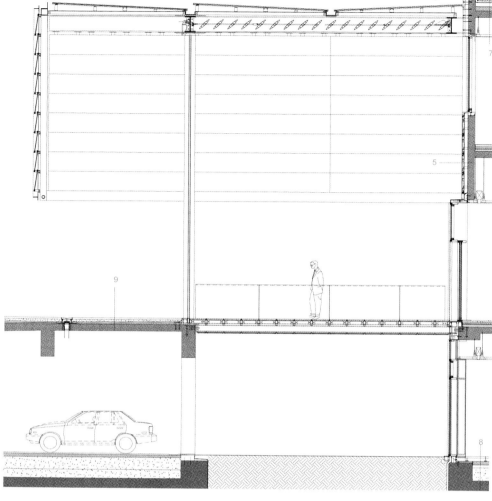

Detail
1. Tempered laminated glass 2x6mm
2. Basic steel structure HEB 500
3. Façade composite panel supporter
4.
- Tempered glass 6mm
- Onyx panel 6mm
- Tempered glass 6mm
5.
- Cement-lime plaster
- Reinforced concrete wall
- Water repellent mineral wool plates
- Ventilated air space
- Wall cladding system with aluminium composite panel
6.
- Earth layer or drainage layers
- Felt, geotexlile
- XPS extruded polystyrene with edge overlaps
- Vapour barrier - polymer welding bitumen strip with aluminium sheet insert laid on cold bitumen primer
- Single layer synthetic waterproof membrane with felt underlayer
- Reinforced concrete wall
- Water repellent mineral wool plates
- Ventilated air space
- Stone cladding fixed with stainless steel brackets
7.
- Soil mix layer for vegetation
- Felt, geotexlile
- Gravel layer
- Felt, geotexlile
- Dimpled plastic membrane
- Single layer synthetic waterproof membrane with felt underlayer
- Vapour barrier, polymer bitumen welding strip with aluminium sheet insert laid on cold bitumen primer
- Concrete base, smoothed, in fall
- Reinforced concrete slab

- Air space, suspended ceiling structure
- Suspended ceilinjg, gypsum boards
8.
- Floating floor layer with bolted stringer system
- Reinforced concrete floor slab
- Polyethylene foil, protection layer
- Sound insulation, elastic expanded polystyrene panel
- Thermal insulation, extruded polystyrene panel
- Compacted gravel layer, installation area
- Reinforced concrete floor slab
- Vapour barrier membrane
- Reinforced concrete foundation plate
9.
- Paving stone layer on dry mortar
- Perforated drainage foil
- Geotextile
- Waterproof thermal insulation, extruded polystyrene panel
- Waterproof membrane, synthetic foil
- Reinforced concrete slab

节点
1. 钢化夹层玻璃 2x6mm
2. 基础钢结构 HEB 500
3. 立面复合板支架
4.
– 钢化玻璃 6mm
– 缟玛瑙板 6mm
– 钢化玻璃 6mm
5.
– 水泥石灰抹面
– 钢筋混凝土墙
– 防水矿物棉板
– 通风空气腔
– 墙面覆盖系统，配铝复合板
6.
– 土层或排水层
– 毛毡，土工布
– XPS 挤塑聚苯乙烯，边缘叠加
– 隔汽层，聚合焊接沥青带，配铝板，铺装于冷沥青底漆上方
– 单层合成防水膜，毛毡衬垫
– 钢筋混凝土墙
– 防水矿物棉板
– 通风空气腔
– 石材覆面，不锈钢支架固定
7.
– 土壤混合层，用于植被种植
– 毛毡，土工布
– 碎石层
– 毛毡，土工布
– 浅凹塑料膜
– 单层合成防水膜，毛毡衬垫
– 隔汽层，聚合焊接沥青带，配铝板，铺装于冷沥青底漆上方
– 混凝土底座，光滑面
– 钢筋混凝土板
– 空气腔，吊顶结构
– 吊顶，石膏板
8.
– 浮式地板层，螺栓串联系统
– 钢筋混凝土楼板
– 聚乙烯膜，保护层
– 隔音层，弹性发泡聚苯乙烯板
– 隔热层，挤塑聚苯乙烯板
– 压实碎石层，隔离区
– 隔汽膜
– 钢筋混凝土地基板
9.
– 干砂浆上方铺路石
– 穿孔排水膜
– 土工布
– 防水隔热层，挤塑聚苯乙烯板
– 防水膜，合成膜
– 钢筋混凝土板

Dingli Sculpture Art Museum
鼎立雕刻馆

Location/ 地点：Quanzhou, China/ 中国，泉州
Architect/ 建筑师：Wang Yan/ 王彦
Project team/ 设计团队：Gao Guangye/ 高广也，Zhang Xu/ 张旭
Structural design/ 结构设计：Shanghai Tong Zhu Structure Design Co.
Photos/ 摄影：Lu Hengzhong/ 吕恒中
Site area/ 占地面积：8,000m²
Built area/ 建筑面积：3,900m²
Cost/ 预算：15 million RMB/1,500 万元人民币
Completion date/ 建成时间：2013

Key materials: Façade – local stone
主要材料： 立面——当地花岗岩

Overview
Chongwu ancient town was built in 1387, the best preserved ancient stone city in China. It locates about 30 minutes drive away from Quanzhou City of Fujian province. City of Chongwu has long history of fame as "Town of Stone Sculpture" in China. Dingli Sculpture Art Museum lies beside the Chonghui Road leading to the ancient stone town.

The museum is facing to the south, lies on the center axis. The new reception centre was built to west and the existing office building to east was recently renovated. 3 buildings have formed a long rectangular entrance square.

Detail and Materials
The building of art museum looks like many huge stones stacking over the other, silent and generous. The Façade implies the function of the building, meanwhile gives out a kind of natural, simple, but strong vision. The folding surfaces of stone wall create vivid shadow and the obtuse angle stone makes the corner look more firm and powerful.

The gallery interior space layout is symmetry, there is a round patio in the center, which surrounded by four exhibition rooms on each floor level. From the terrace on the top, people can enjoy the leisure and overlook the beautiful sea.

We choose a kind of common local stone as façade material. This kind of stone is locally called "G654". It is widely used as stone curving material, meanwhile as well as the basic construction material of Chongwu ancient city. Many traditional houses are made of stacking stones. Thus the museum design concept of stacking stones would arise the thinking of relationship between Dingli Art Museum and local stone tradition. The pure form of stacking stone has a strong contemporary vision as well.

项目概况

崇武古城始建于 1387 年，是中国现存比较完好的明代石头城。这里距福建泉州大约半小时车程，素有"中国石雕之乡"的美誉。而鼎立雕刻馆就位于直达古城的惠崇国道旁。

鼎立雕刻馆座北面南，处于广场中轴位置；东西两侧分别是保留下来的原有办公楼和新建的接待中心。它们与艺术馆形成 U 字型广场，环抱中间水池。

细部与材料

雕刻馆外观像是错位垒叠起来的巨石堆，方正大气，暗示着石雕馆功能特征，同时给人以质朴拙然的视觉感受。众多折面的石材幕墙单元在日光下分出光影，视觉层次丰富；转角处钝角转折处理更增加了浑厚有力的建筑感。

雕刻馆内部空间体现天圆地方的主题，中心圆形内庭空间统摄全局，四隅展厅分别设置石雕作品。顶层露台设置休憩空间，俯瞰南侧广场。

雕刻馆立面选用了当地易采的普通花岗岩，崇武人称它为"G654"，是经常被用作石雕的建筑材料。崇武古城是座明代石头城，传统石房子历来就地取材，用石块垒叠砌筑而成，独具特色。雕刻馆立面巨石垒叠的设计概念不禁让人与崇武当地的"出砖入石"建筑传统联系起来。而简洁纯粹的垒叠形象极富现代感，也抽象地表达了建筑与当地人文历史的联系。

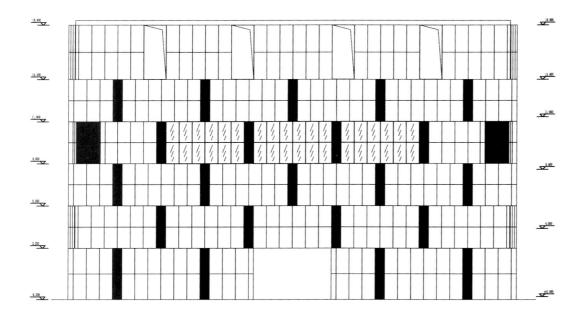

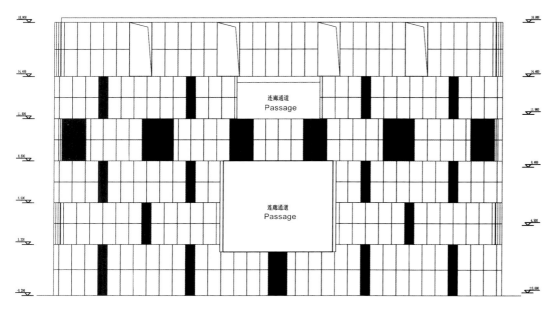

连廊通道
Passage

连廊通道
Passage

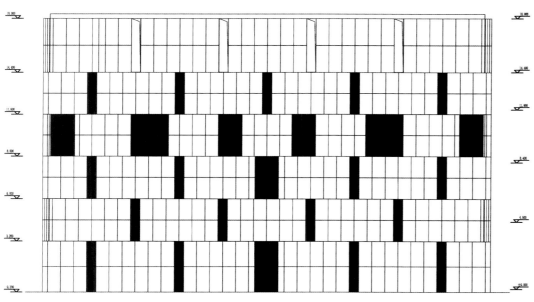

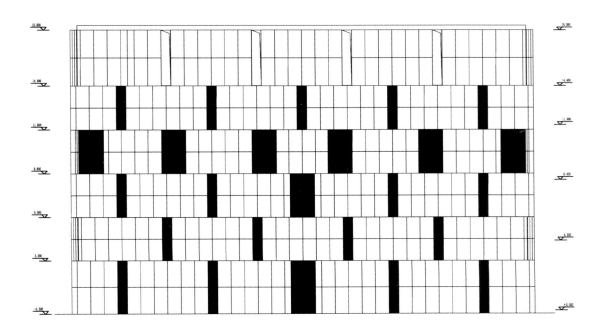

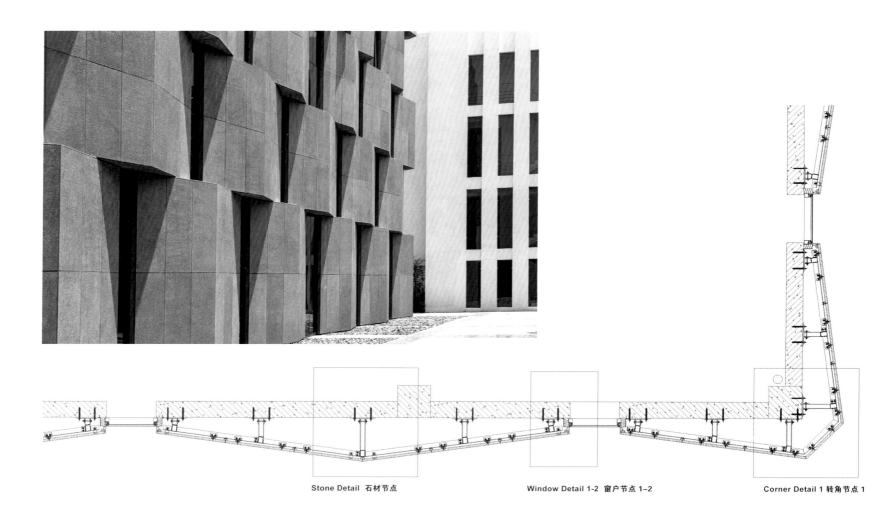

Stone Detail 石材节点   Window Detail 1-2 窗户节点 1-2   Corner Detail 1 转角节点 1

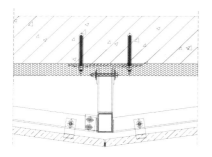

Window Detail 1 窗户节点 1

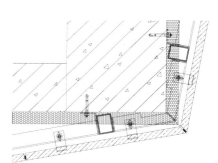

Wall Construction
15mm interior paint
250mm concrete masonry
300mm curtain-wall steel structure
25mm stone-panel (654#)

墙壁结构
15mm 内墙涂料
250mm 混凝土砌体
300mm 幕墙钢结构
25mm 石板（654#）

Window Detail 2 窗户节点 2

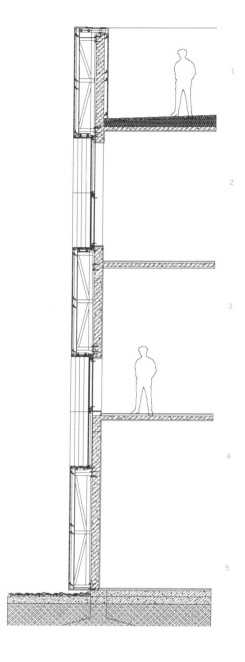
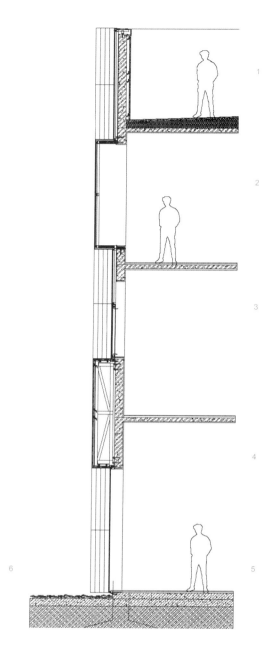

1. Wall construction
   25mm stone-panel (654#)
   300mm curtain-wall steel structure
   250mm concrete masonry
   80mm curtain-wall steel structure
   25mm stone-panel (654#)
2. Roof construction
   20mm stone floor tile
   40mm concrete mortar
   2mm waterproof coil
   200mm expand perlite
   120mm concrete floor slab
   15mm ceiling paint
3. Wall construction
   25mm stone-panel (654#)
   300mm curtain-wall steel structure
   250mm concrete masonry
   15mm interior paint
4. Floor construction
   15mm stone floor tile (654#)
   30mm concrete mortar
   120mm concrete floor slab
   15mm ceiling paint
5. Ground floor construction
   15mm stone floor tile (654#)
   30mm concrete mortar
   150mm concrete floor slab
   Underground structure
6. Ground construction
   100mm pebble stones (654#)
   100mm concrete

1. 墙壁结构
   25mm 石板（654#）
   300mm 幕墙钢结构
   250mm 混凝土砌体
   15mm 内墙涂料
   25mm 石板（654#）
2. 屋顶结构
   20mm 石板地砖
   40mm 混凝土灰浆
   2mm 防水卷材
   200mm 膨胀珍珠岩
   120mm 混凝土楼板
   15mm 天花板涂料
3. 墙壁结构
   25mm 石板（654#）
   300mm 幕墙钢结构
   250mm 混凝土砌体
   15mm 内墙涂料
4. 地面结构
   15mm 石板地砖（654#）
   30mm 混凝土灰浆
   120mm 混凝土楼板
   15mm 天花板涂料
5. 底层地面结构
   15mm 石板地砖（654#）
   30mm 混凝土灰浆
   150mm 混凝土楼板
   地下结构
6. 地面结构
   100mm 卵石（654#）
   100mm 混凝土

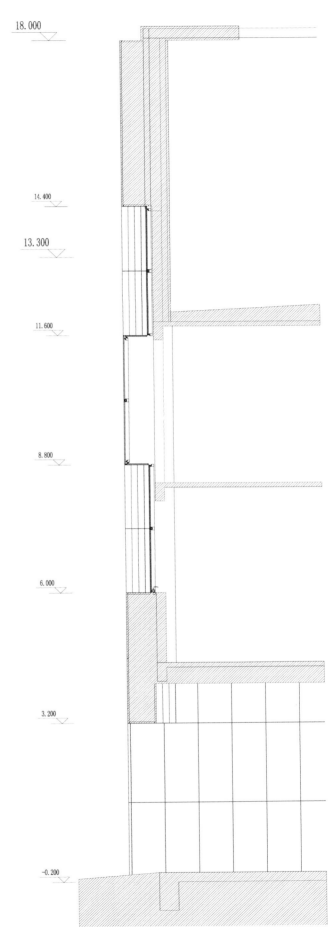

305

Centre for Interpretation of Jewish Culture Isaac Cardoso
犹太文化讲解中心

Location/ 地点: Trancoso, Portugal/ 葡萄牙，特兰科索

Architect/ 建筑师: Arch. Gonçalo Byrne (Gonçalo Byrne Arquitectos, Lda.) www.byrnearq.com, Arch. José Laranjeira (Oficina Ideias em linha – Arquitectura e Design, Lda.) www.oil-arq.com

Project team/ 项目团队: Doriana Reino, Ana Abrantes, Arq.º Tiago Oliveira

Photos/ 摄影: Fernando Guerra | FG + SG Fotografia de Arquitectura | www.ultimasreportagens.com

Area/ 面积: 500m²

Completion date/ 竣工时间: 2012

Key materials: Façade – granite slab
Structure – concrete
主要材料：立面——花岗岩板
结构——混凝土

Overview

The Interpretation Centre was plotted in the dense urban fabric of a medieval fortified village, in an area once referred as the Jewish quarter of Trancoso.

Starting from a ruined allotment, the aim was to re-erect a building that reinforces the corner geometry, still displaying an acute angle on the intersection of two narrow streets, and establishing a symbolic gesture in the context of Jewish urban culture.

The massive character of the building is also reflected on the interior design and "excavated" spaces, like a sequence of voids sculpted from within a large stone monolith. For the exception on this sense of mass, the building is provided with the existence of a large glazing which allows visibility over the Master Pit, a core that enhances all the Jewish culture symbolism with the presence of water.

Detail and Materials

Given the small size of the building, unique geometry and privileged location within the urban medieval tissue of Trancoso, the option pointed towards one outer shell is insulated and coated with granite slabs providing a ventilated façade solution.

The structure of reinforced concrete column/slab, with walls filled with brick masonry is fully lined, on the inside, with walls and ceilings of acoustic control plasterboard.

Altogether, the irregular granite slab stereotomy and tiny fenestrations define the elevation towards the two confining streets. The excavated granite mass, where the openings are also craft-

Floor Plan
1. Lobby
2. Synagogue sefradita model room
3. Exhibition room
4. Projection room
5. Terrace
6. Technical area

平面图
1. 大厅
2. 犹太教会堂样板间
3. 展览室
4. 放映室
5. 平台
6. 技术区

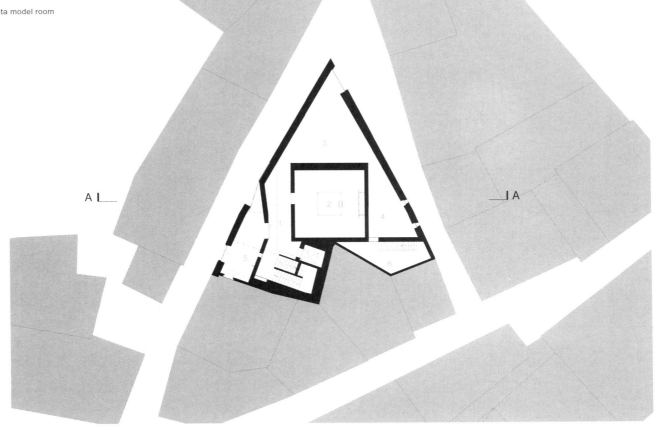

ed with a special plastic approach, prevents overall perception of interior space from the outside, also controlling lighting, recreating and reinterpreting some of the most expressive features of Jewish Architecture in Beira Interior region.

Outside paving and coatings have the same nature, made with regional granite slabs, keeping the colours and textures of the urban environment inside the fortified village and castle guard.

The main room, which refers to the sacred space of the Sephardic Synagogue and the Synagogue of Tomar (also in Portugal) has the most obvious inspiration, rising in the stony mass of the building on all its height, filtering the sunlight to the inside through a ceiling where the complex geometry veils and shapes the perception of all sacred space.

This area differs from the others not only for its size, but also for the lining of the vertical strained panelling in glazed wood, providing an inner atmosphere bathed in golden light. The religious space is dominated by the texture and the sense of rising by the wood cladding of the walls, giving it a temperature and a particular colour and smell.

项目概况

讲解中心坐落在一座中世纪要塞村的中心地带，该地区曾经是特兰科索的犹太人聚集区。

项目从拆迁开始，其设计目标是重建一座能突出尖角地形的建筑，展现两条狭窄街道所成的锐角并打造一个能够体现犹太城市文化的象征性地标。

建筑的体块特色还体现在室内设计和"挖掘"空间上，整座建筑就像是一块大石块上挖掘切割而成。与这种体块感不同的是，建筑拥有大面积的玻璃开窗，可以看到通过水来象征所有犹太文化内涵的"主坑"（Master Pit）。

细部与材料

由于建筑规模较小、造型奇特，并且占据着城市中比较重要的位置，建筑师选择了花岗岩石板作为建筑的外壳，从而保证了建筑的自然通风。

钢筋混凝土支柱和地面以及砖砌墙壁的内部全部配有隔音石膏板内衬。

朝向两条窄街的建筑立面以花岗岩石板的不规则切割形状和小窗口为特色。花岗岩体块与特制的窗口保护室内空间不会过多地显露出来，同时也控制了光照，重现了贝拉因特拉地区犹太建筑最具代表性的特色。

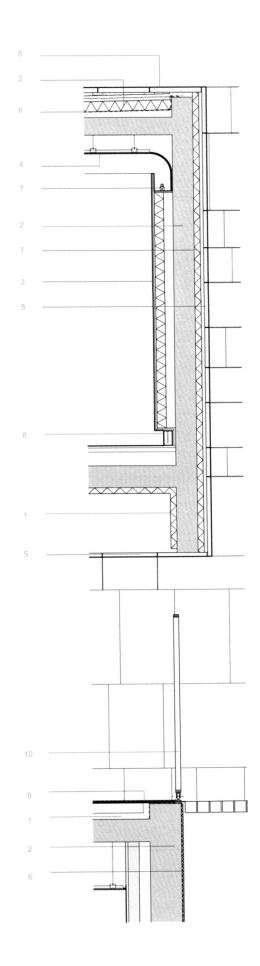
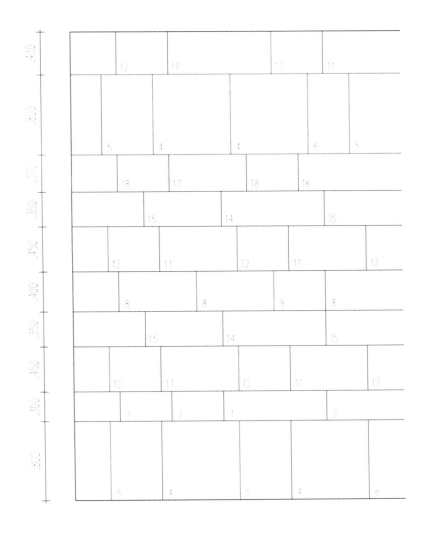

Façade Detail
1. Thermal insulation, extruded polystyrene, e=60mm
2. Concrete structure
3. Dry wall
4. Perforated acoustic gypsum board ceiling
5. Granite slabs, e=30mm
6. Waterproofing system
7. Lighting system
8. Service channel
9. Aluminium profiles rug
10. Exterior gate

立面节点
1. 隔热层，挤塑聚苯乙烯，e=60mm
2. 混凝土结构
3. 干式墙
4. 穿孔隔音石膏板吊顶
5. 花岗岩板，e=30mm
6. 防水系统
7. 照明系统
8. 服务管道
9. 铝型材地块
10. 大门

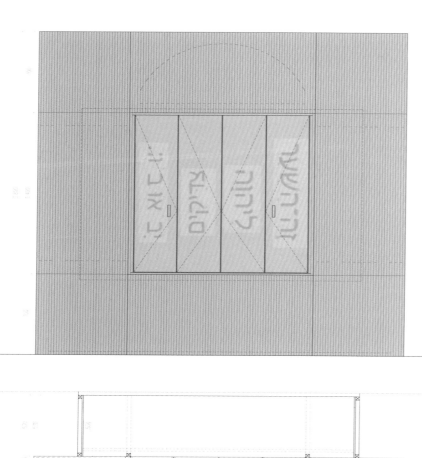

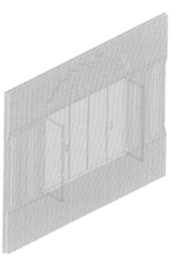
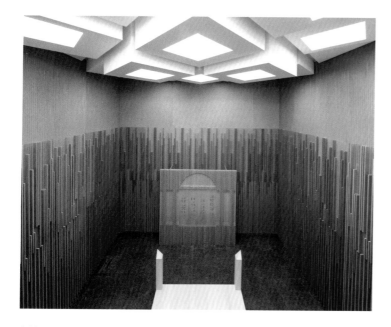

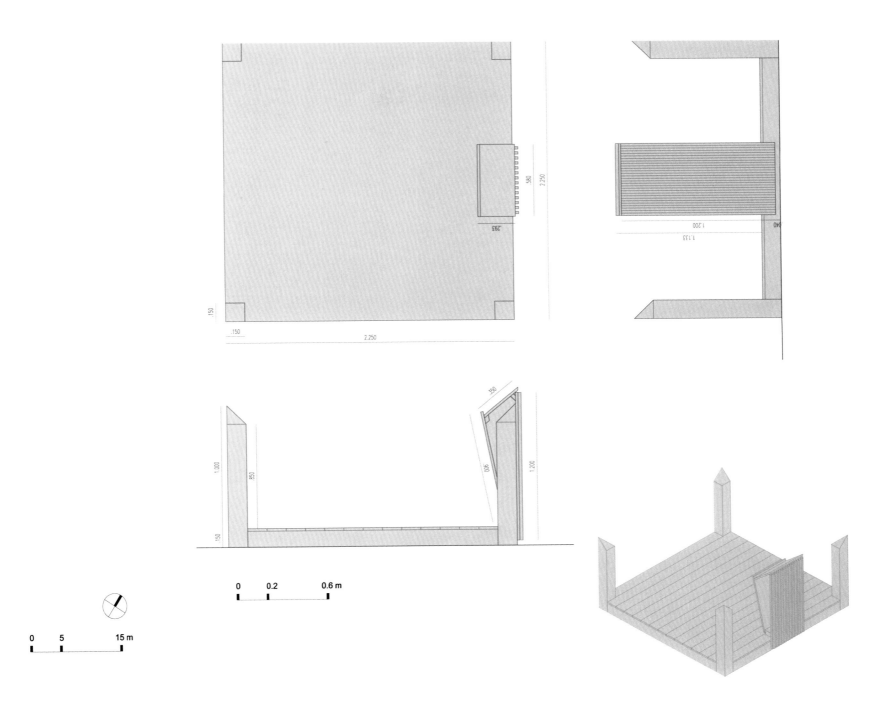

外部的铺装和覆盖层具有同样的特征，全部采用当地花岗岩板建造，保留了要塞村和城堡守卫区的环境特色和材质。

主厅参考了犹太教会堂的空间设计，它的挑高空间与建筑本身等高，通过带有复杂几何图案的半透明天花板让阳光进入室内，赋予了整个空间一种神圣感。

这个区域与其他区域的区别不仅在于尺寸，而且还在于它的内墙被垂直木板条所覆盖，让整个室内氛围沐浴在金色的光芒中。木板墙面的纹理和质感弥漫在宗教空间中，赋予了空间温度与独特的色彩和气味。

Cangzhou Merchant Mansion
沧州市招商大厦

Location/ 地点: Cangzhou, China/ 中国，沧州
Architects/ 建筑师: Daniel Schulz, Zhang Zhuo, Shenyang Jackson Architecture Co., Ltd.
Construction Engineer/ 施工图设计: Norendar International (Group), LTD.
Photos/ 摄影: Shenyang Jackson Architecture Co., Ltd
Site area/ 占地面积: 15,633.4m²
Built area/ 建筑面积: 74,573.3m²
Completion date/ 竣工时间: 2012

Key materials: Façade – granite, glass curtain wall
主要材料: 立面——花岗岩、玻璃幕墙

Overview

Cangzhou Merchant Mansion locates in Cangzhou High-tech Park – the simple yet elegant architecture has proved itself to be one of the emerging landmarks of the high-tech park. With a gross floor area of 74,573.3m² (46943.7m² aboveground, 27629.6m² underground) and a height of 33.9m, main functions of the Mansion include administrative examination and approval, emergency operations center, and e-government center. Efficient spatial organisation and clear function zoning has been achieved – spaces are smoothly connected visually and physically, which provide convenient and efficient experiences to different users.

Detail and Materials

The thermal insulation structure for exterior walls is constructed under supervision of specialized manufacturers of exterior wall thermal insulation materials according to relevant regulations. Surfaces of exterior walls use granite finishing – 30mm stone plate, dowels fixed with epoxy resin adhesive, 6mm seam between stone materials filled with silicone sealant; fix stainless steel accessories and paint 1.2mm polyurethane waterproofing coatings; make level with 20mm 1:3 cement mortar; apply 60mm rock wool board and 15mm adhesive EPS thermal insulation mortar; aerated concrete for exterior walls. Detailed construction method shall be affirmed by specialized manufacturers. Samples of exterior decoration materials shall be provided by the construction company to design company and other relevant parties to confirm texture, size and colour, etc.

The curtain walls adopt grey bridge-cut-off aluminum alloy hidden framing materials; glass structure is 6mm online low-e glass + 12mm air + 6mm transparent tempered glass, air tightness grade not lower than Grade 3.

Thickness of underground concrete exterior wall is detailed in construction drawings and partition wall is constructed with concrete blocks in

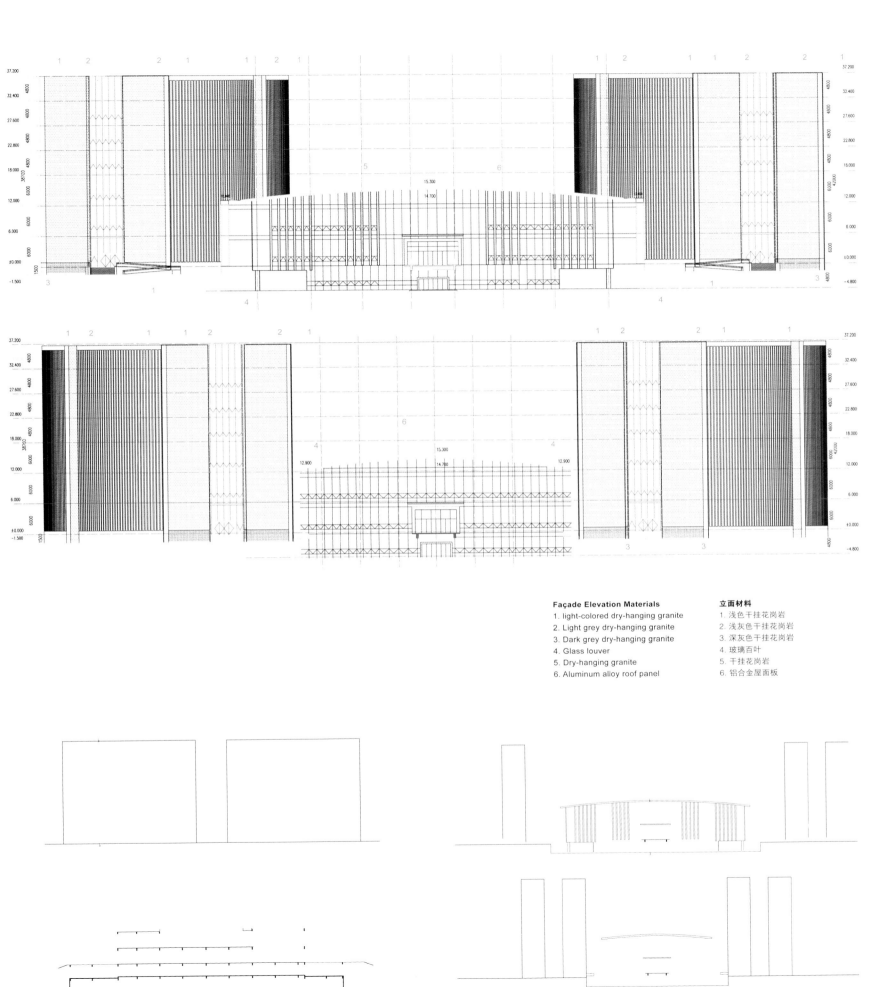

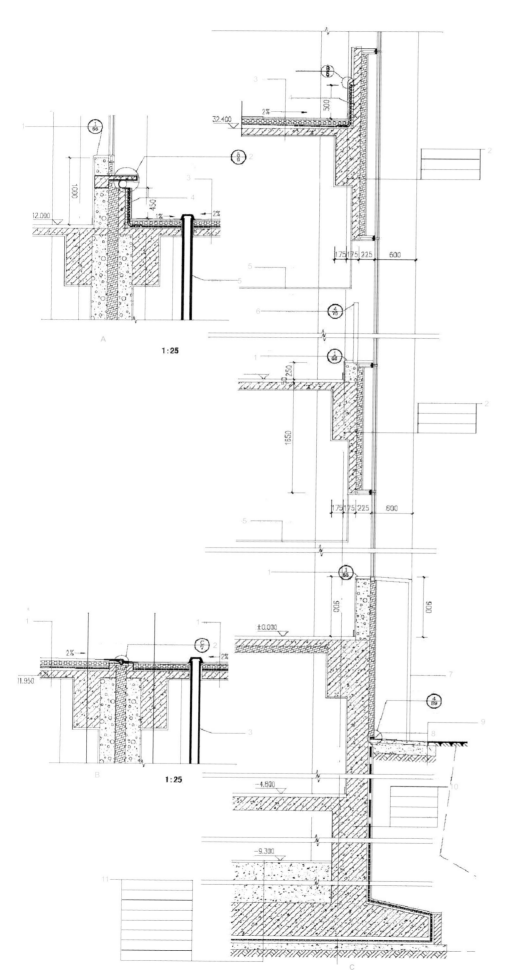

200mm. Walls, without any specific indications, are aligned with column edge or construction axes.

Infilled wall aboveground is constructed with aerated concrete blocks in 200mm or NALC partition wall in 100mm, aligned with column edge or construction axes if not specifically indicated; exterior wall are aligned with column exterior. Above ground part of wallsare constructed with M5 composite mortar, underground part are constructed with M7.5 cement mortar. Dry density of aerated concrete block is B06 and strength is A3.5. Walls are constructed with holes.

项目概况

位于沧州高新区内的沧州市招商大厦造型简洁大气，素雅纯净，是新区的地标性建筑。招商大厦建筑面积74573.3m², 其中地上面积46943.7m², 地下面积27629.6m², 高33.9米，主要功能为行政审批、应急指挥中心和电子政务中心。在建筑空间里，不同属性的功能空间区分明确，且每个空间无论是在物质上还是视觉上，彼此之间的组织和连接都做到了流畅，并在此基础上实现了更高效畅通的空间以供不同的人群使用。

A. Elevation 11.950, parapet deformation joint
1. Window sill
2. Parapet deformation joint
3. Room 1
4. 30mm extruded polystyrene board
5. Internal drainage

B. Elevation 11.950, deformation joint between roofs on same elevation
1. Room 1
2. Deformation joint between roofs on same elevation
3. Internal drainage

C. Walls
1. Marble window sill
2. 6+low-E+12insulating glass/white aluminum composite panel /80cm fire prevention board /concrete beam
3. 30mm extruded polystyrene board
4. Suspended ceiling
5. Window protecting rail
6. Dry-hanging granite
7. Filleting sealed with sealant
8. Aproll
9. HDPEwaterproof roll /3+3SBSwaterproof roll /one layer of primer/make level with 20mm cement mortar /reinforced concrete wall
10. Wearresistant aggregate floor/ Plain concrete/ Reinforced concrete slab/50mm C20 Fine stone concrete protective layer/3+3SBSwaterproof roll/one layer of primer/make level with 20mm cement mortar /100mm C15concrete cushion/ rammed earth
11. Room 2
12. Aluminum alloy suspended ceiling
13. Sunken courtyard
14. Flower beds
15. Underground garage

D. Elevation 37.200, parapet
1. Dry suspending stone curtain wall parapet
2. Room 1
3. 05J5-1

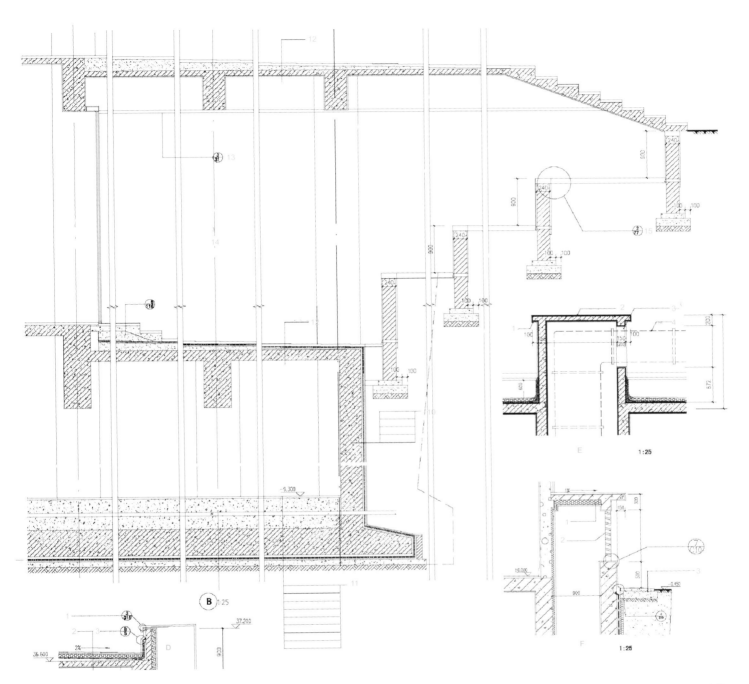

E. Air duct
1. Drip
2. Cement-basedcrystalline waterproofing coating
3. Seam sealed
4. Equipment air duct

F. Air duct
1. 100mm rock wool board
2. Aluminum alloyluover
3. Aproll

A. 标高 11.950 女儿墙变形缝节点详图
1. 大理石窗台板
2. 女儿墙处变形缝
3. 屋一
4. 30 厚挤塑聚苯板
5. 内排水

B. 标高 11.950 同标高屋面变形缝节点详图
1. 屋一
2. 同标高屋面变形缝
3. 内排水

C. 墙身详图
1. 大理石窗台板
2. 6+low-E+12 中空玻璃 / 白色铝塑板 /80cm 防火岩棉 / 混凝土梁
3. 30 厚挤塑聚苯板
4. 吊顶
5. 护窗栏杆
6. 干挂花岗岩
7. 建筑密封胶嵌缝
8. 散水
9. HDPE 防水卷材 /3+3 厚 SBS 防水卷材 / 刷基层处理剂一遍 /20 厚水泥砂浆找平 / 钢筋混凝土墙
10. 特殊耐磨骨料地面 / 素混凝土 / 钢筋混凝土底板 /50 厚 C20 细石混凝土保护层 /3+3 厚 SBS 防水卷材 / 刷基层处理剂一遍 /20 厚水泥砂浆找平 /100 厚 C15 混凝土垫层 / 素土夯实
11. 屋二
12. 铝合金吊顶
13. 下沉庭院
14. 花池
15. 地下车库

D. 标高 37.200 女儿墙节点详图
1. 干挂石材幕墙女儿墙
2. 屋一
3. 05J5-1

E. 风道出屋节点详图
1. 滴水
2. 水泥基结晶型防水涂料
3. 缝隙封堵
4. 设备风道

F. 风道出屋节点详图
1. 100 厚岩棉板
2. 铝合金百叶窗
3. 散水

细部与材料

外墙面保温构造做法由外墙保温材料厂家按相应规范、规定配合施工单位施工。外墙面为花岗岩饰面，参见以下做法，详细做法由有资质的专业厂家进行二次设计：30厚石质板材，用环氧树脂胶固定销钉，石材接缝6mm，用硅酮密封胶填缝；安装与石材配套的不锈钢挂件；刷1.2厚聚氨酯防水涂料；20厚1:3水泥砂浆找平；60厚岩棉板；15厚黏结型胶粉聚苯颗粒保温浆料；加气混凝土外墙。另外，外装修选用的各项材料，其材质、规格、颜色等，均由施工单位提供样板，经建设和设计单位确认后进行封样，并据此施工验收。

幕墙采用灰色断桥铝合金隐框型材，玻璃为6 Low-E（在线）+12空气+6透明钢化玻璃，气密性能等级不低于3级。

墙体材料上，地下部分混凝土墙外墙厚度见结构图纸，隔墙采用混凝土砌块200厚，墙体定位未有特殊注明均齐柱边或居轴线中；地上部分填充墙采用加气混凝土砌块200厚或100厚蒸压轻质加气混凝土板（NALC）隔墙，未有特殊注明均齐柱边或居轴线中，外墙齐柱外皮。墙体地上部分采用M5混合砂浆砌筑，墙体地下部分采用M7.5水泥砂浆砌筑；加气混凝土砌块干密度级别为B06，强度级别为A3.5；同时，墙体留洞。

Index 索引

Abalo Alonso arquitectos
www.abaloalonso.es

AbCT
www.abct.kr

Amas4arquitectura
http://www.amas4arquitectura.com/

André Espinho – Arquitectura
http://www.andrespinho.com/

Archiland
www.archiland.com

ARKIBULLAN architects
www.arkibullan.is

ATR
www.atr-atelier.com

binaa.co
http://www.binaa.co/

Benthem Crouwel Architects
www.benthemcrouwel.nl

Biuro Projektów Lewicki-Łatak
Cracow, Poland

Bruck + WeckerleArchitekten
www.bruck-weckerle.com

Bruno Gaudin Architectes
www.bruno-gaudin.fr

Chalabi Architekten & Partner ZT GmbH
www.chalabi.at

dmvA
www.dmva-architecten.be

D'HOUNDT+BAJART Architects & Associates,
www.dhoundtplusbajart.fr

DMV architects
http://www.dmvarchitecten.nl/

Doojin Hwang Architects
www.djharch.com

EGM architecten
http://www.egm.nl/

EQUIP Xavier Claramunt
http://www.equip.com.es

FAAB Architektura Adam
www.faab.pl

Francisco Berreteaga
www.berreteaga.com

FRES architectes
www.fres.fr

g+f arquitectos
https://gmasfarquitectos.wordpress.com/

Gonçalo Byrne Arquitectos, Lda
www. byrnearq.com

Gramazio & Kohler
http://www.gramaziokohler.com/

HILBERINKBOSCH architects
http://hb-a.nl/

Höweler and Yoon Architecture, LLP
www.hyarchitecture.com

Ingarden & Ewý Architects
www.iea.com.pl

Ignacio Quemada Arquitectos
http://www.ignacioquemadaarquitectos.com/

Jackson Architecture Co., Ltd.
www.promontorio.net

Jiangyin Architectural Design & Research Institute Co., Ltd
http://www.jyadi.com.cn/

Jiran Kohout architekti
www.jkarch.cz

JOHO Architecture
http://www.johoarchitecture.com/

Jo Janssen Architecten
www.ronaldjanssen.eu

K2S Architects Ltd.
www.k2s.fi

Karelse & den Besten, Rotterdam
www.karelse-denbesten.nl

keithwilliamsarchitects
www.keithwilliamsarchitects.com

L'ATELIER
www.atelier.net

Magén Arquitectos
www.magenarquitectos.com

Marlies Rohmer Architects and Planners
www.rohmer.nl

MAYU architects+
http://www.malonearch.com.tw/

Mestura architects
www.mestura.es

NocasaBaumanagement
http://gc_yp60245423939.en.forbuyers.com

O'Donnell + Tuomey
www.odonnell-tuomey.ie

Oficina Ideias em linha – Arquitectura e Design, Lda
www.oil-arq.com

Óscar Pedrós
http://www.oscarpedros.com/

PROMONTORIO
www.promontorio.net

Ramón Fernández-Alonso and associated
http://www.fernandez-alonso.com/

Smart Architecture
http://www.smart-arch.com/

SpreierTrenner Architekten
www.spreiertrenner.de

TRIANERA DE ARQUITECTURA S.LP
www.tridarq.com

UAB "Laimos ir Ginto projektai"
http://www.lgprojektai.lt/

UArchitects
www.uarchitects.com

UNIT architekti,
www.unitarch.eu

University of Zagreb
www.unizg.hr

VIRAI ARQUITECTOS
www.viraiarquitectos.com

ZON-E ARCHITECTS
http://www.zon-e.com/

ZSK Architects
http://www.zsk.hu/

图书在版编目（CIP）数据

建筑材料与细部结构：砖石 /（荷）麦里恩博尔编；常文心译.
— 沈阳：辽宁科学技术出版社，2016.3（2016.12 重印）
ISBN 978-7-5381-9366-4

Ⅰ.①建… Ⅱ.①麦… ②常… Ⅲ.①建筑材料－砖石结构
Ⅳ.① TU36

中国版本图书馆 CIP 数据核字 (2015) 第 178324 号

出版发行：辽宁科学技术出版社
（地址：沈阳市和平区十一纬路 25 号 邮编：110003）
印　刷　者：上海利丰雅高印刷有限公司
经　销　者：各地新华书店
幅面尺寸：245mm×290mm
印　　张：20
字　　数：200 千字
出版时间：2016 年 3 月第 1 版
印刷时间：2016 年 12 月第 2 次印刷
责任编辑：鄢　格
封面设计：何　萍
版式设计：何　萍
责任校对：周　文

书　　号：ISBN 978-7-5381-9366-4
定　　价：368.00 元

联系电话：024-23280367
邮购热线：024-23284502
E-mail: 1207014086@qq.com
http://www.lnkj.com.cn